INNOVATION IN ANESTHESIOLOGY

Clinical and Medical Innovation Series

Series Editors

Adam E.M. Eltorai, MD, PhD
Harvard Medical School, Boston, MA, United States

Nancy Patterson, MBA
President and CEO, Strategy Inc., Austin, TX, United States

Volumes in Series

Innovation in Anesthesiology

Innovation in Nephrology

Innovation in Interventional Radiology

Innovation in Emergency Medicine

Clinical and Medical Innovation Series

INNOVATION IN ANESTHESIOLOGY

Technology, Development, and Commercialization Handbook

Series Editors

ADAM E.M. ELTORAI
Harvard Medical School, Boston, MA, United States

NANCY PATTERSON
Strategy Inc., strategyinc.net, Austin, TX, New York, NY, Research Triangle Park, NC, United States

Volume Editors

ASHLEY ELTORAI
Anesthesiology and Critical Care, Yale-New Haven Hospital; School of Medicine, Yale University, Connecticut, United States

PIYUSH MATHUR
Anesthesia Critical Care Fellowship Program, Anesthesia Institute, The Cleveland Clinic, OH, United States

Academic Press is an imprint of Elsevier
125 London Wall, London EC2Y 5AS, United Kingdom
525 B Street, Suite 1650, San Diego, CA 92101, United States
50 Hampshire Street, 5th Floor, Cambridge, MA 02139, United States

ISBN: 978-0-12-818381-6

For Information on all Academic Press publications
visit our website at https://www.elsevier.com/books-and-journals

Publisher: Mara Conner
Acquisitions Editor: Chris Katsaropoulos
Editorial Project Manager: Fernanda Oliveira
Production Project Manager: Manju Paramasivam
Cover Designer: Miles Hitchen

Typeset by MPS Limited, Chennai, India

Contents

List of contributors

Odmara L. Barreto Chang Department of Anesthesia and Perioperative Care, University of California San Francisco, San Francisco, CA, United States

P.K. Benson Department of Anesthesiology, HRH, Secaucus, NJ, United States

Cristy Berg Codonics, Middleburg Heights, OH, United States

Poorwa G. Bhaskar Department of Health Care Management, The Wharton School, University of Pennsylvania, Philadelphia, PA, United States

Ty L. Bullard Department of Anesthesiology, University of North Carolina School of Medicine, Chapel Hill, NC, United States

Harry Burke Department of Anesthesiology and Perioperative Medicine, University of Pittsburgh Medical Center, Pittsburgh, PA, United States

Brian Coblitz Technology Commercialization Office, The George Washington University, Washington, DC, United States

Ettore Crimi Department of Anesthesiology, Atrium Health Wake Forest Baptist, Wake Forest University Health Sciences Medical Center Boulevard, Winston-Salem, NC, United States; Department of Anesthesiology, Atrium Health Wake Forest Baptist, Wake Forest University Health Sciences, Orlando, NC, United States

Jennifer Dawson University Hospitals Health System, Cleveland, OH, United States

Erin Dengler Department of Anesthesiology, University of North Carolina School of Medicine, Chapel Hill, NC, United States

D. John Doyle Cleveland Clinic Lerner College of Medicine of Case Western Reserve University, Cleveland, OH, United States; Department of General Anesthesiology, Cleveland Clinic, Cleveland, OH, United States

Anterpreet Dua Department of Anesthesiology and Perioperative Medicine, Medical College of Georgia, Augusta University, Augusta, GA, United States

Trent Emerick Department of Anesthesiology and Perioperative Medicine, University of Pittsburgh Medical Center, Pittsburgh, PA, United States

David N. Flynn Department of Anesthesiology, University of North Carolina School of Medicine, Chapel Hill, NC, United States

Zach Frabitore Department of Anesthesiology and Perioperative Medicine, University of Pittsburgh Medical Center, Pittsburgh, PA, United States

Ross Goodman Codonics, Middleburg Heights, OH, United States

Michael P. Harlander-Locke Department of Anesthesiology and Perioperative Medicine, Medical College of Georgia, Augusta University, Augusta, GA, United States

Gary Haynes Department of Anesthesiology, Tulane University School of Medicine, New Orleans, LA, United States

Bernadette Henrichs Nurse Anesthesia Program, Goldfarb School of Nursing, Barnes-Jewish College, St. Louis, MO, United States; CRNA Education and Research, Department of Anesthesiology, Washington University, St. Louis, MO, United States

Joshua Herskovic Baptist Health Bethesda Hospital, Boynton Beach, FL, United States; Veterans Affairs Medical Center, West Palm Beach, FL, United States

James Lee Hill University Hospitals Health System, Cleveland, OH, United States

Rachel Wang Hoffman Clinical & Regulatory Affairs, Medcura, Inc, Riverdale, MD, United States; Syneos Health, Morrisville, NC, United States

Xiaodong Hua Department of Anesthesiology and Perioperative Medicine, Medical College of Georgia, Augusta University, Augusta, GA, United States

Gary Keefe Codonics, Middleburg Heights, OH, United States

Raina Khan Department of Anesthesia and Perioperative Care, University of California, San Francisco, CA, United States

Steven Kubisen InnoComm, LLC, Annapolis, MD, United States

Elisa T. Lund Department of Anesthesiology, University of North Carolina School of Medicine, Chapel Hill, NC, United States

Vittorio Mottini Department of Biomedical Engineering, Michigan State University, East Lansing, MI, United States

Junichi Naganuma Department of Anesthesiology, University of California, San Diego, CA, United States

Jungmin On The Department of Anesthesiology and Perioperative Medicine, Medical College of Georgia, Augusta University, Augusta, GA, United States

Neil Padharia Ray Board Certified Pediatric Anesthesiologist, Raydiant Oximetry, Inc, Sacramento, CA, United States

Bahram Parvinian Lighthouse Regulatory Consulting Group, Rockville, MD, United States

Ioana Pasca Loma Linda University at Riverside University Health System, Moreno Valley, CA, United States; University of California Riverside, Riverside, CA, United States

Nancy Patterson Strategy Inc., strategyinc.net, Austin, TX, New York, NY, Research Triangle Park, NC, United States

Douglas Portnow Douglas Portnow, Schwegman Lundberg & Woessner, P.A., Minneapolis, MN, United States

Aniruddha P. Puntambekar Cincinnati Children's Innovation Ventures, Cincinnati Children's Hospital Medical Center, Cincinnati, OH, United States

Chunyuan Qiu Anesthesiology & Perioperative Medicine, Kaiser Permanente Southern California, Pasadena, CA, United States

Kory Razaghi Aptus Advisors Inc., Laguna Niguel, CA, United States

Lauren Rodriguez Strategy Inc., strategyinc.net, Austin, TX, New York, NY, Research Triangle Park, Durham, NC, United States

Reid Rubsamen Department of Anesthesiology and Perioperative Medicine, University Hospitals Cleveland Medical Center, Cleveland, OH, United States; Case Western Reserve University School of Medicine, Cleveland, OH, United States

Umme Rumana Department of Anesthesiology, University of Central Florida College of Medicine, Orlando, FL, United States; Department of Anesthesiology, University of Central Florida College of Medicine, Winston-Salem, FL, United States

Michael Salomon Department of Anesthesiology, Cleveland Clinic Foundation, Cleveland, OH, United States

Ruth Segall University of Pittsburgh, Pittsburgh, PA, United States

Zuhayr Shaikh School of Medicine, University of Virginia, Charlottesville, VA, United States

Ashish Sinha Loma Linda University at Riverside University Health System, Moreno Valley, CA, United States; University of California Riverside, Riverside, CA, United States

Bryan Smith Department of Biomedical Engineering, Michigan State University, East Lansing, MI, United States

Che Antonio Solla University of Tennessee Medical Center, Knoxville, TN, United States

Zhuo Sun Department of Anesthesiology and Perioperative Medicine, Medical College of Georgia, Augusta University, Augusta, GA, United States

Anna M. Varughese Department of Anesthesiology, Johns Hopkins All Children's Hospital, St. Petersburg, FL, United States; School of Medicine, Johns Hopkins University, Baltimore, MD, United States

Robert P. Walsh Nurse Anesthesia Program, Goldfarb School of Nursing, Barnes-Jewish College, St. Louis, MO, United States; CRNA, HCA Florida Fawcett Hospital, Port Charlotte, FL, United States

Abby V. Winterberg Department of Anesthesiology, Cincinnati Children's Hospital Medical Center, Cincinnati, OH, United States; College of Nursing, University of Cincinnati, Cincinnati, OH, United States

Lucy Xu Department of Biomedical Engineering, Michigan State University, East Lansing, MI, United States

Joshua Younger Henry Ford Hospital, Henry Ford Innovations, Detroit, MI, United States

About the editors

Ashley Eltorai, MD, is a practicing Anesthesiologist in Hartford, Connecticut; she is an Assistant Professor at the University of Connecticut School of Medicine and is also a research fellow at a law firm in New Haven, Connecticut.

Piyush Mathur, MD, FCCM, FASA, is an anesthesiologist and intensivist. He is a thought leader and innovator in perioperative artificial intelligence. He has extensive experience in quality improvement and safety.

About the series editors

Adam E.M. Eltorai, MD, PhD, completed his graduate studies in Biomedical Engineering and Biotechnology along with his medical degree from Brown University. His work has spanned the translational spectrum with a focus on medical technology innovation and development. Dr. Eltorai has published numerous articles and books.

Nancy Patterson, MBA, is a Venture Analyst and President and CEO, of Strategy Inc. www.strategyinc.net, a life science market and financial due diligence company founded in 2000. Strategy Inc. provides market and business analysis services for emerging life science technology entities to inform and drive successful commercialization.

About the authors

Chunyuan Qiu, MD, MS, is an Associate Professor of clinical science at Kaiser Permanente School of Medicine. He is the current Regional Chief of Anesthesiology at Kaiser Permanente Southern California and the Chief and Physician Director of anesthesiology and perioperative medicine at Kaiser Permanente Baldwin Park Medical Center. After earning his medical degree from Beijing University and completing residency training at the University of Maryland Medical Center, his career has been marked by notable achievements in medical education, practice innovation, leadership development, and clinical research. His research, conducted with his team, has been primarily centered around maternal obesity, diabetes, preeclampsia, labor epidural analgesia, and autism spectrum disorder. He has collaborated with universities, research institutes, and private and public companies, contributing to developing medical devices from concept to patents to FDA-approved products. He has a prolific record of scientific publications, patents, and inventions and is a visiting professor at numerous universities worldwide.

Junichi Naganuma, MD, MPA, MBA, MPH, is an Associate Clinical Professor of Anesthesiology at the University of California, San Diego. During his initial clinical training in general surgery and cardiothoracic surgery at the University of Tokyo Hospital in Japan, Dr. Naganuma saw a critical need for physicians capable of understanding the entire perioperative environment. He then completed residency in anesthesiology and perioperative administration fellowship at Massachusetts General Hospital, and held faculty positions and clinical leadership roles at Massachusetts General Hospital and the Johns Hopkins Hospital. In addition to currently being an academic anesthesiologist at the University of California, San Diego, his nonclinical interests include operating room management, patient safety, and resident education. Dr. Naganuma holds an MD from the University of Tokyo, an MPA from Harvard Kennedy School, an MBA from The Wharton School of the University of Pennsylvania, and an MPH from The Harvard T.H. Chan School of Public Health.

Dr. D. John Doyle is a Professor Emeritus at Case Western Reserve University in Cleveland, Ohio, and Consultant Anesthesiologist at the Cleveland Clinic. In addition to his medical qualifications, he has a doctorate in Biomedical Engineering. He has special interests in medical instrumentation and clinical airway management. Further details are available at his personal website @danieljohndoyle.com.

Bahram Parvinian, PhD, is the founder and Principal Consultant at Lighthouse Regulatory Consulting Group. He has been advising manufacturers of medical devices for various phases of medical device design, development, and regulatory approval. Prior to founding Lighthouse, Bahram served at the FDA-CDRH from 2007 to 2019 as a Senior Lead Reviewer and Team Leader of critical care and autonomous drug delivery devices. In his role at the FDA, he authored numerous publications, the FDA guidance document on physiological closed-loop controlled medical systems and white papers related to regulation of critical care and anesthesia medical devices. He holds a BS in biological engineering from the University of Maryland College Park, MS in biomedical engineering from Johns Hopkins University, and PhD in mechanical engineering from the University of Maryland College Park. His graduate research focused on credibility assessment of mathematical physiological models used for design and evaluation of clinical decision support and automated medical devices.

Gary Haynes, MD, PhD is a Professor and the Merryl and Sam Israel Chair in Anesthesiology at Tulane University School of Medicine in New Orleans. In addition to a lifetime career in academic and clinical practice, he serves in leadership positions in professional medical organizations and has presented clinical and practice management topics at national and international conferences. He is a consultant to pharmaceutical and medical device companies, advising on the development and marketing of new products. His undergraduate education was at Illinois College with graduate and medical degrees from Case Western Reserve University and advanced studies at the Wharton School of the University of Pennsylvania and the TH Chan School of Public Health at Harvard University.

Dr. Neil Padharia Ray is a board-certified pediatric anesthesiologist and CEO/Founder of Raydiant Oximetry, Inc. Raydiant

Oximetry is a medical device corporation that focuses on developing medical technologies to improve outcomes for mothers and babies during childbirth. Dr. Neil P. Ray has raised over $15 million dollars in private investments and has been awarded over $5 million dollars of federal funding from the NIH and NSF. He was nominated for physician entrepreneur of the year by the Society of Physician Entrepreneurs in 2021. Dr. Ray completed a residency in anesthesiology at the Brigham and Women's Hospital in Boston, Massachusetts, and a fellowship in pediatric anesthesiology at the Hospital for Sick Children in Toronto, Canada. He also serves as an oral board examiner for the American Board of Anesthesiology.

Douglas Portnow is a registered US patent attorney with nearly 20 years of legal experience focusing on medical and surgical inventions. His practice focuses on counseling early-stage companies and entrepreneurs to develop patent portfolios that provide global protection for inventions. He also works closely with investors and companies during intellectual property due diligence as part of fundraising or an exit strategy. Before entering the legal profession, Doug worked as an Engineer for over a decade designing medical devices. He is an international speaker on intellectual property, has been quoted in the New York Times, and has published numerous articles on intellectual property. In particular, he is a coauthor of the book "IP Strategies for Medical Device Technologies: Be Your Own Incubator." His educational background includes a BS in mechanical engineering from MIT, an MS in bioengineering from the University of Michigan, an MBA in finance from Santa Clara University, and a JD from the University of New Hampshire School of Law.

Nancy Patterson, MBA, is a Venture Analyst and President and CEO, and **Lauren Rodriguez**, Vice President (www.strategyinc.net) of Strategy Inc., a life science market and financial due diligence company, founded in 2000. Over the years, they have leveraged their experience in market and business strategy for emerging life sciences technology to drive the probability of successful commercialization for emerging entrepreneurs seeking to value their technology for funding, acquisition, or licensing, allocate development resources, and confirm strategic alliances. They often support enterprise companies seeking to increase technology adoption in a competitive landscape or identify unmet clinical needs and financiers staging investments with technologies in all product life cycle stages. Both are published authors and frequent presenters on value analysis for commercial adoption and the strategic role of KOLs throughout the product lifecycle. They enjoy supporting life science innovators in a range of clinical areas to understand the US landscape, capture funding, identify expanded indications for market growth, and position their innovation for success.

Dr. Odmara L. Barreto Chang is an Anesthesiologist and Neuroscientist at the University of California San Francisco, with a research focus on translational neuroscience research. She conducts human studies and clinical trials that aim to understand the molecular and behavioral basis of cognitive impairment after surgery. By focusing on the perioperative period, she seeks to improve both the information patients receive during the perioperative evaluation in terms of their brain health assessment and to develop potential neuroprotection approaches during the period of surgery and postoperative recovery. In addition, she studies long-term cognitive trajectories of older adults after surgery and the role of the immune system and inflammation in postoperative neurocognitive disorders.

Rachel Wang Hoffman is a Healthcare Executive specializing in medical device research. Currently, Rachel is the Senior Vice President of Clinical & Regulatory Affairs for Medcura, Inc. Previously, she was the Global Head of the Medical Device & Diagnostics Division at Syneos Health, one of the world's largest Contract Research Organizations (CROs). She has also held senior leadership and consultancy positions at both medical device manufacturers and CROs.

Rachel has an educational background in biomedical engineering and molecular biology; she studied and began her research career at Duke University before moving into the medical technology industry. Rachel's experience spans product development, nonclinical testing, regulatory strategy, and clinical development and operations. Her expertise is in the full medical device development life cycle and providing strategic guidance on how to get products to market. She can advise medical device manufacturers from as early on as the concept or prototype and guide them through the entire testing pathway.

Rachel is a published author and speaker, with experience leading workshops training US Food and Drug Administration

(FDA) reviewers on how to assess both the safety and performance of medical devices with a least burdensome approach. The combination of her engineering background and knowledge of industry positions Rachel to advise on a wide range of aspects of device development, translational research, efficient clinical execution, and regulatory pathway strategies.

Steven Kubisen, PhD, has spent much of his career focusing on the intersections between science, technology, and the business world. He has extensive university technology transfer experience. He spent 8 years at the George Washington University as the Managing Director of technology commercialization, 3 years at the Johns Hopkins University as the Senior Director of ventures and marketing, and 4 years at the Utah State University Research Foundation as the Vice President for technology commercialization. For his contributions to commercializing academic technologies, he was recognized as a National Academy of Inventors Fellow. Dr. Kubisen also has extensive startup and corporate experience, including president or CEO roles at Seguro Surgical and VEC Technology and director/manager at Alcoa Corporate Lab and GE Plastics. Currently, Steven is focused on management consulting and angel investing through InnoComm LLC, where he is the President and Founder.

Brian Coblitz, PhD, is currently the Executive Director of the George Washington University (GW) Technology Commercialization Office. Before that role, he managed the patenting and licensing of GW's life science inventions for 9 years. Dr. Coblitz is a registered US patent agent and an active angel investor in life science startup companies. Dr. Coblitz owned and operated a small business that provided instructional services, District Self-Defense LLC, for 10 years.

Dr. Reid Rubsamen received his undergraduate degree from the University of California majoring in biochemistry and computer science and received an MD and a master's degree in computer science from Stanford. At the beginning of his career, while he was Chief Resident at Massachusetts General Hospital, he was funded by the Harvard NIH Center Grant to study at MIT's Laboratory of Computer Science, leaving after 2 years to found Aradigm Corporation where he served on the board of directors and the IPO transaction team which took the company public in 1996. Dr. Rubsamen was a member of Medical

Anesthesia Consultants, the largest anesthesia group in the San Francisco Bay Area, and was on the transaction team that sold the practice to Sheridan Corporation in 2013. He also served as Medical Director of Surgical Services for John Muir Medical Center from 2004 to 2016. He received a master's degree in healthcare management from the Harvard School of Public Health in 2021 and is now a Senior Attending Physician in the Department of Anesthesia and Perioperative Medicine at University Hospitals, Cleveland Medical Center, and a Clinical Associate Professor at Case Western Reserve University School of Medicine. Dr. Rubsamen is a coauthor of multiple peer-reviewed journal articles, a named inventor on more than 70 issued US patents, and has more than 11,000 citations to his collective work.

Che Antonio Solla is an Anesthesiologist and Pain Medicine Specialist at the University of Tennessee Medical Center in Knoxville, TN. He completed an Anesthesiology Residency at Walter Reed National Military Medical Center and a Pain Medicine Fellowship at Weill Cornell Medical Center. Following his return to practice in Tennessee, Dr. Solla earned his Master of Business Administration degree from the University of Tennessee—Haslam College of Business. Dr. Solla holds several leadership positions including the medical directorship of the hospital's acute pain service, a faculty position as an assistant Professor, and an associate program director role in the hospital's pain medicine fellowship.

Kory Razaghi is a Seasoned Finance Executive and the President of Aptus Advisors, Inc. (http://www.aptuscorp.com). Kory offers executive leadership, financial due diligence, and valuation analysis to identify commercially viable technologies and execute funding strategies to secure debt and equity financings.

Aptus Advisors, Inc., was founded in 2006 in Orange County, CA, and offers executive management, operational turnaround, and corporate finance services. Aptus' clients include medical device and advanced materials technology companies and sector-focused venture and private equity firms. Typical Aptus clients have matured beyond the seed stage and are preparing to secure funding from professional investors.

Kory has received Master of Business Administration degree from the University of California, Irvine; an MS in biomedical sciences with emphasis in neurology; and a BS in chemistry.

Joshua Herskovic, MD, is the founder of Guidance Airway Solutions, LLC. He is an Anesthesiologist in the South Florida area. He received his BS in mechanical engineering from Case Western Reserve University and worked in the medical device arena before pursuing a medical degree from Wayne State University. He completed a residency in anesthesiology at the University of Illinois at Chicago with top national board scores. Guidance Airway Solutions, LLC is a joint venture between Dr. Herskovic and Harter Investment Strategies. Guidance is committed to bringing to market novel devices related to airway management, including the **First**Look LTA and the Guidance Multi-Axis Endoscope.

Trent Emerick, MD, MBA, serves as an Associate Professor in the Departments of Anesthesiology and Perioperative Medicine and Bioengineering at the University of Pittsburgh Medical Center. He serves as the Fellowship Director for the Chronic Pain Fellowship and Medical Director/Associate Chief of the Chronic Pain Division. He is board certified in anesthesiology, chronic pain medicine, and addiction medicine. He has a specific interest in substance use disorders, craniofacial pain, complex abdominal pain, nerve stimulation, entrepreneurship and innovation, and start-up formation.

Bernadette Henrichs, PhD, CRNA, CCRN, CHSE, FAANA, FAAN, is a Professor and Director of CRNA Education and Research at the Department of Anesthesiology in Washington University in St. Louis, Missouri, and a Professor and Director of the Nurse Anesthesia Program at Barnes-Jewish College-Goldfarb School of Nursing.

Robert P. Walsh, PhD, MBA, MS, CRNA, is a staff CRNA at HCA Fawcett Florida in Port Charlotte, Florida. Rob previously worked at a small community hospital in Missouri where he served as the Director and Chief CRNA. Rob also taught the nurse anesthesia students at Barnes-Jewish College during this time.

Dr. Henrichs and Dr. Walsh both received their PhD in educational studies from St. Louis University. Both have a history of teaching nurse anesthesia students, both clinically and in the classroom. They have written articles together regarding MRI patient safety and other anesthesia-related topics. Dr. Walsh and Dr. Henrichs have been professional colleagues for over 20 years.

Abby V. Winterberg, APRN, DNP, is a nurse practitioner, clinical researcher, and intrapreneur at Cincinnati Children's Hospital and an Affiliate Assistant Professor at the University of Cincinnati. She serves as a cochair for the American Nurses Association Innovation Advisory Committee for Technology & Devices. Clinically, Abby works as an APRN in the preanesthesia consult clinic. Her research and innovation work is focused on preoperative anxiety reduction. Abby has received multiple innovation awards, including the first Johnson and Johnson Nurses Innovate Quickfire Challenge. These awards supported the development of a novel breathing-controlled video game application for anxiety reduction prior to surgery. In partnership with Cincinnati Children's commercialization department, Abby has led all phases of product development from design to partnering with a company for commercialization. Her innovation work has led to creating a new program at Cincinnati Children's Hospital to foster and support clinical innovation.

James Lee Hill, MD, MBA, CPE, FASA, FACHE, is a distinguished Senior Healthcare Executive known for his strategic vision and operational excellence. Currently he is the Chief Operating Officer at University Hospitals Parma Medical Center, Dr. Hill has spearheaded the development of new clinical programs, resulting in notable bottom-line improvements, despite the challenges of the COVID-19 pandemic. His leadership has earned recognition by Leapfrog, CMS, and Becker's Healthcare. With a background spanning role such as Chief Medical Officer and Division Chief of Trauma Anesthesiology, Dr. Hill has consistently championed patient safety and healthcare innovation. He is also a cofounder of Hemaptics, LLC, focusing on healthcare IT solutions. Dr. Hill holds Fellowships in the American College of Healthcare Executives (FACHE) and the American Society of Anesthesiologists (FASA), and is a Certified Physician Executive (CPE). He actively contributes to academia, research, and national speaking engagements, embodying a commitment to excellence in healthcare.

Dr. Vittorio Mottini is an esteemed researcher and a PhD candidate at Michigan State University, specializing in biomedical engineering within the Institute for Quantitative Health Science and Engineering. With a rich foundation in materials science and nanotechnology from Politecnico di Milano, Dr. Mottini has seamlessly transitioned to profound biomedical applications, focusing on the development of innovative wearable technologies and advanced manufacturing techniques such as two-photon polymerization (2PP) 3D printing and bioprinting.

Dr. Mottini's research is driven by a passion to enhance human well-being and health, as evidenced by his leadership in projects that span from micro-robotics to fungal bioelectronics. Notable for his contributions to the field, his work includes significant publications in Nature and is actively involved in the ongoing development of soft bioelectronics and precision neuroengineering tools.

Being an advocate for knowledge sharing and mentorship, Dr. Mottini cofounded the Biomedical Engineering Graduate Student Association at Michigan State and has mentored over 10 students, fostering a new generation of engineering talent. His career is a testament to his dedication to improving quality of life through cutting-edge science and technology.

David N. Flynn, MD, MBA is a board-certified anesthesiologist practicing at East Carolina Anesthesia Associates, PLLC, in Raleigh, NC. Previously, he was an Associate Professor of Anesthesiology at the University of North Carolina in Chapel Hill. Dr. Flynn has an interest in finance and entrepreneurship and has served as an advisor, investor, and founding member of multiple healthcare startups. Dr. Flynn completed medical school and anesthesia residency at the University of Pennsylvania. Additionally, he completed an MBA in healthcare management at the Wharton School, where he was selected as a Ford Fellow to conduct research in innovation management. Prior to medical school, he was a Fulbright Scholar at the Pasteur Institute in Paris, France.

Dr. Anna M. Varughese, MBBS, MD, FRCA, MPH, FAAP, is a pediatric anesthesiologist and serves as a Professor, Johns Hopkins Medicine, Associate Chief Quality and Safety Officer and Perioperative Director at Johns Hopkins All Children's Hospital, Johns Hopkins SOM in St. Petersburg, Florida. Dr. Varughese is recognized nationally and internationally for her clinical excellence and commitment to improving patient care through quality improvement and safety programs. Her pioneering work in action-oriented health-services improvement science research has been instrumental in advancing the field of pediatric anesthesia and perioperative care. Dr. Varughese's extensive training and experience have equipped her with a diverse skill set combining pediatrics, anesthesiology, perioperative medicine, and public health. Her academic achievements, including a master's degree in public health from the prestigious Harvard School of Public Health, highlight her commitment to scholarly activities and her dedication to helping anesthesiologists and other clinicians enhance and improve their practices. She is also currently the President of "Wake Up Safe," an internationally known Patient Safety Organization that advances the safety of children undergoing anesthesia and perioperative care.

Aniruddha P. Puntambekar, PhD, MBA, Acceleration Manager, Medical Devices & Diagnostics, brings specialized expertise in early-stage technology commercialization. He previously led a startup based on a microfluidic platform technology developed at the University of Cincinnati and led the technology to successful commercialization and acquisition of the startup. He played a key role in commercialization strategy development including market analysis, product positioning, competitive analysis, and financial analysis that led to over $11 million in equity funding and $3.5 million in nondilutive SBIR funding. This allows him to provide an insight into market-driven innovation and selecting early-stage technologies based on defined user needs and advancing them to value inflections of relevance to investors and prospective acquirers.

Ioana Pasca, MD, is an Associate Professor of Anesthesiology, Critical Care, and Neurocritical Care at Loma Linda University and Associate Program Director at Riverside University Health System where she is the Director of Neuroanesthesia. She is also an Adjunct Associate Professor of Internal Medicine at the University of California Riverside and an Associate Professor of Anesthesiology Diplomate at the University of California San Diego. Dr. Pasca serves in leadership roles within the American Society of Anesthesiologists and California Society of Anesthesiologists. She is passionate about her growing residency and her three children. Her academic work focuses on obstetrics, neurocritical care, and medical education.

Ashish Sinha, MD, PhD, MBA, MSEd, is a Professor at Loma Linda University at Riverside University Health System and University of California Riverside. His clinical focus is anesthesia with focus on care for the patient with obesity. One of the clinical challenges that he tackled was the introduction of a nasogastric tube. He created and patented a nasoenteral introducer to decrease discomfort, bleeding, and infection. As a prolific publisher, with over 150 pieces authored or coauthored, and three textbooks edited, he continues on the path of innovation in academic anesthesiology. An international lecturer, with over 700 lectures delivered all over the world, he is a well-rounded academician. He most enjoys mentoring and teaching of junior colleagues, residents, and medical students.

Zhuo Sun, MD, is a board-certified Anesthesiologist and chronic pain management physician at the Medical College of Georgia, Augusta University. As a physician, he not only finds fulfillment in clinical practice but also harbors a keen interest in conducting research to deepen our understanding of the effects of perioperative hemodynamic changes on postoperative outcomes. Additionally, he has spearheaded several clinical studies aimed at enhancing care for chronic pain patients undergoing nerve modulation treatment. Furthermore, he actively engages in managing complex regional pain syndrome (CRPS), a severely disabling condition that typically affects the extremities following injury or a medical event, such as surgery, trauma, stroke, or heart attack, and has contributed to published chapters on related topics.

Anterpreet Dua, MD, is a distinguished physician specializing in anesthesiology and chronic pain management, holding the position of Associate Professor at the Medical College of Georgia, Augusta. As the Program Director of the Chronic Pain Fellowship Program, he is instrumental in mentoring and shaping emerging experts in pain medicine. He actively engages in groundbreaking clinical research focusing on neuromodulation and functional restoration. His innovative approach to patient care combines his dedication to advancing medical knowledge through clinical trials. Additionally, he has contributed a chapter on geriatric anesthesia, further showcasing his expertise in specialized areas of anesthesiology. He has a passion for music outside his medical profession, reflecting his multifaceted interests and creativity.

Jennifer Dawson, MBA, MSN, RN, NE-BC, is a seasoned professional serving as a Senior Operations Engineer, Lean Six Sigma Black Belt, and High Reliability Medicine Strategist at University Hospitals Health System in Cleveland, Ohio. With two decades of clinical, change management, and program development expertise, she integrates evidence-based practice and lean principles to optimize healthcare resource utilization. Dawson spearheaded the creation of a groundbreaking solution to optimize transfusion practices, mitigate risks, and enhance patient outcomes. Widely implemented across the hospital system, this innovative software reduced transfusion rates while fostering heightened practice awareness. As the Director of Product Development at Hemaptics, a spinoff of UHHS, she continues to pioneer process improvements, overseeing the delivery of the software solution, HemaLogiX, to healthcare institutions nationwide. Her commitment to innovation and efficiency underscores her pivotal role in advancing healthcare excellence.

Dr. Raina Khan is an Associate Professor in the Department of Anesthesia & Preoperative Care at the University of California San Francisco. In addition to her practice as an anesthesiologist, her research focus include non-operating room anesthesia (NORA) safety and quality improvement. In particular, Dr. Khan is interested in the study of NORA adverse events, NORA provider perceptions of safety, and systems-based practice changes in NORA. Additionally, Dr. Khan's clinical interests include neuroanesthesia and critical care anesthesia.

Ross Goodman is currently a Senior Engineering Product Leader and has spent over 35 years leading teams that invent, launch, and produce innovative new global health care businesses. He has spent half of his career at Codonics leading global launches of new medical devices such as the Safe Label System which is the standard of care in over 1000 worldwide hospitals to help prevent the three main medication errors in the operating room, infinity data storage system, integrity data importing system, and Virtua medical disc image creation device. Previously, he worked for Procter & Gamble for 19 years leading cross-functional technical teams and running operations that delivered new brands such as Crest Whitestrips, Crest Mouth Rinse, and Crest Dual Phase Toothpaste leading to Crest achieving global Oral Care Brand market leadership. His educational background includes a BS in mechanical and industrial engineering from Clarkson University.

Cristy Berg, CTSM, serves as the Vice President of Communications at Codonics, where she has dedicated 25 years to developing and overseeing the company's global trade show marketing initiatives. As a key member of the leadership team, Cristy played an integral role in shaping the brand identity for Codonics' Safe Label System product line, contributing to its success in the market. With a yearly portfolio spanning more than 30 global exhibits, Cristy demonstrates her expertise at managing large-scale projects and ensuring their success. Her passion for storytelling and adeptness in digital and print marketing have been instrumental in driving targeted traffic and increasing ROI for Codonics. Cristy's commitment to creating compelling content has enabled Codonics to effectively connect with the healthcare industry and communicate its mission and value. She holds a cum laude degree from the University of Akron in Mass Media/Communications and is a Certified Trade Show Marketer (CTSM), reflecting her dedication to professional excellence. Throughout her tenure, Cristy's contributions have directly impacted Codonics' award-winning recognition for best-in-class innovation and product excellence.

Umme Rumana, MD, DO, MBI, is a physician specializing in medicine, surgery, anesthesia, and biomedical informatics. She is now undergoing anesthesia training at the University of Central Florida College of Medicine in Orlando, Florida. Rumana holds a doctor of osteopathic medicine from the New York Institute of Technology College of Osteopathic Medicine and a master's degree in biomedical informatics, research track, from the University of TX Health Science Center: School of Biomedical Informatics. Rumana has a strong academic background and has experience in developing and testing software solutions for electronic health records as well as conducting big data analytics. Umme Rumana's expertise in healthcare and commitment to advancing the area of anesthesia have significantly shaped her enthusiasm for anesthetic innovations.

Dr. Ettore Crimi is a Professor of Anesthesiology at Wake Forest University. As a cardiac anesthesiologist and intensivist, he is deeply engaged in education and research, particularly focusing on translational studies in endothelial dysfunction and epigenetic changes in critical care diseases. He received training at Massachusetts General Hospital, Stanford University, and the University of Florida, where he completed his Anesthesiology Residency, Cardiac Anesthesia Fellowship, and Critical Care Fellowship, respectively. Dr. Crimi also pursued an Executive MBA from the University of Florida. Through his versatile contributions to academia and medicine, Dr. Crimi remains dedicated to advancing patient care and scientific understanding within the realms of anesthesiology and critical care.

Gary Keefe is the Vice President of New Product Development at Codonics, Inc., and has been with the company for 32 years.

In that role, he helps to identify and define product ideas that align well with the corporate mission and strengths. As Codonics is a company with strong engineering roots and a tight knit senior management, guiding products from idea to proof of concept and finally to a signed off business plan requires in depth knowledge of the business, stakeholders, and product concepts. His journey to this role included 15 years as the Director of Software Engineering at Codonics and many years of domestic and international travel to hospitals, OEMs, and vendors to gain market perspectives and specific business skills that would help shape the products that Codonics eventually developed. Once a product is released, the need for ongoing refinement and adaptation as market needs evolve is the final role he performs to ensure long-term product success.

Zuhayr Shaikh, BS, is an MD/MBA student at the UVA School of Medicine & Darden School of Business in Charlottesville, Virginia, with a certificate in venture creation from VCU 's Da Vinci Center for Innovation from his undergraduate years. He has numerous experience in biomedical innovation from academic, corporate, and regulatory angles and notably led design efforts for VCU's N95 respirator decontamination setup during March 2020. Currently, he is the Founder and Head of Innovations in Access to Care, a nonprofit aimed at promoting healthcare innovation among students to solve health equity issues, and he also leads his peers as both club founder and President of the School of Medicine's AR/VR Club, AI&ML Club, Robotic Surgery Interest Group, and 3D Printing Club, as well as the Generative AI club at the Darden School of Business. He will graduate in May 2025 with internship experience at Pfizer & the FDA.

Michael Salomon, MD, is a Resident Anesthesiologist at the Cleveland Clinic Foundation. He serves on the Resident Wellness Committee for Graduate Medical Education at the Cleveland Clinic. He graduated from medical school at the University of Virginia and was a member of AOA. He is interested in medical device innovation, perioperative safety and quality improvement, perioperative artificial intelligence, physician education, and physician wellness. He serves as a member of the board for Innovations in Access to Care, a nonprofit that he helped cofound in medical school. His undergraduate education was completed at the College of William and Mary where he graduated with a BS in Biochemistry.

1

Introduction—overview and innovation process flow diagram

Michael Salomon[1] and Zuhayr Shaikh[2]

[1]Department of Anesthesiology, Cleveland Clinic Foundation, Cleveland, OH, United States [2]School of Medicine, University of Virginia, Charlottesville, VA, United States

Chapter outline

Abstract

Technology innovation and development is vital to the progress of the field of anesthesiology. A fundamental understanding of basic biodesign principles is paramount to idea development. These principles can be divided into the identify, invent, and implement phase. This chapter explores these concepts.

Keywords: Stakeholders; perioperative care; Stanford biodesign; identify phase; invent phase; implement phase

- Why it Matters?
- A Tour of This Book
- Getting Started
- Pitfalls

Key points

- Anesthesiology is an exciting field of medicine ripe for innovation with the potential to positively impact care for patients across many surgical and nonsurgical disciplines

Innovation in Anesthesiology. DOI: https://doi.org/10.1016/B978-0-12-818381-6.00009-7

- This book takes inspiration from Stanford's famous and well-regarded Biodesign process, and book chapters can be compartmentalized into three phases: identify, invent, and implement.
- Although this book can be read front-to-back, it is also written to double as a reference handbook for each phase of healthcare innovation.

Why it matters

The healthcare industry is a global, dynamic, and complex market composed of private, corporate, and governmental **stakeholders** that influence the physical and financial health and well-being of virtually every person on the planet. The WHO estimates that 10% of global gross domestic product (GDP) is spent on healthcare [1]. A staggering US $8.3 trillion. Operative surgical care, including **perioperative care**, accounts for 40% of total hospital expenses and generates 70% of hospital revenue [2]. In the recent past, we have seen an increase in the number of patients consenting for procedures in the operating room, inpatient, and outpatient settings, demand for intensive care increases, and spending on pain management exceeds $293 B annually [3–5]. With an aging population and higher chronic disease burden, these numbers will likely continue to increase in the future. As a result, greater stakeholder attention and capital will be brought to anesthesiology innovations that continue to improve access and quality of care.

It is these same pressures that helped us advance from Clover's nitrous oxide and ether apparatus (Fig. 1.1) to the modern-day anesthesia machine (Fig. 1.2).

The widely diversified global anesthesia device market size has considerably grown since the field of anesthesia's inception and is predicted to reach approximately $30 billion by 2030 [4]. With billions of lives and trillions of dollars at stake, a complex mire of local, federal, and international regulations have been created to govern every level of healthcare research, development, and business. To navigate this space efficiently and productively, a multidisciplinary team of medical, engineering, and business experts have authored this step-by-step handbook for Innovation in Anesthesiology. This comprehensive resource is written for clinicians, students, researchers, and entrepreneurs to reflect recent trends in industry globalization and value-conscious healthcare. Follow along in this introductory chapter for a tour of thoughtfully chosen book contents that we

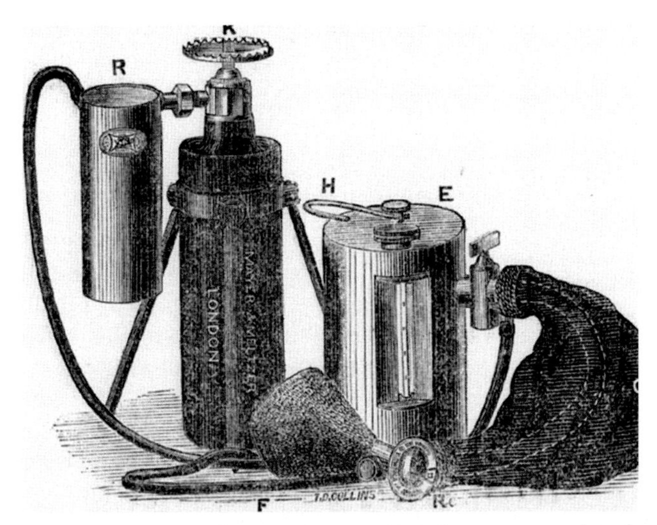

Figure 1.1 Anesthesiology Innovations Then: Clover's nitrous oxide/ether apparatus (1876). Credit: Manufacturer's catalog, p. 325. Public Domain. Credit: Wellcome Collection. CC BY.

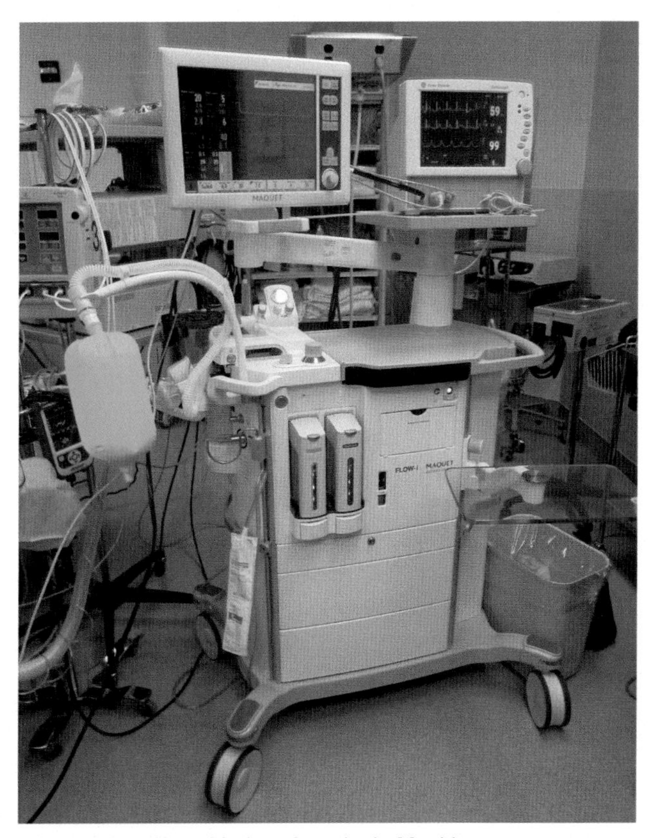

Figure 1.2 Anesthesiology Innovations Now: Modern Anesthetic Machine. Credit: https://commons.wikimedia.org/wiki/File:Maquet_Flow-I_anesthesia_machine.jpg.

hope will empower any aspiring anesthesiology innovator to build into existence impactful solutions that will transform patient care for the better.

A tour of this book innovation in anesthesiology

Technology Development and Commercialization Handbook is a step-by-step guide from identifying a clinical problem to planning a business. Structurally, you'll find a bulleted chapter outline, key point/summary, and introductory "why it matters" section at the opening of each chapter, subsequently followed by chapter contents and closing with actionable "get started" and "potential pitfalls" and reference "Examples," "Resources," and "References" sections. To better understand how this book is thematically organized, it is best to reflect on the **Stanford Biodesign Process** (Fig. 1.3) [6]. The biodesign process is composed of three major phases, each composed of two steps, that outline a systematic method for discovering unmet clinical needs, bringing to live valuable solutions, and creating freestanding commercialized ventures around them [5].

The **Identify phase** focuses on understanding a system, identifying a problem, and understanding that problem in the scientific and financial context of healthcare. Anesthesia has a large but specific scope of practice as shown in Fig. 1.4.

While it is not necessary to be an expert in every aspect of anesthesiology to identify a problem, it is critical to approach the problem objectively, systematically, and comprehensively. Chapters 3–7 explore how to identify and define a problem as well as how to understand the opportunities, markets, **stakeholders**, and barriers to creating a solution. Following the guide provided will help ensure a sound foundation for any project.

The **Invent phase** explores how to create a viable solution and demonstrate its feasibility. Chapters 8–18 provide details

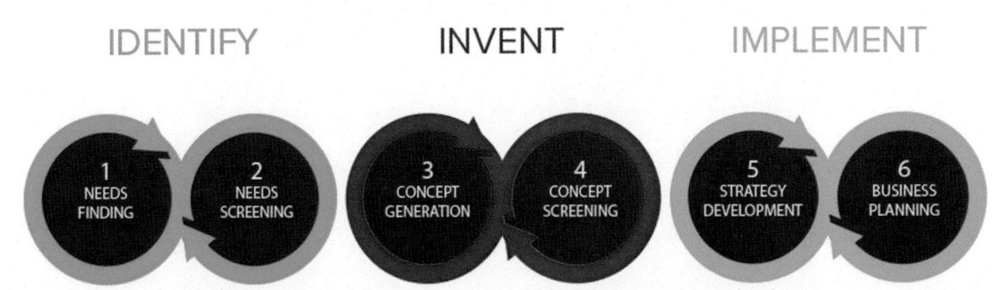

Figure 1.3 The Stanford Biodesign Process [6]. Credit: Reproduced with permission from Biodesign: The Process of Innovating Medical Technologies.

ABA Definition of Anesthesia as a Specialty:

1. Assessment of, consultation for, and preparation of patients for anesthesia.
2. Relief and prevention of pain during and following surgical, obstetric, therapeutic, and diagnostic procedures.
3. Monitoring and maintenance of normal physiology during the perioperative period.
4. Management of critically ill patients including those receiving their care in an intensive care unit.
5. Diagnosis and treatment of acute, chronic, and cancer-related pain.
6. Management of hospice and palliative care.
7. Clinical management and teaching of cardiac, pulmonary, and neurologic resuscitation.
8. Evaluation of respiratory function and application of respiratory therapy.
9. Conduct of clinical, translational, and basic science research.
10. Supervision, instruction, and evaluation of performance of both medical and allied health personnel involved in perioperative or periprocedural care, hospice and palliative care, critical care, and pain management.
11. Administrative involvement in health care facilities and organizations, and medical schools.

Figure 1.4 The scope of anesthesia. Credit: American Board of Anesthesiology Booklet of Information, 2013.

for generating successful solutions by identifying solution requirements, understanding shortcomings, and creating prototypes for proof of concept. Like the Identify phase, this is a dynamic process that unfolds over time. Insight generation may occur in a cyclical nature and failure is expected and encouraged to reach an optimal solution.

The **Implement phase** involves strategic solution development and business modeling. Chapters 19–30 provide walkthroughs for how to build and fund a company, how to navigate regulatory bodies, and how to launch a successful business. Chapters 31–37 cover many of the later-stage essentials necessary to maintain and perpetuate successful healthcare ventures. The team of medical, business, and engineering experts has carefully crafted detailed and digestible explanations of these business-related topics. As in other chapters, case studies and explicitly defined action items enhance reader understanding and success.

Get started

- Take some time to familiarize yourself with book contents, noting what parts you feel are of particular interest.
- If you already have an idea for an anesthesia innovation, consider checking out the intellectual property section for helpful tips and tricks to properly safeguard your idea.
- Take notes!

Potential pitfalls

- Only read this book front-to-back. It is not necessary to read this guide in its entirety before beginning or continuing a project.
- Giving up! Failure is an inherent necessity of the process and presents valuable learning opportunities.
- Attempting to do everything on your own—just like patient care, healthcare innovation is significantly improved by collaboration.

Resources

Biodesign: The Process of Innovating Medical Technologies Book by Paul Yock, Stefanos A. Zenios, and Todd J. Brinton

References

[1] Alhat A, Sumant O. Anesthesia devices market size: key analysis: forecast – 2030. Allied Market Research. 2022, January. Retrieved July 14, 2022, from https://www.alliedmarketresearch.com/anesthesia-devices-market.

[2] Kinderscher J., Rockford M. Chapter 99: Operating room management. Miller's anesthesia. 7th edition, 2010, page 3023.

[3] Lunsford, L. ICU statistics during COVID drove pandemic response. SOC Telemed. 2021, July 20. Retrieved July 15, 2022, from https://www.soctelemed.com/blog/covid-icu-statistics/

[4] Dobson GP. Trauma of major surgery: a global problem that is not going away. Int J Surg 2020;81:47–54.

[5] Gaskin DJ, Richard P. The economic costs of pain in the United States. J Pain 2012;13(8):715–24.

[6] Zenios SA, Makower J, Yock PG. Biodesign: the process of innovating medical technologies. Cambridge: Cambridge University Press; 2010.

2

Technology in anesthesia and perioperative medicine: exploring opportunities and challenges

D. John Doyle[1,2]

[1]Cleveland Clinic Lerner College of Medicine of Case Western Reserve University, Cleveland, OH, United States [2]Department of General Anesthesiology, Cleveland Clinic, Cleveland, OH, United States

Chapter outline

Abstract

There is substantial agreement that over the past several decades continuing advances in physiological monitoring and clinical practice protocol development have improved perioperative outcomes. Technical developments such as pharmacokinetically based drug delivery systems ("smart pumps"), electroencephalographic

Innovation in Anesthesiology. DOI: https://doi.org/10.1016/B978-0-12-818381-6.00006-1

depth of anesthesia monitoring, bedside ultrasonography systems, advanced neuromuscular monitoring devices, and other important advances have all contributed to patient safety, as have advances in "systems" thinking that has led to innovations such as perioperative care guidelines and safety checklists.

These encouraging facts notwithstanding, significant room for improvement remains, such as the need to improve equipment that suffers from poor software design or offers a confusing user interface. Additionally, publishing user reports on the usability of anesthesia equipment would be helpful to assist individuals seeking to purchase new anesthesia equipment.

Keywords: Anesthesia technology; artificial intelligence in medicine; clinical simulation; closed-loop anesthesia delivery; early warning score; patient safety; ultrasound scanners; Ventrain patient ventilation system

Introduction

Recent technological advances have led to notable enhancements in perioperative and critical care equipment, resulting in significant improvements in patient safety [1–4]. Developments such as pharmacokinetically based drug delivery systems [5], electroencephalographic depth of anesthesia monitoring [6], bedside ultrasonography systems [7], advanced neuromuscular monitoring devices [8], and related advances have all contributed to safer, easier delivery of patient care. Similarly, advances in "systems" thinking have led anesthesiologists worldwide to embrace perioperative care guidelines and standards that include the use of safety checklists[1] as well as the real-time perioperative monitoring of hemodynamics, oxygenation, ventilation, urine production, body temperature, and other physiological parameters[2]; these advances have also contributed to patient safety in important ways. As an example, in a study by Haynes et al. [9] the introduction of a surgical checklist reduced the surgical death rate from 1.5% to 0.8%, with the inpatient complication rate similarly dropping from 11% to 7%.

[1]A copy of the WHO Surgical Safety Checklist is available online at https://www.who.int/patientsafety/safesurgery/checklist/en.

[2]For information on perioperative monitoring equipment standards and other practice guidelines applicable to American physicians, visit https://www.asahq.org/standards-and-guidelines, where one can also find helpful guidelines for perioperative fasting, acute and chronic pain management, managing the patient with obstructive sleep apnea, and much more.

Another important development in "systems" thinking has been the development of clinical early warning algorithms, which have proven to be effective in detecting the early onset of clinical deterioration. For example, the Modified Early Warning Score uses systolic blood pressure, heart rate, respiratory rate, temperature, and neurological responsivity information to identify patients in need of urgent clinical review [10]. In a study with 334 patients, the authors found that the Modified Early Warning Score with a threshold of four or more as a trigger for action was 75% sensitive and 83% specific for predicting which patients required transfer to a higher level of care. Several other clinical early warning systems have also been described, such as pediatric systems and obstetrical early warning systems [11].

In the domain of perioperative cardiac monitoring, the use of conventional and 3D echocardiography based on advanced ultrasonography now allows for real-time monitoring of valvular function, ventricular filling, cardiac contractility, and other cardiac parameters. Similarly, hand-held ultrasound scanners (Fig. 2.1) are now available so that clinicians can augment the traditional patient physical exam with a focused ultrasonic examination; this is expected to facilitate perioperative patient management by providing information such as cardiac function and gastric volume and facilitate the early detection of adverse clinical situations.

In the realm of clinical airway management, advances such as airway management algorithms, video laryngoscopy, airway introducers, extubation catheters, and advanced supraglottic airway devices are improving patient safety [12–15]. The Ventrain system, which allows patients to be ventilated using ultra-narrow ventilation catheters, is one notable recent innovation [16,17] (Fig. 2.2).

Ergonomic design issues

Poor ergonomic design of medical equipment, producing a confusing and cumbersome user interface, can result in patient morbidity and mortality. For example, patient-controlled analgesia (PCA) machines are used extensively for postoperative pain and can be programmed to give an opioid analgesic every time the patient pushes the demand button. Unfortunately, such devices sometimes have confusing user interfaces. As an example, flaws in the design of the user interface for the Abbott Lifecare 4100 PCA Plus II machine have been deemed responsible for many deaths [18,19]. Here is an explanation of the design flaw [20]:

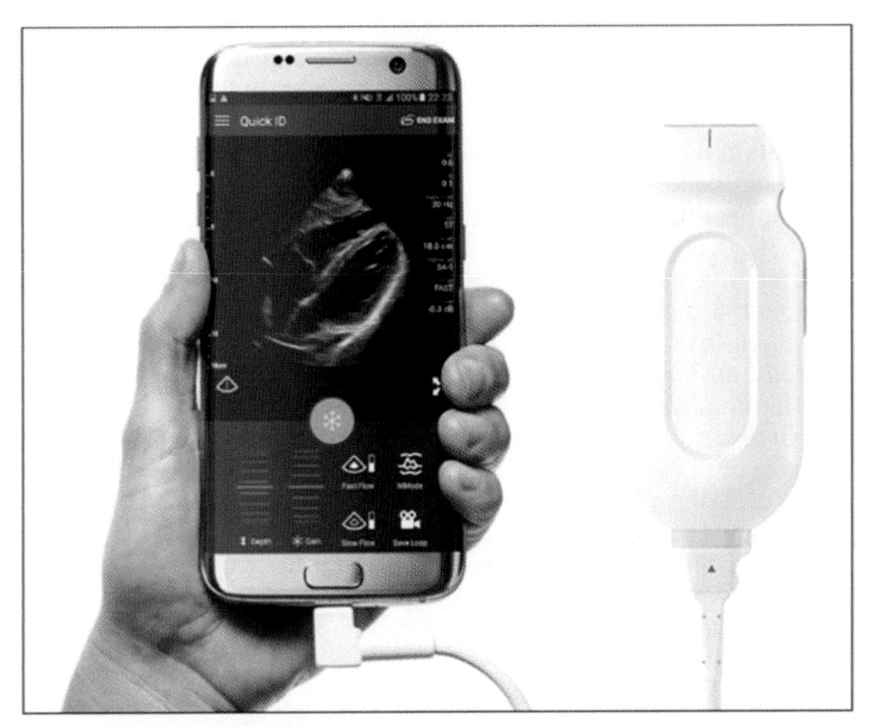

Figure 2.1 Ultrasound machines for applications such as echocardiography, regional anesthesia, or central line placement have now evolved to the point that they can connected to a smartphone or tablet. From Michard F. Smartphones and e-tablets in perioperative medicine. Korean J Anesthesiol. 2017;70(5):493–9. https://doi.org/10.4097/kjae.2017.70.5.493. PubMed PMID: 29046768; PubMed Central PMCID: PMC5645581. Image used under the terms of the Creative Commons Attribution Non-Commercial License, which permits unrestricted non-commercial use, distribution, and reproduction in any medium, provided the original work is properly cited.

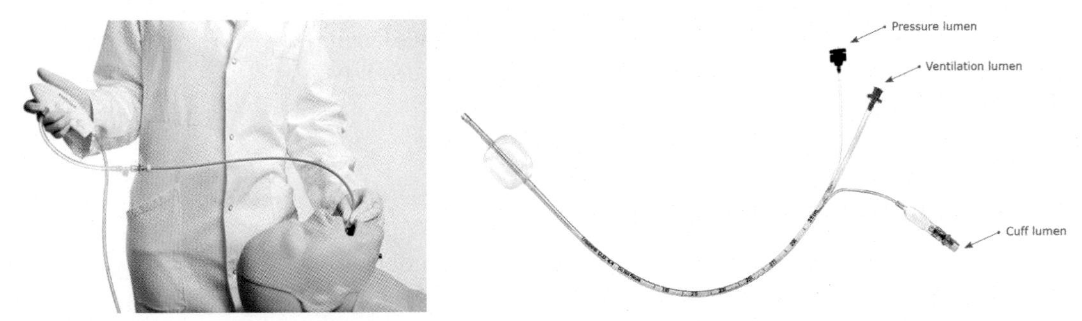

Figure 2.2 The Ventrain (left) is a simple-to-operate manual ventilator used in conjunction with small-bore airway catheters such as those used to exchange endotracheal tubes or those used to assist with tracheal extubation. The unit can also be used in conjunction with the Tritube (right), a narrow-bore cuffed endotracheal tube with inner and outer diameters of 2.4 mm and 4.4 mm, respectively. The Ventrain features active expiration based on the Bernoulli Principle. In addition to inspiratory (positive pressure) and expiratory (negative pressure) modes of operation, an equilibration (safety) mode is available where no significant positive or negative pressure is present at the tip of the attached ventilation catheter. A pressure measurement lumen in the Tritube permits continuous intratracheal pressure measurements. This device may be useful in both elective and emergency airway settings. Image and text used with permission of Ventinova Medical, Eindhoven, The Netherlands.

When nurses program the drug concentration the display shows a particular initial concentration (e.g., 0.1 mg/mL). Nurses can either accept this initially displayed value or modify it using the controls. The critical flaw in the design is that in this situation the Lifecare 4100 offers a minimal drug concentration as the initial choice. If nurses mistakenly accept the initially displayed minimal value (e.g., 0.1 mg/mL) instead of changing it to the correct (and higher) value (e.g., 2.0 mg/mL), the machine will "think" that the drug is less concentrated than it is. As a result, it will pump more liquid, and thus more opioids, into the patient than is desired. A safer design would have the unit offer a much higher drug concentration than the initial offering so that if nurses mistakenly accept this value, the result would be an underdose rather than an overdose.

Robotically assisted anesthesia

Advocates of robotically assisted clinical care expect that future technological developments will eventually help human practitioners improve rendered care by increasing their productivity, assisting with precision tasks, improving clinical attentiveness, and freeing up clinicians so they can focus on complex procedures and empathic bonding instead of mundane, repetitive tasks. Robotic assistance is also being explored for much more complex tasks, such as airway management for anesthesiologists [21–24].

The untested assumption of these advocates is that robotic clinical assistance will make clinical care delivery safer and more enjoyable, eliminating menial repetitive tasks that machines do so well, automating clinical documentation, and allowing clinicians to focus on the patient rather than equipment and paperwork. However, excessive reliance on technology, and especially on software, carries drawbacks such as the need for complex repairs when equipment fails and extremely impaired workflow in the event of outages. The notion that clinical technology will inevitably improve workflow and patient care has not always been borne out by experience.

Artificial intelligence in anesthesia and perioperative medicine

Artificial intelligence (AI) is the capability of a computer to respond intelligently to assigned tasks. Computer capabilities considered AI include competing at championship levels in games such as chess and Go, "understanding" human verbal requests ("Alexa, find me some music by The Beatles"), driving autonomous automobiles, or even recommending cancer treatment protocols[3].

While the so-called "expert systems" approach[4] remains an option for some AI platforms, as a practical matter most practical AI systems utilize the "machine learning" approach, and nearly all machine learning platforms are built on "neural network" technology. One key difference between these two AI platforms is that expert systems can, at least in principle, explain their reasoning (e.g., "a first-degree heart block is present because the PR interval on the ECG exceeds 200 ms"), while AI systems based on machine learning cannot explain their "thought processes" in terms meaningful to humans.

Some important anesthesia-related technical initiatives that draw on AI methods include attempts at "closed-loop" drug delivery for anesthesia, such as the automatic control of depth of anesthesia based on electroencephalographic monitoring, or the automatic delivery of muscle relaxants based on neuromuscular blockade monitoring. Other initiatives center around "smart alarm" systems that combine data sources to reduce the rate of false alarms, or offer some helpful diagnostic hints in addition to sounding an alarm (e.g., dual hypotension and bradycardia alarms might be a result of excessive beta-blocker administration or perhaps a spinal anesthetic that has traveled to an excessively high neuraxial level).

As another example of AI application in perioperative medicine, the Food and Drug Administration (FDA) has approved an AI algorithm from GE Healthcare that examines chest X-rays for a possible pneumothorax (https://www.genewsroom.com/press-releases/ge-healthcare-receives-fda-clearance-first-artificial-intelligence-algorithms). The algorithm was trained using a database of thousands of X-rays that were known to either have or not have a pneumothorax. This would be useful in the postoperative care of thoracic surgical procedures, among others.

[3]IBM's **Watson for Oncology** can explore a patient's electronic medical record to identify cancer treatment options and recommend therapy. For more information see https://www.ibm.com/us-en/marketplace/clinical-decision-support-oncology and https://www.theatlantic.com/magazine/archive/2013/03/the-robot-will-see-you-now/309216/.

[4]In an "expert system" AI platform, the knowledge of a subject matter expert is coded as a series of software rules (typically in the form of IF-THEN-ELSE statements), and these rules are used for decision making.

Anesthesiology, artificial intelligence, and the technological singularity

Many individuals ask whether developments in robotics and AI will result in clinicians losing their jobs as a result of a hypothetical future "technological singularity" where uncontrolled and irreversible technological growth renders them obsolete. In one variant of this hypothetical scenario, an "intelligence explosion" would result in an all-powerful intelligent machine network that would greatly surpass all human capability and thereby make human employment unnecessary. Discussing this matter in the context of the possibility that anesthesiology jobs might vanish, Atchabahian and Hemmerling [23] have noted:

> *While the broad consequences of such an event are unpredictable and beyond the topic of this editorial, this would make human anesthesia providers redundant; however, that would be true of most other sectors of human activity. Ultimately, we might lose our jobs, but so will everyone else.*

It appears implausible that developments in AI will eliminate the need for clinical anesthesiologists at any time soon. While the practice of anesthesiology is a mix of cognitive and mechanical skills, future AI developments will predominantly assist with the cognitive aspects of anesthesia practice—that is, aid in clinical decision-making—but not replace it, nor will they assist with most procedural tasks or the provision of kindness and empathy to patients.

Regulatory hurdles

Some well-established anesthesia technologies are not available in the United States because of insurmountable regulatory hurdles. As an example, one form of "smart" drug infusion pumps, termed "Target Controlled Infusion" (TCI) pumps, uses population-based pharmacokinetic modeling to provide an innovative form of intravenous anesthesia drug delivery. While this technology is widely available in Europe and elsewhere, and has proven to be both safe and reliable, it has not been approved by the FDA and has been deemed unlikely to gain approval under the current generation of FDA officials [25].[5] Similarly, the FDA is unlikely to offer approval

[5]The German physicist Max Planck is said to have quipped that "science advances one funeral at a time," although apparently his exact words actually were: "A new scientific truth does not triumph by convincing its opponents and making them see the light, but rather because its opponents eventually die, and a new generation grows up that is familiar with it."

for closed-loop anesthesia delivery systems at any time in the foreseeable future, despite numerous technical successes. On this matter, Atchabahian and Hemmerling [23] have noted:

> *In the light of the recent FDA approval of the rather controversial Sedasys system, a semi-automated propofol delivery system that will be used by non-anesthetic health care providers (gastroenterologists) controversial because, at this stage, a machine is unable to adjust the level of sedation to the anticipated discomfort of discrete parts of the procedure, or to the specific skill and speed of a given operator, but also because, if a patient gets overly sedated, as is possible even with an experienced practitioner, Sedasys will not lift the jaw or ventilate the patient using a face mask, and there is no reversal agent for propofol. One might question the validity of NOT making automated anesthesia delivery systems, tested in thousands of patients, available for use by anesthesiologists, who are experts in anesthesia delivery. Regulatory agencies need to reassess their attitude towards robotic or automated anesthesia systems.*

Remaining challenges

Despite the many advances described above, substantial challenges remain. One lies in the realm of alarms; audio alarms whose source remains completely mysterious despite a thorough search quickly lead to "alarm fatigue," whereby alarms genuinely indicating a problem no longer receive adequate attention from the clinician, and can also produce fixation errors where focus upon troubleshooting an unnecessarily sounding alarm results in clinician failure to attend to actual imperative tasks or changes in patient condition. Problems related to poor software or user interface designs can also result in patient harm and may be more likely when biomedical engineers are not fully aware of the nuances of the clinical environments for which they are designing products. It has been previously recommended for clinical equipment evaluation reports to be widely available, perhaps in a format similar to the popular Consumer Reports magazine for household products [26]. Structured evaluations of new equipment would allow for better decision-making for purchasers within medical facilities and also facilitate further medical device innovations (Tables 2.1 and 2.2).

Table 2.1 A list of some important advances in clinical practice development that have been particularly beneficial to patients undergoing anesthesia and surgery.

- Electronic medical record-keeping systems (EMRs)
- Early warning advisory systems to help detect clinical deterioration
- Surgical checklists (e.g., WHO Surgical Safety Checklist)
- Enhanced recovery after surgery (ERAS) patient management protocols
- Airway management algorithms (e.g., can't intubate, can't oxygenate)
- Perioperative monitoring standards and other practice guidelines
- Cognitive aids (e.g., management of intraoperative anaphylaxis)
- Tools to facilitate perioperative pain assessment (e.g., pain scores)
- Clinical simulation (e.g., PC-based and mannequin-based systems)

Early warning systems provide clinical teams with information to assist with the early recognition of clinical deterioration and trigger early clinical intervention. An example is the Modified Early Warning Score that charts systolic blood pressure, heart rate, respiratory rate, temperature, and neurological responsivity (alert, reacting only to pain, etc.) and where patients with a score over four were referred for urgent clinical evaluation and possible ICU admission. A copy of the WHO Surgical Safety Checklist is available online at https://www.who.int/patientsafety/safesurgery/checklist/en/. For information on perioperative monitoring equipment standards and other practice guidelines applicable to American physicians, visit https://www.asahq.org/standards-and-guidelines, where one can also find helpful guidelines for perioperative fasting, acute and chronic pain management, managing the patient with obstructive sleep apnea, and much more. A great collection of cognitive aids is available from the UK Association of Anaesthetists at https://anaesthetists.org/Home/Resources-publications/Safety-alerts/Anaesthesia-emergencies/Quick-Reference-Handbook. For an example of a modestly priced software-based anesthesia simulation trainer, visit https://anesoft.com/demos/anesth6/index.html.

Table 2.2 Some important advances in anesthesia technology have been of assistance to anesthetists caring for patients undergoing anesthesia and surgery.

- Smart alarms/intelligent alarms
- Pharmacokinetic-based anesthesia infusion pumps
- Advanced supraglottic airways (e.g., LMA Gastro)
- Video laryngoscopy (e.g., GlideScope Titanium)
- Video-capable endotracheal tubes containing an embedded camera
- Blood flow monitoring applied to goal-directed fluid therapy
- Depth of anesthesia monitoring (e.g., Bispectral Index, Patient State Index)
- Systaltic pressure variability monitoring applied to goal-directed fluid therapy
- Advanced neurological monitoring (e.g., motor evoked potentials)
- Perioperative echocardiography (transthoracic and transesophageal variants)
- Barcode readers to assist with patient and drug identification

Some examples: endotracheal tubes containing an embedded camera can allow the clinician to continuously view the position of the endotracheal tube relative to the carina; video laryngoscopy provides a superior view of the airway structures compared to conventional (direct) laryngoscopy, making intubation easier, especially in "difficult airway" patients; perioperative echocardiography, which allows clinicians to look for blood around the heart or elsewhere in trauma patients (known as "FAST," or Focused Assessment with Sonography in Trauma).

Conclusions

Advances in technology and clinical protocol development have improved outcomes for surgical patients. Bedside ultrasonographic imaging systems, "smart pumps," and countless other innovations have contributed to improved patient safety, as have advances in "systems" thinking leading to innovations such as safety checklists. Nevertheless, room for improvement always remains. Ongoing challenges include the need to improve software designs and confusing user interface designs in medical equipment. Finally, making summative reports on medical equipment usability widely available would be ideal.

Acknowledgments

Not applicable.

Declarations

Ethics approval and consent to participate

Not applicable.

Consent to publish

Not applicable.

Availability of data and materials

Not applicable (manuscript contains no original data).

Competing interests

The author declares that he has no academic or financial conflicts of interest with respect to this manuscript.

Funding

Not applicable

References

[1] Barrington MJ, Uda Y. Did ultrasound fulfill the promise of safety in regional anesthesia? Curr Opin Anaesthesiol 2018;31(5):649–55. Available from: https://doi.org/10.1097/ACO0.0000000000000638.

[2] Beck S, Reich C, Krause D, Ruhnke B, Daubmann A, Weimann J, et al. For beginners in anaesthesia, self-training with an audiovisual checklist improves safety during anaesthesia induction: a randomised, controlled two-centre study. Eur J Anaesthesiol 2018;35(7):527–33. Available from: https://doi.org/10.1097/EJA0.0000000000000781.
[3] Parotto M, Cooper R. Recent advances in laryngoscopy in adults F1000Res 2019;8:pii: F1000 Faculty Rev-797. Available from: https://doi.org/10.12688/f1000research.18544.1.
[4] Birnbach DJ, Bateman BT. Obstetric anesthesia: leading the way in patient safety. Obstet Gynecol Clin North Am 2019;46(2):329–37. Available from: https://doi.org/10.1016/j.ogc.2019.01.015.
[5] Dave S, Shriyan D, Gujjar P. Newer drug delivery systems in anesthesia. J Anaesthesiol Clin Pharmacol 2017;33(2):157–63. Available from: https://doi.org/10.4103/joacp.JOACP_63_16.
[6] Koch S, Spies C. Neuromonitoring in the elderly. Curr Opin Anaesthesiol 2019;32(1):101–7. Available from: https://doi.org/10.1097/ACO.0000000000000677.
[7] Bainbridge D, McConnell B, Royse C. A review of diagnostic accuracy and clinical impact from the focused use of perioperative ultrasound. Can J Anaesth 2018;65(4):371–80. Available from: https://doi.org/10.1007/s12630-018-1067-5.
[8] Hund HC, Rice MJ, Ehrenfeld J. An evaluation of the state of neuromuscular blockade monitoring devices. J Med Syst 2016;40(12):281.
[9] Haynes AB, Weiser TG, Berry WR, Lipsitz SR, Breizat AH, Dellinger EP, et al. Safe Surgery Saves Lives Study Group A surgical safety checklist to reduce morbidity and mortality in a global population. N Engl J Med 2009;360(5):491–9. Available from: https://doi.org/10.1056/NEJMsa0810119.
[10] Gardner-Thorpe J, Love N, Wrightson J, Walsh S, Keeling N. The value of Modified Early Warning Score (MEWS) in surgical in-patients: a prospective observational study. Ann R Coll Surg Engl 2006;88(6):571–5. Available from: https://doi.org/10.1308/003588406X130615.
[11] Doyle DJ. Clinical early warning scores: new clinical tools in evolution. Open Anesth J. 2018;12:26–33. Available from: https://doi.org/10.2174/2589645801812010026, https://benthamopen.com/FULLTEXT/TOATJ-12-26.
[12] McNarry AF, Patel A. The evolution of airway management – new concepts and conflicts with traditional practice. Br J Anaesth 2017;119(suppl_1):i154–66. Available from: https://doi.org/10.1093/bja/aex385.
[13] Gómez-Ríos MA, Gaitini L, Matter I, Somri M. Guidelines and algorithms for managing the difficult airway. Rev Esp Anestesiol Reanim 2018;65(1):41–8. Available from: https://doi.org/10.1016/j.redar.2017.07.009.
[14] Bjurström MF, Bodelsson M, Sturesson LW. The difficult airway trolley: a narrative review and practical guide. Anesthesiol Res Pract 2019;2019:6780254. Available from: https://doi.org/10.1155/2019/6780254.
[15] Edelman DA, Perkins EJ, Brewster DJ. Difficult airway management algorithms: a directed review. Anaesthesia. 2019;74(9):1175–85. Available from: https://doi.org/10.1111/anae.14779.
[16] de Wolf MW, Gottschall R, Preussler NP, Paxian M, Enk D. Emergency ventilation with the Ventrain(®) through an airway exchange catheter in a porcine model of complete upper airway obstruction. Can J Anaesth 2017;64(1):37–44. Available from: https://doi.org/10.1007/s12630-016-0760-5.
[17] Wahlen BM, Al-Thani H, El-Menyar A. Ventrain: from theory to practice. Bridging until re-tracheostomy. BMJ Case Rep 2017;2017:bcr-2017–220403. Available from: https://doi.org/10.1136/bcr-2017-220403.

[18] Lin L, Vicente KJ, Doyle DJ. Patient safety, potential adverse drug events, and medical device design: a human factors engineering approach. J Biomed Inf 2001;34(4):274–84.
[19] Doyle D. Ergonomics, patient safety, and engineering ethics: a case study and cautionary tale. J Long Term Eff Med Implant 2007;17(1):27–33.
[20] Doyle DJ, Vicente KJ. Patient-controlled analgesia. CMAJ. 2001;164 (5):620–1.
[21] Tighe PJ, Badiyan SJ, Luria I, Lampotang S, Parekattil S. Robot-assisted airway support: a simulated case. Anesth Analg 2010;111(4):929–31. Available from: https://doi.org/10.1213/ANE0.0b013e3181ef73ec.
[22] Hemmerling TM, Taddei R, Wehbe M, Zaouter C, Cyr S, Morse J. First robotic tracheal intubations in humans using the Kepler intubation system. Br J Anaesth 2012;108(6):1011–16. Available from: https://doi.org/10.1093/bja/aes034.
[23] Atchabahian A, Hemmerling TM. Robotic anesthesia: how is it going to change our practice? Anesth Pain Med 2014;4(1):e16468. Available from: https://doi.org/10.5812/aapm.16468.
[24] Wang X, Tao Y, Tao X, Chen J, Jin Y, Shan Z, et al. An original design of remote robot-assisted intubation system. Sci Rep 2018;8(1):13403. Available from: https://doi.org/10.1038/s41598-018-31607-y.
[25] Egan TD, Shafer SL. Target-controlled infusions for intravenous anesthetics: surfing USA not!. Anesthesiology. 2003;99(5):1039–41.
[26] Doyle DJ, Dahaba AA, LeManach Y. Advances in anesthesia technology are improving patient care, but many challenges remain. BMC Anesthesiol 2018;18(1):39. Available from: https://doi.org/10.1186/s12871-018-0504-x.

3

Specific opportunities for innovation in anesthesia

Bernadette Henrichs[1,2] and Robert P. Walsh[1,3]

[1]Nurse Anesthesia Program, Goldfarb School of Nursing, Barnes-Jewish College, St. Louis, MO, United States [2]CRNA Education and Research, Department of Anesthesiology, Washington University, St. Louis, MO, United States [3]CRNA, HCA Florida Fawcett Hospital, Port Charlotte, FL, United States

Chapter outline

Abstract

The healthcare industry, including anesthesiology, is constantly changing. Exciting new ways of caring for a patient are being brought into the operating room to improve healthcare and make it safer. From the days when pulse oximetry and capnography brought positive changes to anesthesia, technology has continued to grow and expand. New drugs are being discovered and telemedicine is being expanded to the operating room from the intensive care unit and outpatient clinics.

Current anesthesia practice still has limitations. Innovations could make vital sign monitoring wireless, laboratory results available immediately without drawing blood, and monitoring for ischemia of the brain and other organs much simpler. Although we live in exciting times where advanced technology is used for entertainment, that technology must also be

Innovation in Anesthesiology. DOI: https://doi.org/10.1016/B978-0-12-818381-6.00005-X

available in healthcare so that the safety and quality of patient care can continually be improved.

Keywords: CRNA; nurse anesthetist; anesthesia; nurse anesthesia; anesthesiology; technology in anesthesia; innovative trends; telemedicine; future trends in anesthesia

Trends in anesthesia care

Enhanced recovery after surgery (ERAS) protocols, a growing trend in hospitals and surgery centers, promote early recovery from surgery by aiming to preserve preoperative organ function and attenuate the dramatic stress response that surgery induces. From the anesthetic standpoint, pharmacological agents designed to allow the patient to awaken quickly with minimal residual anesthetic effect are warranted. Minimal use of opioids and attention to administration of agents to reduce postoperative nausea and vomiting (PONV) are mainstays. Regional anesthesia also plays a large role in ERAS, as it can desensitize the specific part of the body about to undergo surgery through the targeted administration of local anesthetics and, thus, decrease the stress response once surgery occurs. With regional anesthesia, anesthesia providers can decrease or minimize the use of general anesthetics during surgery, as well as decrease the amount of opioids required for pain control after surgery, all of which shorten recovery time. Recovery is impeded when opioids saddle the patient with side effects such as respiratory depression, postoperative nausea and vomiting, confusion, sedation, severe itching (pruritus), and constipation—in addition to their potential for addiction. Nonnarcotic analgesics such as intravenous (IV) nonsteroidal anti-inflammatory drugs (NSAIDs), acetaminophen, and continuous lidocaine and magnesium infusions during surgery have been integrated into the multimodal analgesic delivery model, with varying degrees of success toward the goal of perioperative opioid reduction.

Often, powerful muscle relaxant/paralytic drugs designed to prevent spontaneous or reflexive body movement are part of an anesthetic. These medications, such as rocuronium and vecuronium, provide surgeons with "optimal" surgical conditions, meaning. no patient body movement to complicate delicate maneuvers required by the surgeon during the operation. Unfortunately, these medications are slowly metabolized by the body. Quite often the powerful paralytic medication continues to be active in the body at the time of anesthesia emergence from surgery, delaying or even preventing emergence until its effects are reversed. A new paralytic reversal agent, sugammadex, has been approved by the FDA for use in rapidly inactivating the circulating paralytic agent more powerfully and better than the other + reversal agents available. This new reversal agent has positively impacted ERAS protocols by totally inhibiting the deleterious effects of circulating paralytic agents at the end of surgery.

Airway management is the core expertise of the anesthesia care provider. A multitude of devices have been developed to reliably, consistently, and effectively access and maintain a "difficult" airway. Obesity has become a worldwide epidemic and is often associated with a difficult airway. Direct laryngoscopy with the use of a handheld instrument allows direct visualization of the glottic opening for endotracheal intubation. However, this is old technology—similar to a surgeon looking into an "eyepiece" through a rigid cystoscope to diagnose bladder issues. New advancements in video technology have facilitated the trend toward video laryngoscopy. In addition to improving visualization for intubation, an advantage of video laryngoscopy is the ability to photographically "document" the difficult airway and allow other members of the anesthesia care team to see exactly what the primary care provider is viewing, to better assist with current and future tracheal intubations.

Ultrasound technology has provided the "launch pad" for the expansion of regional anesthesia techniques. Over the last 10 years, advances have provided practitioners with the means to accurately and consistently perform peripheral nerve blocks and field blocks that desensitize specific areas of the body to minimize pain.

The neurophysiological science underlying the anesthetic state is still somewhat of a mystery. We do not fully understand how anesthetic drugs create and maintain the state of anesthesia. A minute percentage of patients experience intraoperative awareness, which has been described as a memory of conversations or perceiving pain without the ability to respond during surgery. Several products (Bispectral Index, Entropy) have been developed to monitor the depth of the anesthetic state through the interpretation of EEG (electroencephalogram) signals in the brain. The anesthesia provider uses this information to appropriately titrate anesthetic drugs during the procedure. An additional benefit of this technology is to assess whether the anesthetic state is too deep at the time the procedure ends, which may result in a longer recovery time. Many facilities are adopting this technology to optimize the anesthetic experience and surgical recovery.

Issues and unmet needs in anesthesia care

Sedasys and gastroenterology procedures

When the Sedasys was approved by the FDA in 2013, it was doomed to fail. Sedasys was a computer-assisted personalized sedation system that was used by a handful of gastroenterologists in the United States. It was approved for moderate to light sedation for gastrointestinal procedures [1]. It was a closed-loop system that delivered propofol to the patient. The machine assessed the responsiveness of the patient by testing his or her ability to squeeze a handheld device in response to a stimulus. It also monitored the patient's vital signs (blood pressure, electrocardiogram/ECG, and pulse oximetry), and it used this data to titrate the propofol infusion. Some of the reasons for failure included: per the package insert, propofol should only be administered by people trained in cardiopulmonary resuscitation and airway management; patients preferred deep anesthesia rather than light to moderate sedation for gastrointestinal procedures; and anesthesia providers felt the company was trying, albeit unsuccessfully, to replace them with a machine. Eventually, Sedasys was pulled from the market in the United States in 2016 [1].

Wireless electrocardiogram monitoring

ECG monitoring of the heart, always conducted in the operating room, alerts anesthesia providers if the ST segments become elevated or depressed, signaling cardiac ischemia. Electrocardiogram monitoring is done by placing 3–12 leads on the patient. This can be troublesome when the leads get tangled and accidentally removed while the patient is moved onto the OR table or positioned for surgery. In addition, sometimes the leads need to be placed in unusual places on the chest to avoid lines traversing the surgical field. Wireless monitoring of the electrocardiogram would remove the burden of tangled, malpositioned, or displaced leads.

Noninvasive glucose monitoring

Frequent blood glucose monitoring is performed for diabetic patients under anesthesia. This requires the anesthesia provider to draw blood from the patient. Although the fingers are typically used for testing outside the operating room, sometimes they are not available during surgery if they are tucked and wrapped to prevent being in the surgeon's way of operating. The provider's options are then narrowed to accessing a toe or drawing blood from a vein or an existing intravenous or arterial line to check the glucose. Noninvasive blood glucose monitoring available for widespread perioperative use would be an excellent advancement.

Adjusting anesthesia medications based on genetic variations

Wouldn't it be helpful to obtain genetic information specific to an individual surgical patient that would enable anesthesiologists or CRNAs to tailor the anesthetic to the needs of that specific patient? The genetic makeup is very similar among humans with only about 0.1% variability in our DNA differentiating us from one another [2]. Still, these variations result in humans responding differently to medications in terms of sensitivity and specific drug response. Under an anesthetic, this could translate into a person requiring, for instance, a more volatile anesthetic gas to achieve full general anesthesia. If anesthesia providers knew the genetic variations of the surgical patient, then they could tailor the anesthetic to those variations and improve patient safety and patient outcomes. One example, according to a study by Liem et al. [3], is that patients with red hair need more of the inhaled anesthetic desflurane compared to females with dark hair [3]. It was found that those with red hair were either homozygous or compound heterozygous for mutations in the melanocortin-1 receptor genes, contributing to the sensitivity differences.

Up to 86% of patients experience moderate to severe pain after major surgery, according to a study by Aroke and Kittelsrud [4]. Genetic variations in patients can influence pain perception and the patient's response to pain medications. Genes that may affect postoperative pain management include the opioid ɥ1 receptor (OPRM1), cytochrome P450 (CYP) enzymes, catechol O-menthyl transferase (COMT) enzyme, and adenosine triphosphate-binding cascade (ABCB1) transporter. The effects may be related to the actual activity at the receptor site regarding analgesic efficacy and adverse effects of pain medications, or they may be related to metabolism variations of certain opioids. Nonsteroidal anti-inflammatory drugs can be metabolized differently based on genetic variants. Studies are needed to develop "genotype-guided therapeutic guidelines" for medications used for postoperative pain management [4].

Many questions remain unanswered in anesthesia. Why does propofol cause muscle twitching in some patients but not in others? Why do some patients need very little amounts of medication to cause sedation while others need far greater amounts? It would be helpful to know the genetic variations of each surgical patient, so the anesthetic could be tailored to their specific needs, improving patient safety.

Eliminating malignant hyperthermia

Malignant hyperthermia (MH), although very rare, is a life-threatening complication of anesthesia that can be deadly if not diagnosed and treated quickly and appropriately. When dantrolene, a medication used for the treatment of MH, was discovered in the 1970s, it was predicted that no one would ever die from this syndrome. Nonetheless, people continue to die from MH each year. What do we need to help us quickly recognize MH so treatment can be started immediately? Can machine learning and improved technology help anesthesia providers quickly and correctly diagnose MH so that treatment can be given, and survival will improve? Can death from MH be eliminated with the help of machine learning?

Machine learning and telemedicine

Each year, over 300 million people worldwide and over 40 million people in the United States undergo surgery [5,6]. Despite advances in the delivery of anesthesia and improved patient safety, the risk of morbidity and mortality in surgical patients persists. Approximately 4.2 million people worldwide die within 30 days after surgery each year [7]. This accounts for 7.7% of all deaths globally, making it "the third greatest contributor to deaths, after ischemic heart disease and stroke" [7]. One in every 10–20 patients having surgery will die in the following year [8–11], and 10%–20% will experience a major perioperative complication such as heart attack, chronic pain, infection, or blood clots following their surgical procedures [12–14]. Mortality is significantly higher in the elderly group, specifically those over the age of 80 where mortality is more than twice that for patients 65–69 years of age [15,16].

Telehealth involves monitoring from a distance and provides a "second pair of eyes" on any given situation. While this does not replace the vigilant activities of the anesthetist in the room providing one-on-one care of the perioperative patient, observation, and feedback from another anesthesia provider may enhance the safety aspect of the anesthetic experience by providing additional support via observation and treatment recommendations.

Can anesthesia become safer using telemedicine? Washington University Department of Anesthesiology is conducting a randomized controlled trial to investigate the utility of a telemedicine-based control center similar to telemedicine in intensive care units.

Those involved in the study are assessing the risk of surgery and anesthesia, negative patient trajectories, and implementation of evidence-based anesthesia practice from the Anesthesiology Control Tower (ACT). The single-center, randomized control trial aims to enroll 10,000 patients annually over 4 years. The study will be expanded in the near future when patients in the postoperative care area will also be included.

The primary objective of the trial is to determine whether the ACT can prevent clinically relevant adverse postoperative outcomes, including 30-day mortality, delirium, respiratory failure, and acute kidney injury. Secondary objectives are to determine whether the ACT improves the perioperative quality of care, including the management of temperature, mean arterial pressure, mean airway pressure with mechanical ventilation, blood glucose, anesthetic concentration, antibiotic redosing, and efficient fresh gas flow.

In the Anesthesiology Control Tower each day, there are anesthesia providers who monitor the operating rooms from computers located in the tower. The team consists of an anesthesiologist, a CRNA, a resident, a nurse anesthesia student, and several research physicians. The operating rooms are randomized. For those in the study group, the providers in the tower can reach out to the anesthesia provider in the operating room for any alerts so that they can be addressed. This is done through a phone call or a chat message through the electronic health record. Those in the control group are not contacted by the alerts unless the intervention unless it is a life-threatening situation. An example is this: Those in the tower are alerted that a specific patient has not received a preoperative antibiotic and the surgeon just made the incision (Appendices A). The team can then alert the anesthesia provider in the operating room and remind them to give the antibiotic and chart it in the patient's record.

Telehealth for physician office visits has become increasingly popular due to the COVID-19 pandemic. However, it has not been used in surgical patients. If the study shows positive effects on decreasing morbidity and mortality in surgical patients, then the use of telemedicine for surgical patients may become widespread.

Summary

Overall, surgery and anesthesia are safer today than in past eras due to continuing advances in science and technology. A continuous influx of pharmacologic and technical advances, as well as the advent of telemedicine, are all promising factors in continuing to improve perioperative safety and quality.

Appendix A

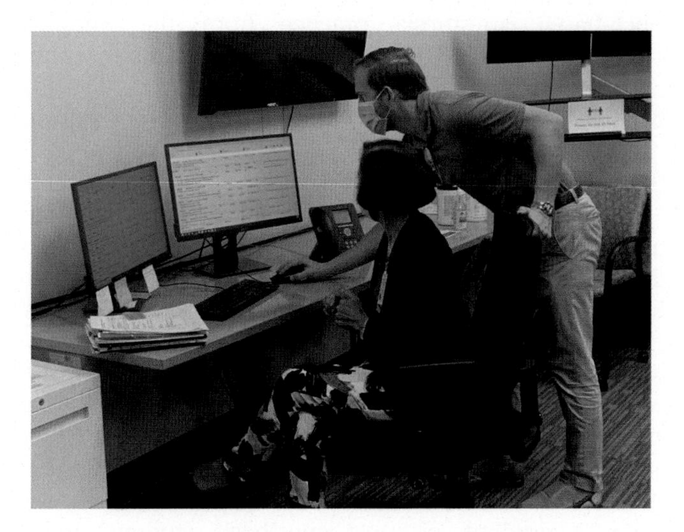

Bernadette Henrichs, CRNA, and Thaddeus Budelier, research physician, monitoring anesthetics in an Anesthesiology Control Tower.en.

References

[1] Goudra B, Singh PM. Failure of Sedasys: destiny or poor design? Anesth Analg 2017;124(2):686–8.

[2] Panditrao MM, Panditrao MM. Anaesthesia and genetics: still, the uncharted territories!! JACCOA 2016;4(1):16–19.

[3] Liem EB, Lin CM, Suleman MK, Doufas AG, Gregg RG, Veauthier JM, et al. Anesthetic requirement is increased in redheads. Anesthesiology 2004;101 (2):279–83.

[4] Aroke EN, Kittelsrud JM. Pharmacogenetics of postoperative pain management: a review. AANA J 2020;88(3):229–36.

[5] Fritz BA, Chen Y, Murray-Torres TM, Gregory S, Abdallah AB, Kronzer A, et al. Using machine learning techniques to develop forecasting algorithms for postoperative complications: protocol for a retrospective study. BMJ Open 2018;8:e020124. Available from: https://doi.org/10.1136/bmjopen-2017-020124.

[6] Gregory S, Murray-Torres TM, Fritz BA, Abdallah AB, Helsten DL, Wildes TS, et al. Study protocol for the Anesthesiology Control Tower-Feedback alerts to supplement treatments (ACTFAST-3) trial: a pilot randomized controlled trial in intraoperative telemedicine (version 2; peer review: 2 approved). F1000 Res 2018;7:623 Last updated: 17 May 2019.

[7] Nepogodiev D, Martin J, Biccard B, Makupe A, Bhangu ANational Institute for Health National Institute for Health Research Global Health Research Unit on Global Surgery. Global burden of postoperative death. Lancet 2019;393:401.

[8] Kertai MD, Palanca BJ, Pal N, et al. Bispectral index monitoring, duration of bispectral index below 45, patient risk factors, and intermediate-term mortality after noncardiac surgery in the B-Unaware trial. Anesthesiology 2011;114:545–56.
[9] Monk TG, Saini V, Weldon BC, Sigl JC. Anesthetic management and one-year mortality after noncardiac surgery. Anesth Analg 2005;100:4–10.
[10] Visser BC, Keegan H, Martin M, Wren SM. Death after colectomy: it's later than we think. Arch Surg 2009;144:1021–7.
[11] Kertai MD, Pal N, Palanca BJ, et al. Association of perioperative risk factors and cumulative duration of low bispectral index with intermediate-term mortality after cardiac surgery in the B-Unaware trial. Anesthesiology 2010;112:1116–27.
[12] Healey MA, Shackford SR, Osler TM, Rogers FB, Burns E. Complications in surgical patients. Arch Surg 2002;137:611–17 discussion 7–8.
[13] Turrentine FE, Wang H, Simpson VB, Jones RS. Surgical risk factors, morbidity, and mortality in elderly patients. J Am Coll Surg 2006;203:865–77.
[14] Hamel MB, Henderson WG, Khuri SF, Daley J. Surgical outcomes for patients aged 80 and older: morbidity and mortality from major noncardiac surgery. J Am Geriatr Soc 2005;53:424–9.
[15] Finlayson EV, Birkmeyer JD. Operative mortality with elective surgery in older adults. Effective Clin practice: ECP 2001;4:172–7.
[16] Howes TE, Cook TM, Corrigan LJ, Dalton SJ, Richards SK, Peden CF. Postoperative morbidity survey, mortality and length of stay following emergency laparotomy. Anaesthesia 2015;70:1020–7.

4

Validate, refine, and define the problem to be solved

Raina Khan
Department of Anesthesia and Perioperative Care, University of California, San Francisco, CA, United States

Chapter outline

Abstract

Now that you've identified an unmet need, you can expand your analysis into the problem and why it exists. While the end goal may provide value and/or an innovative solution, it may be wise to spend your initial time and energy not on the solution, but on the problem. This means confirming the specific problem and how it manifests in your clinical procedures.

Keywords: Validate; refine; define; idea validation; ideal refinement; MVP; minimum viable product

Now that you've identified an unmet need, you can expand your analysis into the problem and why it exists. While the end goal may provide value and/or an innovative solution, it may be wise to spend your initial time and energy not on the solution, but on the problem. This means confirming the specific problem and how it manifests in your clinical procedures.

Innovation in Anesthesiology. DOI: https://doi.org/10.1016/B978-0-12-818381-6.00033-4

Don't be surprised if the answer to this question evolves as more information emerges. In fact, after finishing your research, the problem you try to solve may end up being quite divergent from the initial problem identified. During idea validation, refinement, and defining the problem to be solved, the ultimate solution is validated and verified.

Idea validation

So you have identified an unmet need where a clinical solution would provide value. The purpose of idea validation is to frame your "problem" in the big picture outlining the benefits achieved.

Idea Validation is a multistep process, which doesn't have to be performed in a particular order but has necessary components that must be addressed [10,11]. In anesthesia there are many ways to put a patient to sleep—in product development, there are several ways to conduct idea validation. We've created a mnemonic to help get started.

VALUE—D A T E

V or Value—Does this problem have value?

D—Data and Design—What data is out there? What is your plan?

A—Assumptions—Where is your knowledge deficient?

T—Trials and Test—How will you trial your idea? What will you consider a successful test?

E—Explore and Enhance -Never stop exploring. Continue to search for ways to enhance your product.

Value

Value requires evaluation—Start by asking yourself the following questions:

-What is the problem? [2]

-Why does it exist?

-Is this a problem worth solving? [1]

To get started, you want to not only ask yourself "what is the problem?" but also create a working definition of it—a hypothesis [1,3].

For example, at your hospital, the laryngeal mask airways (LMA) used do not seat properly in many patients regardless of their airway anatomy and habitus. This often leads anesthesiologists to switch to an endotracheal tube (ETT) when intubation otherwise could have been avoided. This poor choice of LMA is part of the hospital's initiative to cut costs. You feel, however, that the extra time spent by the anesthesiologist to remove the LMA and place an ETT, with the associated potential morbidity such as airway edema or bleeding or hypoxia, warrants a closer look and a resolution.

Looking back at the three questions above, converting theory to fact will require researching question #2—why does this problem exist?

You suspect that these specific LMAs are older prototypes with lower quality material, where the LMA often doesn't hold additional air and is not very flexible. You require additional data to obtain verification.

Now, step #3, is this a problem worth solving?

We may think that wasting extra time trying out an LMA and failing is an annoying issue—and one that needs to be resolved. You know (or assume [4]) that the problem is because the hospital is selecting to purchase reduced quality and older model LMAs. If they would upgrade the equipment, it could be possible to reduce efforts performing unnecessary intubations—but is this truly a problem worth solving?

You should determine whether hospitals will pay for an LMA upgrade and whether a safety issue that warrants this update truly exists. Some safety events with LMAs may occur not due to poor LMA seal or fit, but because an ETT, a more secure airway, should have been elected over an LMA anyhow. Will newer, more enhanced features and sizing in LMAs solve the problem—or is it something intrinsic to all LMAs? Are we sure that more education regarding the LMAs currently in use will not fix the problem? Conversely, if the LMA was placed as the next step along the difficult airway algorithm in a cannot intubate/cannot ventilate situation, where hypoxic cardiac arrest is looming, then a high rate of first-placement LMA failures is certainly a serious safety concern. If an increased range of LMA sizes becomes available (such as half or quarter sizes), then would clinical outcomes be better, and how so? And how will you compare and obtain verified data that new and improved LMA is safer and/or providing better seating?

The point is to expand the analysis in a rigorous scientific fashion to determine what may be contributing to this "problem." Some ideas and solutions may initially appear positive but may not provide the desired outcome.

After answering all those questions and more, if you still believe that is a problem worth solving, create a working definition or hypothesis regarding the problem. This hypothesis will be your starting point for further investigation, which leads us to our next section—what data is available?

Data and design

Data

Ask what data [2] supports your working hypothesis. In the case above, this may include past research regarding LMA failure rates/incidences where a switch to ETT is required. Also this involves researching currently available LMAs and other competitive technology. Identifying such data enables you to start the process of *idea refinement* [5].

Design

After you are satisfied with your initial data gathering, you will want to start thinking about what your plan [5] or "design" [6] is. Is it to find a new supplier of LMAs with better quality (e.g., more durable plastic in cuff)? Switch to a different brand? Or disrupt the current market and create your own LMA—new and improved, fulfilling current needs for form, function, and ease of placement. The design part of validation often includes a more formalized and evolved hypothesis regarding the problem, to accurately address how to solve it.

Assumptions

So far we have discussed the potential problem. and why the data might outline this as a problem. If you've gotten this far, it's time to take a closer look at your hypothesis and review some of the critical assumptions [3]—"guesses" that, if wrong, could derail your goals. It is often recommended to start with the most critical assumptions first. For example, would people pay for a new and expensive, but theoretically easier to place, LMA? What kind of evidence or convincing would be required for most to make this change?

Remember, the process of idea validation is meant to expose an idea to reality [3] in the prospective environment. At this point in the process, a good next step is to start talking to people and collecting feedback. This can be done in any number of ways, including surveys [5], interviews, focus groups, and social media [12].

Examples of feedback [7] we would seek in our LMA scenario:

Do you find the current LMAs problematic? (why/why not)

What improvements do you think can be made to the current product?

Are there alternative LMAs you prefer? (why/why not)
If there was a product that improved upon the hiccups of the current models, is there a price point?

It may seem counterintuitive [5] to survey people about an idea you do think is revolutionary and profitable—you will want to try to protect it. In reality, however, until you survey the future market [4] (e.g., anesthesia providers at your hospital, administrators, buyers and suppliers, etc.), you will not be able to define the target customer [1], and their price sensitivity [3]. Interviews and surveys will help provide direct feedback and refine the market [1, 5, 12]. In addition to local users, you will want to expand your data sample. At the same time, it is prudent to contact vendors and seek their industry research [1, 5, 12]. Attending conferences or trade shows [12] featuring product exhibits may also provide an avenue to expand current industry research and product availability.

Trial and test

Trial

So the field research shows there is a demand for this potential product–How will you trial this new product? Often this involves creating a prototype—and the simplest version of your product that possesses its core features is known as an MVP, or *minimum viable product* [8]. Whether the prototype is physical or not, you will eventually need an MVP that you can test with real users.

In our LMA example, we would plan to design and create a physical prototype that has the core qualities we desire—e.g., easy to place (flexible but structured), cuffless (no need to inflate or deflate for a good seal), adequate sizes. Since this is just a prototype, perhaps we would provide just two basic adult sizes (4 and 5), etc.

Once the prototype [6, 9] has been created and we have all the appropriate and necessary approvals, we will plan to trial this product with a small group of users, seeking direct feedback and adjusting the design as necessary. This early trial will be our pilot study and help assess how the product does in the real world with real users.

Once the initial responses are obtained, scientific and objective investigation is required to determine what modifications may yield improvements. It is important to differentiate between an essential versus luxury recommendation; a "must have" improvement versus an "it would be nice to have" recommendation. Before adjusting the prototype (which may be expensive and time-consuming), it may be worthwhile to conduct a few more different users (e.g., trauma hospitals, community hospitals, and emergency departments) in different locations to expand the feedback. You may be able to use the company or hospitals where you've initially trialed the prototype as references for other trial sites, allowing you to continue to trial this prototype on a wider scale and gather more information.

Test

Now that your prototype has been created and enhanced, your next step is to determine what denotes a successful pilot study. It is important to first start by setting up specific success criteria [1]. Questions you will want to ask include:

Did this prototype solve the problem it sought to fix?

Did you reach the set target for user satisfaction?

Would testers purchase this product if it were available?

You can continue testing this prototype at the locations you have already set up for the initial trials, but again, it may serve well to find new places to trial it

Explore and enhance

Assuming that your prototype has had a successful testing period with favorable responses from potential customers, it is now time for another critical evaluation. It is important to continue exploring the "problem"—both refining and redefining what you think it is based on your product research, prototype feedback, and market analysis of the products currently available [5, 12].

The goal is to continue to enhance the product until it is ready for market, but also to see if this product will be scalable with an increase in demand. This leads us to our next chapter, Market Analysis.

References

[1] Idea Validation: Process, Tools, and Tips (2024) Spdload. Available from: https://spdload.com/blog/idea-validation/

[2] A Unique Way to Problem Identification and Validation Joshi. Available from: https://medium.com/startupsco/first-step-to-validate-your-business-idea-e24d357768ee.

[3] Kylliäinen J. Idea Validation: steps and tools for testing your idea. Viima Solutions Oy. Available from: https://www.viima.com/blog/idea-validation

[4] Problem Validation. How to Know if the World Really Needs Your Cool Idea. Rev ventures;1. Available from: https://www.rev1ventures.com/blog/problem-validation-world-cool-idea/.
[5] 7 Ways to Refine your Business Idea Huber. Available from: https://medium.com/swlh/7-ways-to-refine-your-business-idea-fc8bb7b74a4f.
[6] Bland DJ, et al. Testing Business Ideas. Hoboken, N.J: Wiley; 2020. Print.
[7] Holstein M. Idea validation. iPhone App Design for Entrepreneurs. Berkeley, CA: Apress; 2019. Available from: https://doi-org.ucsf.idm.oclc.org/10.1007/978-1-4842-4285-8_3.
[8] LeMay M. What Is Minimum Viable Product? 1st edition O'Reilly Media, Inc; 2018. Film.
[9] Seiden J, Jeff G. Minimum Viable Products and Prototypes. 1st edition O'Reilly Media, Inc; 2017. Print.
[10] Asmar L, et al. Structuring framework for early validation of product ideas. Int. J. Integr. Eng 2021;13(2).
[11] Ries E. The Lean Startup. Brussels, Belgium: Primento Digital,; 2012. Print.
[12] Brown, M. How to Refine Your Business Idea. (2010). inc.com.. Available from: https://www.inc.com/guides/2010/08/how-to-refine-your-business-idea.html

5

Market analysis—the anesthesia market landscape

Nancy Patterson and Lauren Rodriguez
Strategy Inc., strategyinc.net, Austin, TX, New York, NY, Research Triangle Park, NC, United States

Chapter outline

Abstract

The anesthesia market landscape covers important factors influencing the current and emerging anesthesia landscape in the operating room, ambulatory surgery centers, and other surgical locations. The analysis includes advances in technology innovation, market size and trends, procedural site of service progress, surgical volumes, reimbursement, the adoption process and the effect of population demographics. The insight will support entrepreneurs developing technology and investors in this expanding and profitable space. The global anesthesia market expected to reach US $401M by 2030 with expanding growth at a CAGR of 5.2% from 2021 to 2030.

Keywords: Cost-effectiveness analysis; due diligence; market analysis; market opportunity; market segment; market size; value analysis

Innovation in Anesthesiology. DOI: https://doi.org/10.1016/B978-0-12-818381-6.00016-4

Key takeaway points

- Executing a market analysis at the earliest stage in concept formulation is critical to deliver insight into the potential for successful commercialization and to determine the potential market.
- Market size for anesthesia technology is driven by the number of hospitals and ambulatory surgery center operating rooms and the overall number of procedures and less by the incidence and prevalence of specific diseases.
- Best practices dictate that the market analysis is performed before a prototype is developed, before animal or clinical testing, and before significant expenditures have been extended.
- Market segmentation to identify allied life science segments that have yielded significant returns can positively benchmark and thus influence the levels of investment, timelines, and portend potential merger, acquisition, and exit timelines.
- Understanding the competitive market landscape where technology will be positioned is critical to ensure the market space and level of need are accurately confirmed.
- Strategic competitiveness, with an emerging technology, benefits from an objective, dynamic market assessment
- The innovation must deliver evidence-based and value-driven clinical results to ensure adoption. The increased emphasis on determining the comparative value of the technology or service is driven by the escalation of cost pressures for all healthcare delivery systems including institutions, outpatient services, group practices, and individual practitioners.
- Market trends, both positive and negative, influence the potential for an emerging innovation to realize successful commercialization.

Key term definitions

1. **Cost-effectiveness analysis**: An analytic tool that examines both the costs and health outcomes against alternate clinical solutions to compare therapeutic alternatives. Cost-effective clinical outcomes of a novel technology are often compared against the current standard of care.

2. **Disease burden**: The social, political, environmental, and economic factors of a disease measured by epidemiology, mortality, morbidity, financial cost, or other indicators.
3. **Due diligence**: A comprehensive and iterative strategic investigation and appraisal of a technology or potential investment to evaluate and determine commercial potential.
4. **Group purchasing organization** (GPO): An organization that provides healthcare providers savings by leveraging aggregated purchasing volume to negotiate discounted pricing between healthcare providers and manufacturers, distributors, and other vendors. Over 97% of US hospitals are affiliated with the three largest GPOs: Vizient, Premier, and Health Trust Purchasing.
5. **Incidence**: Occurrence rate of a disease or condition within a population, over a given time, usually annually. Disease incidence indicates the number of newly diagnosed cases.
6. **Market analysis**: A multi-step, iterative process to determine the probability of successful commercialization for a technology, pharmaceutical, service, or application within a specific market that involves qualitative and quantitative market assessment including competitive landscape and trend analysis. Important to include a range of target users and influencers in the analysis.
7. **Market opportunity**: The potential size of the market segment is anticipated to be captured over time by the innovation under analysis.
8. **Market segment**: A group or category of customers that share measurable characteristics or traits such as product requirements, needs, or specialization; used to define the relevant market.
9. **Market size**: The volume/value of products or services purchased by target acquirers, and can be stated over a defined time, often annually.
10. **Prevalence**: Rate of the total number of cases existing within a population at a given location within a specific period, usually expressed as a percentage of a population.
11. **Value analysis**: A systematic assessment performed by diverse and experienced healthcare teams to confirm the comparative cost and clinical effectiveness of a product under consideration for purchase. The bar for acceptance can be significant, as healthcare providers and executive administrators hold each purchase to specific standards, including a comparison to current and competitive offerings.

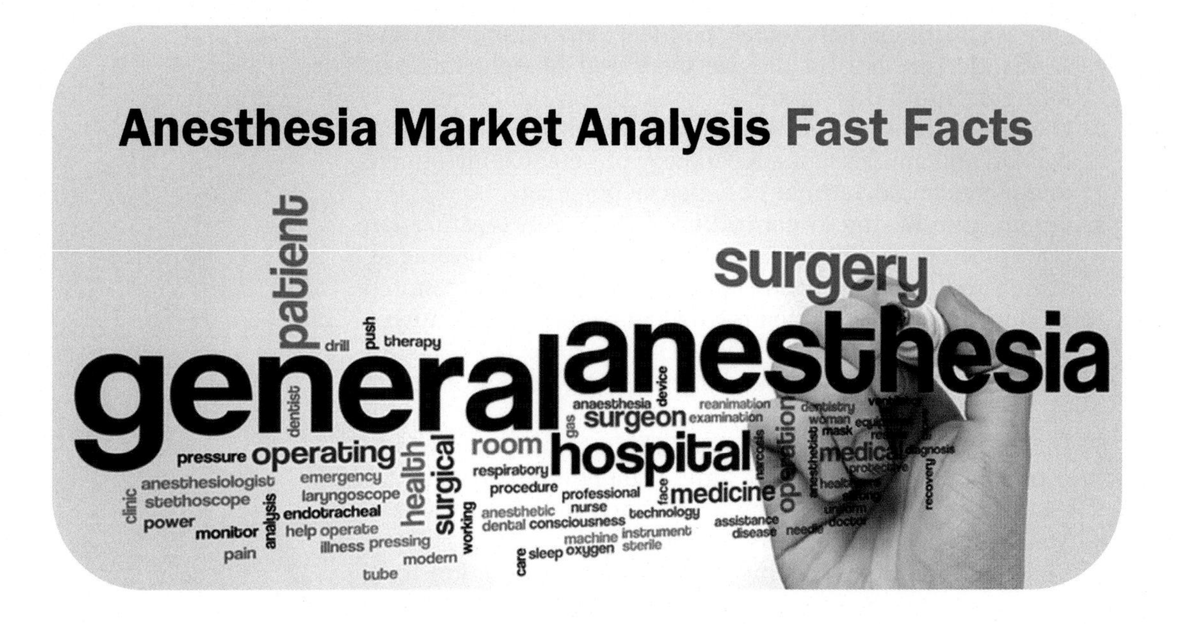

Anesthesia is a clinical process to manage pain through the regional administration of drugs or inhalation of gases to induce a temporary state of pain control by suppressing a patient's inhibitory systems or central nervous system. Anesthesia services include care in the operating room, ambulatory surgery centers (ASCs), and other surgical locations and encompass preoperative evaluation and postoperative management. Development of innovation in this clinical area requires understanding the anesthesia market size, various sites of service, growth, and trends.

The following leading Summary section includes five Fast Fact segments with top-level data to explore the market dynamics in anesthesia. The Market Analysis chapter follows to expand comprehension of the positive and negative influencers, present current procedural data for comprehension of market segments, and explore several components of the anesthesia market important for considerations for innovators desiring to enter this growing market. This analysis will outline areas of growth to support comprehension of different markets where innovation is anticipated to be well received.

Fact 1. Not all information defines the market the same [1,2] (Figs. 5.1–5.3).

Proprietary review and calculations of a significant extent of market data along with an understanding of market

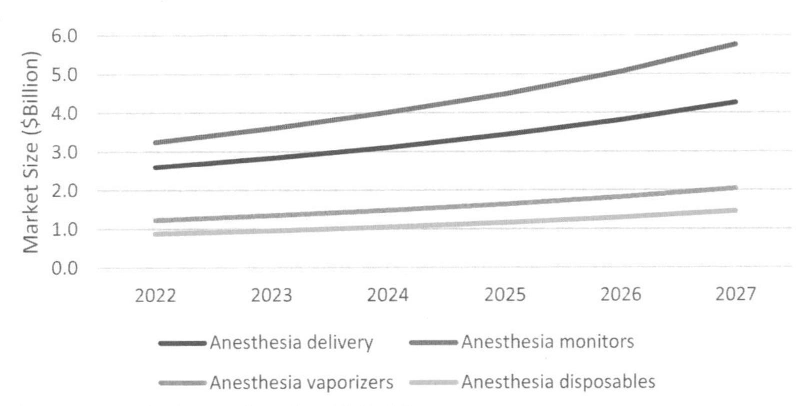

Figure 5.1 Global anesthesia device market size, 2022–27.

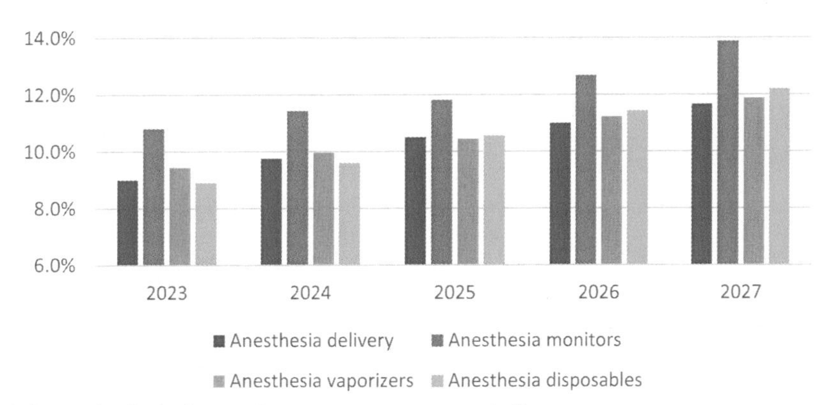

Figure 5.2 Global anesthesia device market year over year growth %.

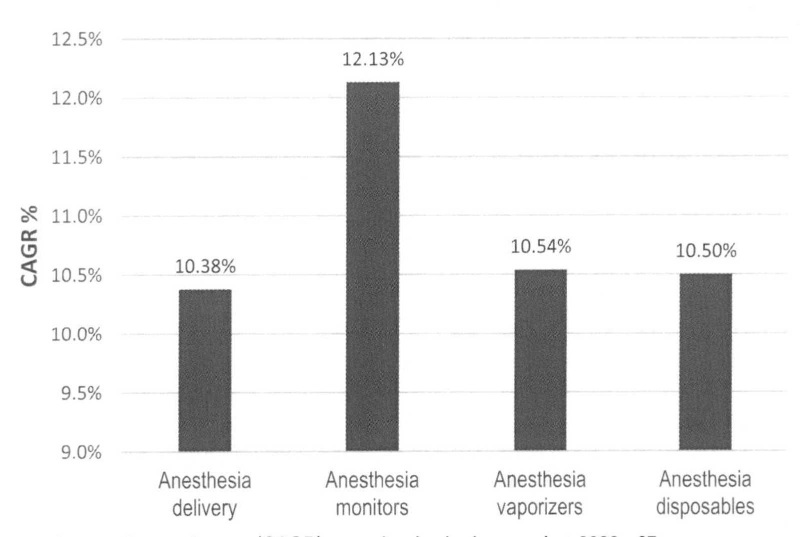

Figure 5.3 Compound annual growth rate (CAGR) anesthesia device market 2022–27.

dynamics and procedures used to prepare the top-level projections.

The **global anesthesia devices market**, including anesthesia delivery, monitors, vaporizers, and disposables, is estimated at US $7.96B in 2022, growing at an average compound annual growth rates (CAGR) of 10.88% to reach US $13.51B in 2027.

The **global anesthesia monitoring devices** market is the most rapidly growing segment and is projected to grow from US $3.25B in 2022 to US $5.76B in 2027, with a CAGR of 12.1%. The US anesthesia monitoring devices market is currently estimated at US $480M.

The **depth of anesthesia monitoring devices** market size is expected to reach US $401M by 2030 with expanding growth at a CAGR of 5.2% from 2021 to 2030.

Fact 2. Anesthesia market driven by the number of operating rooms (ORs) and procedure volumes over disease incidence or prevalence [3,4].

There are an estimated 242,000 ORs in US hospitals and ASCs. In 2016, there were roughly 224,000 US hospital-based ORs, an average of 44 ORs per hospital. In 2022 there were 6087 ASCs with between 1 and 7 + ORs for an estimated 18,739 ORs, a 14% increase since 2014. In 2022, 54% of ASCs have 1–2 ORs and 31% have 3–4 ORs.

The top ASCs by specialty type in 2022 are Orthopedic 36.7%, Other 36.1%, and Pain 35.5%. Other represents General Surgery, Cardiology, Trauma, Neurosurgery, Vascular, Urology, and Pediatrics.

In 2019 there were 28.37M hospital-based inpatient procedures, 50% performed in 10 states.

Nongovernment, not-for-profit hospitals make up 61% of hospitals, and account for 72% of procedures, while for-profit hospitals account for 19% of hospitals and account for 14% of surgeries. Finally, governmental and nonfederal hospitals account for 20% of hospitals and account for 14% of surgeries.

Over 50 million surgical procedures are performed annually in the United States. In 2019, 68.5% of surgeries were performed in an outpatient setting, with 31.5% being performed inpatient.

Fact 3. Anesthesia teams are changing [1,3,5–7].

In 2021 the total number of US anesthesiologists was 31,130 with the five states performing 30.6% of surgeries (New Jersey, New York, California, Florida, and Texas) reporting the highest number of anesthesiologists.

Between July 2017 and July 2021, the number of anesthesiologists, certified nurse anesthetists, and anesthesiologist assistants grew by 12%, 18%, and 39%, respectively.

The addition of Certified Anesthesiologist Assistants (CAAs) is an emerging United States trend in anesthesiology staffing, to manage costs. 1800 US CAAs practice in 20 US states, an increase of 39% from 2017 to 2021. There are now 15 US CAA programs that require about 2 years of postbaccalaureate study.

There are 56,000 Certified Resident Nurse Anesthetists (CRNAs), 73% of those currently employed.

There are 31,130 Anesthesiologists, an ↑ of 12%, with surgeries where the highest number of anesthesiologists are from the 5 states of New Jersey, New York, California, Florida, and Texas encompassing 30.6% of procedures. The number of candidates matching for anesthesiology residency has ↑ increased by 15% between 2017 and 2021. Residents increasingly choose subspecialization.

Differing average salaries: CAA, $160K, CRNAs, $211K, Anesthesiologists $361–$472K, obstetrics up to $582K.

Fact 4. The aging US population is anticipated to drive surgical volume, type, and OR technology needs [8].

The US population was 333.2M as of July 2022. There were 55M Americans >65 years old in 2020, and are forecasted to represent 21% of the national population by 2030.

Fact 5. Inpatient Medicare DRGs (Diagnostic-Related Groups) reveal procedure trends and top surgical volumes.[1]

In 2019 there were 4.27M inpatient Medicare discharges.

The top 20 DRGs make up 45.4% of procedures. Those included: Major hip + knee joint replacement w/o MCC 672.3K discharges (15.7%).

68% of Medicare fee-for-service inpatient discharges with DRG procedures performed in three major diagnostic categories: Musculoskeletal and Connective Tissue (CT) 37%, Circulatory 22%, and Digestive 8%.

There were 431,950 Medicare Cardiovascular Surgical Inpatient discharges in 2022 in the United States. This number supports market size calculations for cardiovascular procedures.

[1]Analysis of data from the Centers for Medicare and Medicaid Services (CMS) MedPAR Inpatient Services 2017–2019.

2017–19 Highest number of procedure changes (DRGs) discussed below:
Increased ↑: The following are the increases in Inpatient Medicare Surgical Discharges: Combined anterior/posterior spinal fusion w/o CC/MCC, +25.6K (321.2%); Combined anterior/posterior spinal fusion w/CC, +22.4K (252.7%); Kidney + ureter nonneoplasm w/CC +21.0K (235.5%).
Decreased ↓: The following are the decreases in Inpatient Medicare Surgical Discharges: Hip + knee joint w/o MCC, −82.9K (−11%); Spinal fusion except cervical, −34.7K (−31.2%); Transurethral w/CC, −10.0K (−53.4%).

Market analysis in anesthesia

In the commercialization of emerging innovation, in anesthesia as well as other markets, it is important to understand more than just the unmet clinical needs. The market analysis process determines a technology's probability for successful commercialization within the target market and to what extent that potential success is possible given the market size, trends, constraints, influencers, and competitors, among other factors. This chapter aims to introduce the important market factors influencing the anesthesia market and its anticipated growth in coming years, provide insights into the anesthesia market size projections, and overview market segments with emerging innovation.

This **Market Analysis** explores the current landscape of Anesthesia to support entrepreneurs considering developing technology to address the clinical needs of patient management for anesthesia. The market dynamics require understanding the various sites of service for anesthesia the trends in identified market segments, and the adoption process for new medical device technology. The market influencers include population trends and their effect on procedures, evolving budgetary constraints of target customers, the direction of procedure numbers, healthcare spending trajectory, competitive product offerings, target users, purchasing decision process, and more (Fig. 5.4).

Understanding the anesthesia market

The market for anesthesia innovation is not clearly defined due to market fragmentation with varying CAGR, the highest for anesthesia monitors. The different types of anesthesia include general, IV-monitored sedation, and regional and local anesthesia. According to NACOR, the National Anesthesia Clinical Outcomes

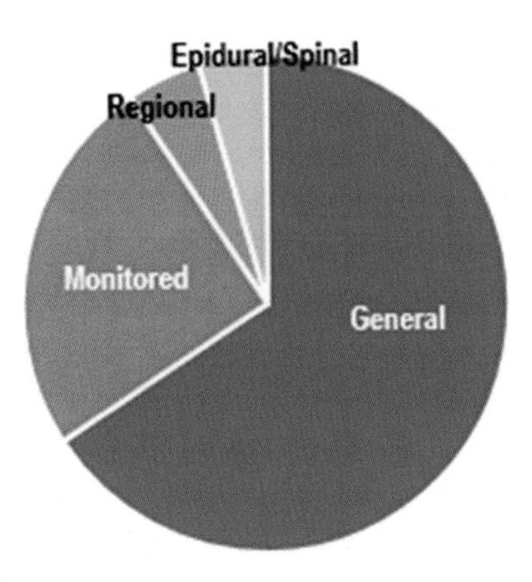

Figure 5.4 US anesthesia types 2019.

Registry, in 2019 anesthesia in the United States included 65.8% of cases for general anesthesia, 24.9% for monitored anesthesia care, 4.9% for regional anesthesia, and 4.4% for epidural/spinal. Anesthesia is delivered in diverse healthcare sites of service including hospitals, outpatient clinics, ASCs, and physician and dental offices, each site with different requirements for technology along with various purchasing processes and decision-makers. In addition, the anesthesia market is a highly competitive market, with significant and increasing market entrants. Having stated the above, the cost-effective user-friendly technology that consistently provides accurate value-based results is most often well received, even in a cost-constrained environment.

Market size and growth often mirror population trends. The US population was 333.2 million as of July 2022 [8]. By 2030 the US population is predicted to increase by 40.9 million people (12.3% growth), primarily driven by births and international migration. There was a reported decline in the US population growth rate during the pandemic years, dropping to 0.35% in 2020 and then to 0.16% in 2021, a historic low supposed from the combination of COVID-related deaths, birth rate declines and pandemic-related limits on immigration [9].

There were 55 million Americans aged 65 years and older in 2020, and this demographic is forecasted to represent 21% of the national population by 2030 [9–11].

Anesthesia market influencers

One important step in the anesthesia market analysis is to consider those aspects that influence the market. The anesthesia market has a significant influence on market dynamics, both

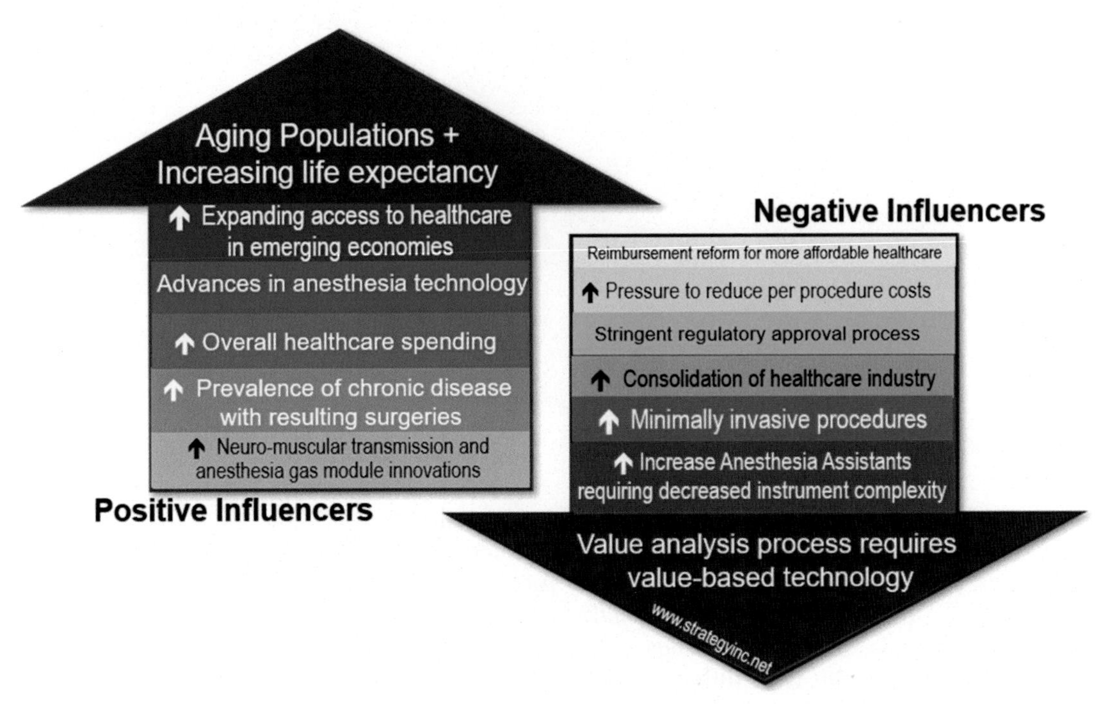

Figure 5.5 Positive and negative influencers of the anesthetic market.

positive and negative. Fig. 5.5 highlights top-level influencers that can both drive the adoption of emerging innovation along those that can prevent achieving the desired return on investment. In addition to identifying these influencing factors, it is important to also understand how their influence may specifically impact innovative technology adoption. The positive and negative influencers should be updated iteratively when making decisions to initiate the development of anesthesia technology, as the influencing forces are continually changing.

Positive influencers

Increase in the aging population

Significant drivers for the increase in anesthetic devices include the increase in aging populations with a concurrent increase in the prevalence of chronic disease and the resulting requirements for surgical procedures. According to the US Census Bureau, individuals 65 years and older will make up almost a quarter of the US population by the year 2030. The growing geriatric population is predisposed to degenerative diseases and chronic illnesses, resulting in an increased reliance on surgical and therapeutic interventions. Such procedures require continuous patient monitoring and thus are expected to propel the anesthetic equipment market forward. According to the American Cancer Society Cancer Statistics 2023, the annual incidence of new US cancer cases is predicted at 1.9M. Out of the total cancer cases for 2023, the projected incidence of breast cancer (women and men) is estimated at 300,590, colorectal at 52,550, and urinary bladder cancer at 16,710. The current standard treatment of these cancers, and others, requires surgeries that utilize anesthesia devices, thereby driving increasing the anesthesia devices market.

Advances in technology

There are significant and increasing examples of engineering advances in technology positively influencing the emergence of new technology in the market [12]. An example of such technological advances includes fully noninvasive assessment of cardiac output using a blood pressure cuff applied to the finger, such as in the CNAP system (CNSystems Medizintechnik GmbH, Graz, Austria) or the ClearSight system [13] (Edwards Lifesciences, Irvine, CA, United States). An additional technology driven by engineering advances is integrated neuromuscular transmission modules that can confirm neuromuscular blocks [14] while anesthesia gas module innovations provide information on patient conditions and for anesthesia system management and ventilation available as integrated modules or stand-alone.

Emerging countries expanding access to healthcare

There is a growing trend of expanding access to healthcare in emerging economies, often referred to as BRICS, an acronym for Brazil, Russia, India, China, and South Africa. As these populous countries continue to build their economies and healthcare systems, market opportunities for medical technology will improve. Innovations strategic for low entry requirements or barriers to use will increase the likelihood of adoption in these countries with significant potential.

Negative influencers

Economic factors encompass most of the negative drivers, such as pressures to reduce the per-procedure cost. Hospitals and health systems are under immense pressure to improve

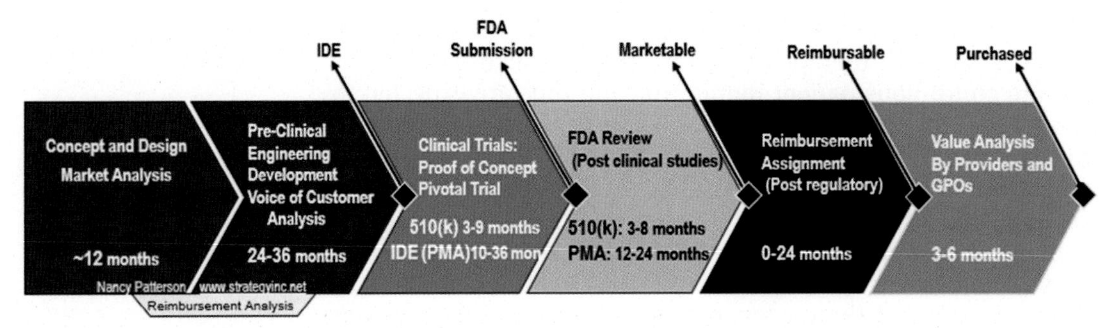

Figure 5.6 Medical device development timeline.

efficiency and reduce their healthcare costs as payers decrease claims reimbursement rates and hospital consolidation continues to grow [15].

Rigorous approval process by regulatory agencies in the United States and most all developed nations. The FDA's Center for Devices and Radiological Health (CDRH) is responsible for regulating firms that manufacture, repackage, relabel, and/or import medical devices sold in the United States. In addition, the CDRH regulates radiation-emitting electronic products such as lasers, X-ray systems, and ultrasound equipment. As outlined in Fig. 5.6, the entire timeline ranges from 4 to 11 years, including the regulatory approval process following clinical trials, which can take from 3 months to 2 years. There are significant differences based on the technology complexity, the predicate devices that can be leveraged for regulatory clearance, and the clinical trial strategy.

Consolidation of the healthcare industry

The COVID-19 pandemic led to decreased procedures and delayed use of healthcare services to lower the spread of the disease. This decrease in the use of healthcare services resulted in a decline in revenue while concurrently being challenged with increased costs due to the pandemic, some in a sustained fashion even with federal assistance. Given the uncertain knowledge of when procedure levels would return to previous levels, some providers have been forced into consolidation. This follows a trend that was already in effect before the pandemic, however, the pandemic accelerated the consolidations [16].

Value analysis

One of the biggest current hurdles in facility adoption of life science innovation is the value analysis processes. Securing healthcare facility acceptance to make your product available for institutional formulary use requires more than just a desirable medical technology able to consistently, and even cost-effectively, deliver results to patients. Medical device and biotech manufacturers must undergo a Value Analysis Committee (VAC) evaluation process at each healthcare institution or GPO where they attempt to sell the innovation and prove the value of their technology, device, biologic, or service. This includes a systematic, detailed process for providing evidence-based documentation to validate the technology value, including both safety and efficacy, overall costs as compared to current and emerging competitive offerings, and the documentation of healthcare user preferences. Members of VAC are multidisciplinary and include physicians, nurses, administrators, supply chain specialists, purchasing agents, and risk-mitigating specialists.

The bar for acceptance can be significant, as healthcare providers and executive administrators hold each purchase to specific standards, including a comparison to current and competitive offerings. In addition to improving the quality of care, VAC members are pressured to reduce the overall costs of care by 10% across the board which drives significant scrutiny in product direct costs and determined use costs. Medical device manufacturers will need to understand the VAC requirements to help develop, manufacture, and conduct clinical studies for validation. VACs review the metrics driving selection, and evidence presented by competitors. Developing evidence-based documentation for a value-based innovation will ensure technology adoption. Such a Value Analysis strategy can help determine potential hurdles and navigate this important gatekeeping review process with efficiency.

Determining procedure-driven market size

The strategy for confirming the specific market size for anesthesia technology will vary by device use and will require expanded information on technology pricing and the competitive landscape. It may require confirming the number of surgeries and OR use. Once the specific market segment where a potential technology will be compared is identified, it is important to determine both the direct and indirect competitors. While an in-depth review of specific technologies is beyond the scope of this analysis, high-level information on the number of procedures, ORs, and national discharge data has been provided to support the market analysis for entrepreneurs. Determining

Medical Records Databases	Government / Society References	Published Journal Databases
HCUP: The Healthcare Cost and Utilization Project includes a collection of longitudinal hospital care data in the United States.	CDC - Center for Disease Control and Prevention Clinicaltrials.gov a database of privately and publicly funded clinical studies conducted around the world	PubMed >28 million citations for biomedical literature from MEDLINE, life science journals, and online books
NASS: From the HCUP tools, the Nationwide Ambulatory Surgery Sample (NASS) database of all-payer major ambulatory surgeries performed in United States hospital-owned facilities.	US Census Bureau	Embase a global, multipurpose and up-to-date biomedical database with literature dating back to 1947.
NAMCS/NHAMCS: The National Ambulatory Medical Care Survey data on the use of ambulatory medical care services in the United States. The **National Hospital Ambulatory Medical Care Survey** data on the utilization of ambulatory care services in hospital emergency and outpatient departments and ambulatory surgery locations.	World Health Organization (WHO) Kaiser Family Foundation American Society of Anesthesiologists (ASA) American Society of Regional Anesthesia and Pain Medicine (ASRA)	Cochrane Collection This collection is an essential source of high-quality healthcare data, not only useful for providers and patients but also for those responsible for researching, teaching, funding and administrating at all levels of the medical profession.
NHDS/NHCS: The National Hospital Discharge Survey inpatient discharge data from non-Federal hospitals in the United States. The National Hospital Care Survey integrates NHDS and NHAMCS data	Society for the Advancement of Patient Blood Management (SABM)	MEDLINE A database of full text for biomedical and health journals indexed in *MEDLINE*. Many are available to doctors, nurses, health professionals and researchers.
NACOR: The National Anesthesia Clinical Outcomes Registry, clinical data warehouse (Clinician access)		Google Scholar A web search engine that indexes scholarly literature across an array of publishing formats and disciplines.
MEPS: The Medical Expenditure Panel Survey data from surveys of families, individuals, medical providers, and employers across the United States.	Created by Strategy Inc. www.strategyinc.net	

Figure 5.7 Validated sources for anesthesia market data.

the market for a specific anesthesia technology requires a detailed market analysis applicable to that technology. Expanded investigation of market size and segments can be supported by a range of validated sources, several outlined in Fig. 5.7.

Surgical volume rates and trends

Identifying procedure volumes comes from a range of sources. The ASA 2021 Anesthesia Almanac provides a compilation of perioperative data for the American Society of Anesthesiologists and includes information about surgical volume anesthesia utilization trends and anesthesia workforce characteristics [3]. Surgical procedures are often classified by site of service and hospital length of stay. The AHA Annual Survey of Hospitals defines inpatient surgical operations as counting each patient undergoing surgery as one surgical operation, regardless of the number of surgical procedures that were performed while the patient was in the operating or procedure room. Outpatient surgical operations are recorded as operations performed on patients who do not remain in the hospital overnight regardless of procedure location.

The American College of Surgery states that 15 million Americans have some kind of surgery annually. The NIH states that 60,000 Americans have surgery under general anesthesia every day [17]. Another report estimates that more than 50 million surgical procedures are performed annually in the United States, however not all require anesthesia [1]. Globally, according to a calculation based on data from 56 member states of the World Health Organization, 230 million major surgical procedures are being carried out under anesthesia worldwide every year [18].

Hospitals and ambulatory surgery centers

Market size projections for anesthesia technology could also be built from the estimated number of ORs. Understanding hospital and ASC trends will impact market potential as well as inform where technology adoption could be greatest. According to the American Hospital Association, there were 6093 US hospitals in 2019, where 90% of hospital-based surgeries occurred in 53% of those hospitals. Of the total number of surgeries, 50% of hospital-based surgeries occurred in just 14% of hospitals. The number of hospitals supporting more than 20,000 surgeries increased from 228 in 2015 to 273 in 2019, and substantial growth in outpatient surgeries helped these large programs become even larger [3].

In 2019, 68.5% of surgeries were performed in an outpatient setting, with 31.5% being performed inpatient. Between 2015 and 2019, these figures showed a 4% increase in outpatient procedures and concurrent 4% decrease in inpatient procedures. Furthermore, in 2019, 88% of hospitals offered a patient-controlled analgesia program and 67% offered a pain management program [3].

The top 10 states for hospital-based surgeries have remained the same since 2015. Ohio moved ahead of Pennsylvania in 2019. Variability in hospital surgical rates reflects differences in population demographics, patient health, the availability of

Rank	State	Hospitals	Hospital-based Surgeries, 2019	
			Number	% of Total
1	California	339	2,271,164	8.0
2	Texas	349	2,199,992	7.8
3	New York	158	2,038,217	7.2
4	Florida	188	1,521,034	5.4
5	Ohio	179	1,363,510	4.8
6	Pennsylvania	170	1,355,075	4.8
7	Illinois	177	1,041,487	3.7
8	Michigan	138	987,262	3.5
9	North Carolina	105	882,221	3.1
10	Georgia	130	784,374	2.8
Top 10 states		**1,933**	**14,444,336**	**50.9**
All other and D.C.		2,578	13,929,286	49.1
TOTAL		**4,511**	**28,373,622**	**100.0**

Figure 5.8 Operating room–based market size estimations by state [3].

# of ORs in ASCs in 2022	
Small (1-2)	3289 (54%)
Medium (3-4)	1881 (31%)
Large (5-6)	575 (9%)
Mega (7+)	342 (6%)

Figure 5.9 Number of operating rooms (ORs) in ambulatory surgery centers (ASCs) in 2022 [4].

freestanding surgery centers, practice patterns, and patient migration (Fig. 5.8).

Nongovernment, not-for-profit hospitals make up 61% of hospitals, and account for 72% of procedures, while for-profit hospitals account for 19% and account for 14% of surgeries. Finally, governmental, nonfederal hospitals account for 20% of hospitals and account for 14% of surgeries (Fig. 5.9).

As of 2016, there were approximately 224,000 US hospital-based ORs with an average of 44 ORs per hospital. The US hospital with the highest number of ORs is the Cleveland Clinic, with 85 ORs. In 2022 there were also 6087 US ASCs (an increase of 9% since 2014) with an estimated total of 18,739 ORs. More than half (54%) of ASCs had one to two operating rooms.

Therefore, the total number of US ORs is estimated at 243,000. Surgical procedures take place in each of these ORs and require anesthesia technology.

In 2019 the top three most frequently performed inpatient surgical procedures included musculoskeletal and CT, circulatory, and digestive operations. They comprised 37%, 22%, and 8% of all hospital-based inpatient surgeries, respectively. Previously in 2014, the highest number of procedures in hospital-affiliated ambulatory surgical settings were musculoskeletal, optical, and digestive procedures, representing 28%, 13%, and 13%, respectively, indicating a rise in circulatory procedure volume and a decline in inpatient digestive procedures over 5 years.

US procedure numbers under Medicare, 2019

A total of 4.27 million inpatient Medicare discharges assigned to a DRG for surgical procedures were performed in 2019, excluding medical procedures. Since 27.3% of procedures for 2019 were inpatient, this projects that a total of 15.4 million Medicare surgical procedures (both inpatient and outpatient) were performed. In 2020 there were 62.6 million people enrolled in the Medicare program, about 18.4% of Americans [19].

Twenty DRGs account for 46% of all Medicare fees for service inpatient Medicare surgical discharges. Based on analysis of data from the Centers for Medicare and Medicaid Services MedPAR Inpatient Services for 2017–19, Fig. 5.10, lists the specific DRGs, the Major Diagnostic category, and the number and percent of total discharges.

Focused needs screening

Looking at the DRGs with the largest change in inpatient discharges provides insights into the market trends of the procedural support required. Fig. 5.11 shows a significant increase in several procedures. The number of kidney and ureter procedures rose to the 45th most common Medicare fee-for-service inpatient procedure type in 2019, when in 2017 it was the 148th most common [3]. Specific spinal procedures had a 300% increase over the same period. In addition, a 50% increase in percutaneous intracardiac and cardiac procedures is noted.

Noting trends such as DRGs with increasing inpatient discharges should motivate an in-depth review of specific procedures in these clinical areas to uncover potential unmet needs or areas where technological advancement would be welcome. The initial research includes anatomical and physiological reviews to provide the foundation. Following, as possible, it is

Rank	Diagnostic Related Groups (DRGs)[i]		Major Diagnostic Category	2019 Discharges Number	Percentage of Total
1	470	Major hip and knee joint replacement or reattachment of lower extremity w/o MCC	Musculoskeletal & CT	672,272	15.7
2	247	Perc cardiovasc proc w/ drug-eluting stent w/o MCC	Circulatory	141,822	3.3
3	853	Infectious and parasitic diseases with OR proc w/ MCC	Infectious & Parasitic Disease & Disorders	139,290	3.3
4	481	Hip and femur proc except major joint w/ CC	Musculoskeletal & CT	123,019	2.9
5	483	Major joint/limb reattachment proc of upper extremities	Musculoskeletal & CT	105,755	2.5
6	330	Major small and large bowel proc w/ CC	Digestive	81,423	1.9
7	246	Percutaneous cardiovascular proc w/ drug-eluting stent w/ MCC or 4+ arteries or stents	Circulatory	80,035	1.9
8	460	Spinal fusion except cervical w/o MCC	Musculoskeletal & CT	76,413	1.8
9	329	Major small and large bowel proc w/ MCC	Digestive	55,885	1.3
10	252	Other vascular proc w/ MCC	Circulatory	50,727	1.2
11	981	Extensive OR proc unrelated to principal diagnosis w/ MCC	Not Assigned	49,344	1.2
12	331	Major small and large bowel proc w/o CC/MCC	Digestive	45,378	1.1
13	480	Hip and femur proc except major joint w/ MCC	Musculoskeletal & CT	43,732	1.0
14	854	Infectious and parasitic diseases w/ OR proc w/ CC	Infectious & Parasitic Disease & Disorders	41,364	1.0
15	253	Other vascular proc w/ CC	Circulatory	41,335	1.0
16	469	Major hip and knee joint replacement or reattachment of lower extremity w/ MCC or total ankle replacement	Musculoskeletal & CT	41,263	1.0
18	274	Percutaneous intracardiac proc w/o MCC	Circulatory	39,970	0.9
17	243	Permanent cardiac pacemaker implant w/ CC	Circulatory	39,743	0.9
19	267	Endovascular cardiac valve replacement w/o MCC	Circulatory	37,619	0.9
20	467	Revision of hip or knee replacement w/ CC	Musculoskeletal & CT	36,810	0.9
Top 20 diagnostic related groups				**1,943,199**	**45.5**
All other diagnostic related groups				2,327517	54.5
TOTAL				**4,270,716**	**100.0**

CC=complications or comorbidities; CT=connective tissue; MCC=major complications or comorbidities

Figure 5.10 Top 20 inpatient Medicare DRG discharges 2019 [3].

advantageous to participate in a clinical observation of aligned surgical procedures to determine where technology could solve or reduce clinical challenges. This screening can assist in identifying detailed specifications of the clinical characteristics, market dynamics, competitors their current solutions, and stakeholder requirements. This is the first stage of determining the needs statement in a single sentence. Disciplined disease state research and therapeutic clinical path review are critical for innovators to validate unmet needs.

Diagnostic Related Groups[ii]		Major Diagnostic Category	INCREASE in Inpatient Discharges: 2017–2019	
			Number	Percentage
455	Combined anterior/posterior spinal fusion w/o CC/MCC	Musculoskeletal & CT	25,635	321.2
454	Combined anterior/posterior spinal fusion w/CC	Musculoskeletal & CT	22,408	252.7
660	Kidney and ureter proc for non-neoplasm w/CC	Kidney & Urinary Tract	21,071	235.5
483	Major joint/limb reattachment proc of upper extremities	Musculoskeletal & CT	19,673	22.9
853	Infectious and parasitic diseases w/ OR proc w/MCC	Infectious & Parasitic Disease & Disorders	19,372	16.2
661	Kidney and ureter proc for non-neoplasm w/o CC/MCC	Kidney & Urinary Tract	17,599	356.6
274	Percutaneous intracardiac proc w/o MCC	Circulatory	13,863	53.1
854	Infectious and parasitic diseases w/OR proc w/CC	Infectious & Parasitic Disease & Disorders	13,138	46.5
267	Endovascular cardiac valve replacement w/o MCC	Circulatory	12,163	47.8
246	Percutaneous cardiovascular proc w/ drug-eluting stent w/MCC or 4+ arteries or stents	Circulatory	11,858	17.4
673	Other kidney and urinary tract proc w/MCC	Kidney & Urinary Tract	10,679	90.5
659	Kidney and ureter proc for non-neoplasm w/MCC	Kidney & Urinary Tract	10,036	266.6
215	Other heart assist system implant	Circulatory	7,305	138.0
247	Perc cardiovascular proc w/ drug-eluting stent w/o MCC	Circulatory	6,767	5.0
023	Craniotomy w/ major device implant or acute complex CNS PDX w/MCC or chemo implant or epilepsy w/ neurostimulation	Nervous	6,389	44.6

Figure 5.11 2019 DRGs with the biggest increase in inpatient discharges [3].

Conversely, monitoring DRGs with significant decreases in inpatient discharges can preempt clinical shifts in therapeutic trends, and increasing caution in development efforts is advised. While musculoskeletal and CT procedures are still in the top three, they saw large Medicare fees for service inpatient declines, either moving to outpatient settings or being controlled medically. In another example, inpatient transurethral procedures declined by 53% (Fig. 5.12).

COVID-19 influence on number of surgeries

In March 2020 the World Health Organization declared SARS-CoV-2 a pandemic. As of May 2021, the highly infectious upper respiratory disease COVID-19, caused by SARS-CoV-2, infected 154.1 million people worldwide, with 3.2 million deaths reported [20]. The US-led globally with approximately 20% of all reported infections and deaths. A year prior, many states suspended elective procedures following guidelines released by the Centers for Medicare and Medicaid Services (CMS), the CDC, and the American College of Surgeons. These recommendations to postpone elective procedures were an effort to mitigate transmission of the virus, prepare for an influx of patient care, and attempt to conserve crucial medical supplies. The impact of implementing these safeguards resulted in a mean loss of $51 billion per month of hospital revenue between March and June 2020. The results of COVID-19 may affect the anesthesia

Diagnostic Related Groups[a]		Major Diagnostic Category	DECREASE in Inpatient Discharges: 2017–2019	
			Number	Percentage
470	Major hip and knee joint replace or reattach of lower extremities w/o MCC	Musculoskeletal & CT	-82,863	-11.0
460	Spinal fusion except cervical w/o MCC	Musculoskeletal & CT	-34,655	-31.2
166	Other resp system OR proc w/MCC	Respiratory	-11,752	-29.7
669	Transurethral proc w/CC	Kidney & Urinary Tract	-10,040	-53.4
708	Major male pelvic proc w/o CC/MCC	Male Reproductive	-9,923	-43.9
473	Cervical spinal fusion w/o CC/MCC	Musculoskeletal & CT	-8,952	-30.5
249	Perc cardiovascular proc w/ non-drug-eluting stent w/o MCC	Circulatory	-6,366	-56.4
167	Other resp system OR proc w/CC	Respiratory	-4,753	-26.6
248	Percutaneous cardiovascular proc w/ non-drug-eluting stent w/MCC or 4+ arteries or stents	Circulatory	-3,939	-51.5
464	Wnd debrid and skn grft exc hand, for musculo-conn tiss dis w/CC	Musculoskeletal & CT	-3,763	-22.2
570	Skin debridement w/MCC	Skin, Subcutaneous Tissue & Breast	-3,597	-39.2
330	Major small and large bowel proc w/CC	Digestive	-3,349	-4.0
328	Stomach, esophageal and duodenal proc w/o CC/MCC	Digestive	-3,085	-19.9
520	Back and neck proc exc spinal fusion w/o CC/MCC	Musculoskeletal & CT	-2,707	-19.4
459	Spinal fusion except cervical w/ MCC	Musculoskeletal & CT	-2,679	-31.6

Figure 5.12 2019 DRGs with the biggest decrease in inpatient discharges [3].

market in coming years, thus diligent review should be continued to determine the influences.

Changes in the anesthesiology care team

Driven by the desire to reduce labor costs, facilities are adding anesthesiologist assistants to the anesthesiology care team to more cost-effectively provide continuity of anesthetic care during procedures and the postoperative recovery period. This trend will influence the required reduction in the complexity of anesthesia equipment.

Anesthesiology care team transitioning

While assessing the specific market size for anesthesia technology requires confirming the number of surgeries and the number of ORs, it is also helpful to understand the dynamics of the anesthesia team. In 2021 the total number of US anesthesiologists was 31,130 [5] with an average age of 52.6 years. The five states (New Jersey, New York, California, Florida, and Texas) with the highest number of anesthesiologists perform 30.6% of surgeries in the United States. The growth of anesthesiologists has considerably outpaced the growth of other clinical specialties with a 9% increase from 2016 to 2020 (Fig. 5.13).

State	Anesthesiologists
New Jersey	2870
New York	2860
California	2560
Florida	1650
Texas	1510

Figure 5.13 Five states with the highest number of anesthesiologists—2021.

From July 2017 to July 2021 there was an increase in the number of anesthesiologists, certified nurse anesthetists, and anesthesiologist assistants, which grew by 12%, 18%, and 39%, respectively. The number of candidates matching for anesthesiology residency has also increased by 15% between 2017 and 2021. Residents are increasingly choosing subspecialization [21].

The significant increase in CAA in 5 years warrants further understanding of this area of specialization. What will this mean for the development of new technology in anesthesiology? CAAs are highly trained master's degree level nonphysician anesthesia care providers that can practice in 20 states and Washington, DC in 2022. CAAs work under the direction of licensed Physician Anesthesiologists to implement anesthesia care plans. The specialty was initiated in 2016, and there are currently 15 anesthesiologist assistants training programs of 24–28 months in length throughout the United States, graduating about 225 new CAA annually. There are a total of 1800 practicing CAAs in the United States in 2022.

Certified Registered Nurse Anesthetist (CRNA) programs initiated the first formal program in 1909, awarding an MS in Nurse Anesthesia. In 2007 a recommendation was made to move to a Doctor of Nursing Anesthesia Practice, (DNAP) that requires one additional year of online course work. This will formally go into effect in 2025. Nurses with other Master's degrees can earn a DNP, while only CRNAs can earn DNAP. There are 128 US nurse anesthesia programs, and 88% are approved to award doctoral degrees. There are 56,000 CRNAs, with more than 40,679 employed in 2022 and around 2400 CRNAs graduate each year.

The stratification of education, responsibilities, and salaries of the anesthesiologist care team means that entrepreneurs will need to develop technology with reduced operator complexity and training going forward, and such features will drive technology acquisition. The difference in the average salaries between the Anesthesia team members is one of the big drivers in a cost-constrained hospital environment. The average salaries listed above: CAA $160K, CRNAs $211K, Anesthesiologists $361–$472K, obstetrics up to $582K.

Understanding diverse technology players

Each anesthesia market segment, from the larger anesthesia monitors to the smaller anesthesia disposables includes four to nine divergent players which leads to significant business requirements to capture and maintain market share in a cost-constrained healthcare market. Most markets are expanding through internal product portfolio diversification and expansion, and through external acquisition of smaller emerging technology that often provide a paradigm shift. Eventually that growth leads to acquisition and less frequently by merger or strategic alliance, though usually later in the commercialization process. Such diversification means that entrepreneurs must create revolutionary solutions that are cost-effective, intuitive to use, and aligned with clinical procedures. The cost of change is so significant, that the technology must deliver significant clinical advantages. Changes in the clinical process through the use of software applications, AI management, and significant cost upheaval can be ways to mitigate.

Expanded anesthesia technology categories

Segmenting the market into technology categories can aid in competitive analyses and determining market opportunities. The categories below are defined to outline where an anesthesia technology can be compared to similar technologies (Fig. 5.14).

Fig. 5.15 shows the categories and technologies with expanded information about each category including examples of global technologies highlighting emerging concepts.

Considering the cardiac anesthesia market

Developing technology to support cardiac anesthesiology requires a full understanding of a long list of clinical, technical, and management issues. Fortunately, there is an excellent resource for developing entrepreneurs entitled Cardiac Anesthesiology Made Ridiculously Simple by Art Wallace, MD, PhD. He is the Vice-Chair of Anesthesiology and perioperative care at the University of California, San Francisco, and Chief of the Anesthesia Service at the Veterans Affairs Medical Center in

	Categories	Definition
1	Airway Devices	Oral and nasal pharyngeal devices to facilitate upper airway patency and the delivery of oxygen or anesthetic gases
2	Depth of Anesthesia Monitoring/ Neuromonitoring	Detects and monitors the degree of CNS depression by a general anesthetic agent
3	Cardiovascular Monitoring Systems	Systems for continuous or intermittent documentation of cardiac observation and documentation of the heart rhythm, either intraprocedural or periprocedural
4	Non-Cardiovascular Monitoring Systems	Noninvasive medical monitoring systems to determine and deliver consistent measurements
5	Drug Delivery Systems	Formulation or device that enables the introduction of a therapeutic substance through various routes of administration for enhancing the efficacy as well as reducing side effects
6	Fluid/Blood Management	Maintenance of normal fluid volume throughout the perioperative period to maintain adequate tissue perfusion. Patient blood management seeks to maximize the use of a patient's blood and avoid unnecessary blood transfusions during surgery.
7	Imaging	Point of care ultrasound and other imaging systems for guided nerve blocks, cardiac assessment, and patient diagnosis to facilitate treatment selection.
8	Regional Anesthesia/Pain Management	Relief of both acute and chronic pain while managing prescriptive opioid utilization with the use of regional and local infusions often with the use of ultrasound-guided peripheral nerve block
9	Software	Software to manage the infusion of anesthetic agents for clinical decision support

Figure 5.14 Anesthesia technology categories.

San Francisco. His publication covers the patient's clinical needs, the anesthesiologist's management, detailed comprehension of the process, medication, the recommended form of anesthesia, induction and intubation, cannula placement, hemodynamics, fluids, planning for early extubation, anticoagulation, checklist for going on and off bypass, cardioplegia, inotropes and vasoactive compounds, laboratory values, and even mid-CABG or Off-Pump CABG. Required reading for the development of **Cardiovascular Monitoring Systems**.

The market size for cardiovascular anesthesia technology can be determined using a surrogate metric of the number of Medicare inpatients discharged, with the DRGs for major cardiovascular procedures. They are listed below, with the 2022 discharges, including the average payment, to provide a metric of the facility's financial amounts paid for such procedures. There were 431,950 Medicare Cardiovascular Surgical Inpatient discharges for 2022 for major surgical involvement in the United States. This number supports market size calculations for cardiovascular procedures (Fig. 5.16).

CMS Medicare Inpatient Reimbursement—Cardiovascular: Abbott [22].

	Categories	Technologies	Example Technologies (Company – Product)
1	**Airway Devices**	• Airway Management Systems • Laryngoscopes • Capnographic Analysis / Continuous respiratory management • Carbon dioxide absorber • Oxygen Delivery Systems • Scavenger Systems	**Micropore Inc** (USA) – SpiraLithCa CO2 Absorber **Ventinova Medical** (Netherlands) – **Ventrain** ventilation systems for obstructive airways **Verathon Inc.** (USA) – **GlideScope** video laryngoscope and **GlideScope Core** airway visualization system **Vyaire Medical** (USA) - **SuperNO2VA** nasal continuous positive airway pressure device
2	**Depth of Anesthesia Monitoring/ Neuromonitoring**	• Neuromuscular transmission monitoring systems / Neuromuscular blockade monitors • Bispectral index (BIS) • Electroencephalographic (EEG)	**Blink Device Company** (USA) – **TwitchView** Quantitative electromyography (EMG) based neuromuscular monitoring of depth of neuromuscular block **BrainStem Biometric** – **Tremor Monitor Unit** non-invasive disposable sedation monitoring system **Quantum Medical** (Spain) - **qCON – NOX** for advanced EEG depth of anesthesia and nociception level monitoring of a patient during general anesthesia in surgery or intensive care
3	**Cardiovascular Monitoring Systems**	• Blood Pressure Monitors • Non-invasive hemodynamic monitoring • Cardiac output monitoring • TEE (Transesophegeal echocardiogram) • Electrocardiogram • Central Venus Catheter • Pulmonary Arterial Catheter	**CNSystems** (Austria) – **CNAP System** for non-invasive continuous cardiac output and hemodynamic monitoring **Deltex Medical** (UK) - **CardiacQ** esophageal Doppler monitoring **Hemosonics** – Quantra Hemostasis Analyzer ultrasound technology for point of care clot elasticity measurement **Retia Medical** – Argos Cardiac Output (CO) Monitor
4	**Non-Cardiovascular Monitoring Systems**	• Respiratory System Monitoring • Pulse Oximeters • Critical Care Monitoring • Accelerometers / Acceleromyographic (AMG) monitors • Magnetometers • Gyroscopes	**Nonin** - noninvasive pulse oximeters **QuSpin** – Optically Pumped Magnetometer (OPM) sensors **Xavant Technology** (South Africa) - **Stimpod NMS450X NMT monitor** Dual-Sensor Neuromuscular Patient Monitor

Figure 5.15 Example emerging anesthesia technologies.

	Categories	Technologies	Example Technologies (Company – Product)
5	**Drug Delivery Systems**	• Automated medication delivery systems • Target-controlled infusion pumps	Acromed (Switzerland)– **Volumed** target-controlled infusion in anesthesia and sedation Conan Argus (Switzerland) - infusion technology systems Infinium Medical (USA) - ADSII Anesthesia Delivery System
6	**Fluid/Blood Management**	• Blood/Fluid Warming Systems • Fluid management systems • Blood Salvage Systems	Ace Medical (Korea) – **Mega Acer Kit** heat and moisture exchanger circuit system Getinge (Sweden) – Perioperative blood flow monitoring for fluid management iSep (France) – **Same** ™ technology Smart autotransfusion for me Intraoperative cell salvage technology
7	**Imaging**	• Imaging Systems • Point of Care Ultrasound • Intraoperative Ultrasound	Butterfly Network (USA) – beside imaging for Ultrasound-guided peripheral nerve blocks line placement, gastric volume and cardiac assessment Clarius (USA) - Ultra-portable wireless ultrasound for the nimble regional anesthesiologist Sonoscanner (France) – portable and stationary ultrasound for anesthesia
8	**Regional Anesthesia/Pain Management**	• Pain Management	Anutra Medical (USA) - Anutra Local Anesthetic Delivery System Avanos (USA) – **ON-Q Pain Relief System** with Select-A-Flow is a non-narcotic elastomeric pump that automatically and continuously delivers local anesthetic Intelligent Ultrasound (UK) - **ScanNav Anatomy Peripheral Nerve Block** real-time AI assistant for ultrasound-guided regional anesthesia …
9	**Software**	• Infusion Software • Anesthesiology Information Management Systems (AIMS) • Clinical Decision Support Systems (CDSS)	AneScan – Anesthesia EHR to digitize the entire perioperative process Demed (Belgium) – **Rugloop II** windows-based TCI infusion control program and general data management program iMD-Soft (Israel) - MetaVision Anesthesia, anesthesia information management systems

Figure 5.15 *(Continued)*

Technology	MS-DRG	Description	Severity	FY 2022	
				Payment	Medicare Inpatient Discharges
Surgical Valves	216	Cardiac valve & other major cardiothoracic procedures with cardiac catheterization	MCC	$66,202	7,517
	217		CC	$42,754	2,610
	218		None	$40,286	479
	219	Cardiac valve & other major cardiothoracic procedures without cardiac catheterization	MCC	$53,134	15,810
	220		CC	$35,644	15,227
	221		None	$30,201	2,486
Congenital Defects-Ventricular Septal	228	Other cardiothoracic procedures	MCC	$35,149	2,747
	229		CC-16, None-17	$22,692	3,515
TEER and TAVR	266	Endovascular Cardiac Valve Replacement & Supplement Procedure	MCC	$46,476	20,611
CABG	231	Coronary bypass with PTCA	MCC	$57,475	1,060
	232		None	$39,261	720
	233	Coronary bypass with cardiac catheterization	MCC	$52,242	13,684
	234		None	$35,187	15,313
	235	Coronary bypass without cardiac catheterization	MCC	$40,252	11,858
	236		None	$27,017	20,092
	301		None	$4,900	7,630
ICD Systems and CRT-D	222	Cardiac defibrillator implant with cardiac catheterization with AMI/HF/shock	MCC	$52,431	2,081
	223		None	$38,237	404
	224	Cardiac defibrillator implant with cardiac catheterization without AMI/HF/shock	MCC	$49,583	2,119
	225		None	$37,045	1,629
	226	Cardiac defibrillator implant without cardiac catheterization	MCC	$43,291	5,317
	227		None	$34,370	4,202
Pacemaker Systems; CRT-P	242	Permanent cardiac pacemaker implant	MCC	$24,581	19,301
	243		CC	$16,608	24,932
	244		None	$13,606	14,466
ICDs	245	AICD generator procedures	NA	$35,726	1,889
	265	AICD lead procedures	NA	$22,193	580
Pacemaker Generator Replacement	258	Cardiac pacemaker device replacement	MCC	$20,891	592
	259		None	$13,777	857
Pacemaker Revision and ICMs Implant	260	Cardiac pacemaker revision except device replacement	MCC	$23,524	2,707
	261		CC	$13,148	3,619
	262		None	$11,251	1,354
	274		None	$21,673	26,476

Figure 5.16 2022 Abbott CMS-inpatient-reimbursement-prospectus.

MCC Major Complications and Comorbidity CC Complications and Comorbidity
None No Complications or Comorbidity

Left Ventricular Assist Device (LVAD)	001	Heart Transplant or Implant of Heart Assist System	MCC	$190,661	2,236
	002		None	$98,716	122
Acute Mechanical Circulatory System (MCS)	003	ECMO or Tracheostomy with MV >96 Hours or PDX Except Face, Mouth and Neck with major O.R. procedure	NA	$125,986	14,274
	215	Other heart assist systems implant	NA	$69,625	4,409
Major Chest	163	Major chest procedures	MCC	$33,016	11,069
	164		CC	$17,512	15,841
	165		None	$12,639	8,164
Aortic Heart Assist	268	Aortic and heart assist procedures except pulsation balloon	MCC	$45,918	3,672
	269		None	$28,455	16,310
Aortic Heart Assist	286	Circulatory disorders except AMI, w cardiac catheterizations	MCC	$14,087	50,172
	287		None	$7,353	51,797

Figure 5.16 *(Continued).*

References

[1] Anaesthetic Equipment Market Size, Share, Growth Analysis, 2022, Facts and Factors, Anesthesia Devices Market: Forecast and Analysis 2023–2027, Technavio, Infiniti Research Limited, 2023, Anesthesia and Respiratory Devices Global Markets, Aug 2022, BCC Publishing, Boston, MA, Anesthesia Devices Market – Growth, Trends, Covid-19 Impact and Forecasts 2023–2028, Modor Intelligence, Hyderabad, India.

[2] Vision Research Reports. Depth of anesthesia monitoring devices market size, share, growth, trends, consumption, revenue, company analysis, regional insights and forecast 2021–2030; 2022.

[3] The ASA 2021 Anesthesia Almanac, American Society of Anesthesiologists; September 2021. Data in the almanac were compiled from the American Hospital Association (AHA) Annual Survey, the Healthcare Cost and Utilization Project Nationwide Inpatient Sample (HCUP-NIS), CMS National Plan and Provider Enumeration System (NPPES), National Residency Match Program (NRMP), State Ambulatory Surgery and Services Databases and State Inpatient Databases, CMS National Downloadable File (NDF) Dataset, Physician/Supplier Procedure Summary (PSPS) Files and Provider of Service Files, RAND Corporation, ASA Member Services, and the National Anesthesia Clinical Outcomes Registry (NACOR). Calculations were performed by the ASA Analytics and Research Services Department.

[4] ASC Data. Industry overview_Q2 2022, <https://ascdata.com/>; n.d. [retrieved 14.03.23].

[5] US Bureau of Labor Statistics. Labor statistics. <https://www.bls.gov/>; n.d. [retrieved 13.03.23].

[6] Patel AY, Eagle KA, Vaishnava P. Cardiac risk of noncardiac surgery. J Am Coll Cardiol 2015;66(19):2140–8.

[7] Payscale. Payscale index US. <https://www.payscale.com/>; n.d. [retrieved 13.03.23].

[8] United States Census Bureau. Quick facts. <https://www.census.gov/quickfacts/fact/table/US/PST045222>; n.d. [retrieved 09.03.23].
[9] Frey W. New census estimates show a tepid rise in U.S. population growth, buoyed by immigration. Brookings.edu; 2023. Available from: https://www.brookings.edu/research/new-census-estimates-show-a-tepid-rise-in-u-s-population-growth-buoyed-by-immigration/.
[10] Meola A. The aging US population is creating many problems—especially regarding elderly healthcare issues. Insider Intelligence; 2023. Available from: https://www.insiderintelligence.com/insights/aging-population-healthcare/.
[11] Kilduff L. Which U.S. states have the oldest populations? PRB; 2021. Available from: https://www.prb.org/resources/which-us-states-are-the-oldest/.
[12] Seger C, Cannesson M. Recent advances in the technology of anesthesia [version 1; peer review: 2 approved] F1000Res 2020;9(F1000 Faculty Rev). Available from: https://doi.org/10.12688/f1000research.24059.1.
[13] ClearSight system. A noninvasive solution that enables clinical decision support to help optimize patient perfusion. Edwards. Available from: https://www.edwards.com/healthcare-professionals/products-services/hemodynamic-monitoring/clearsight-system.
[14] Mindray North America. Mindray introduces NMT module for passport OR monitors at ASA 2017; 2017. Available from: https://www.mindraynorthamerica.com/mindray-introduces-nmt-module-passport-monitors-anesthesiology-2017/.
[15] Chernew M., Cutler D., Shah S. Reducing health care spending: what tools can states leverage? The Commonwealth Fund; 2021. Available from: https://www.commonwealthfund.org/publications/fund-reports/2021/aug/reducing-health-care-spending-what-tools-can-states-leverage.
[16] Gaynor M. 'Examining the impact of health care consolidation' statement before the Committee on Energy and Commerce, Oversight and Investigations Subcommittee, U.S. House of Representatives. SSRN; 2018. Available from: https://papers.ssrn.com/sol3/papers.cfm?abstract_id = 3287848.
[17] National Institutes of Health. Waking up to anesthesia. National Institutes of Health, part of the U.S. Department of Health and Human Services; 2011. Available from: https://newsinhealth.nih.gov/2011/04/waking-up-anesthesia.
[18] Weiser TG, Regenbogen SE, Thompson KD, Haynes AB, Lipsitz SR, Berry WR, et al. An estimation of the global volume of surgery: a modelling strategy based on available data. Lancet 2008;372(9633):139–44.
[19] Statista Research Department. Medicare – statistics & facts. Statista; 2023. Available from: https://www.statista.com/topics/1167/medicare/#dossierContents__outerWrapper.
[20] American Hospital Association. Guides/reports; 2020. Available from: https://www.aha.org/.
[21] Rock-Klotz JA, Miller TR. Anesthesiology residency matches hit record high for fifth year in a row. ASA Monitor 2021;85:40–1.
[22] Abbott. Medicare inpatient reimbursement prospectus; 2023. Available from: https://www.cardiovascular.abbott/content/dam/bss/divisionalsites/cv/cv-live-site/hcp/reimbursement/heart-failure/CMS-Inpatient-Reimbursement-Prospectus.pdf.

6

Competitive analysis

David N. Flynn, Erin Dengler and Elisa T. Lund

Department of Anesthesiology, University of North Carolina School of Medicine, Chapel Hill, NC, United States

Chapter outline

Abstract

One effective model for competitive analysis is Michael Porter's Five Forces. Each of the Five Forces is described, and advice on identifying, analyzing, and comparing competitors is provided. The information, advice, and tools in this chapter will help readers perform a thorough competitive analysis and shape their competitive strategy.

Keywords: Competitor analysis; competition; SWOT analysis; competitive strategy

Abbreviations

IV intravenous
GPO group purchasing organization
SWOT strengths, weaknesses, opportunities, and threats

Innovation in Anesthesiology. DOI: https://doi.org/10.1016/B978-0-12-818381-6.00003-6

Introduction

Every business starts with an idea. The idea may be a technological breakthrough that could revolutionize an industry, an incremental design improvement to an existing technology, or a service that helps clients save time or money.

Unfortunately, most businesses fail. Even an excellent idea can fail in the marketplace because founders did not understand the dynamics of their industry or adequately study their competitors. In this chapter, we first describe a simple framework that can be used to analyze the competitive forces that shape your industry. We then discuss strategies to help you identify and better understand your competitors. The insights gained through these exercises will help you understand the industry foundations of a successful business strategy.

The competitive arena—industry analysis through Porter's Five Forces

Before launching a business, you must understand the economics of your industry. Do current companies enjoy high margins or has competition pushed the firms to accept razor-thin margins? How price-sensitive are your customers? Can you manufacture your product or provide your service at a low enough cost to turn a profit? How intensely do existing companies compete and how will they respond to your entry into the market? If you successfully launch your business, what is the probability that new businesses will enter the market and capture your customers?

In his 1979 groundbreaking paper [1] and best-selling book [2], Harvard Business School Professor Michael Porter described the 5-forces that shape industry profitability. These are commonly referred to as "Porter's Five Forces":

1. Bargaining power of supplies
2. Bargaining power of buyers
3. Threat of new entrants
4. Threat of substitute products or services
5. Rivalry among existing firms

Porter's Five Forces framework provides a structured approach to industry analysis, which helps determine if the market for your business warrants further investment of time and capital. If so, evaluation of each of the forces generates valuable insights to guide business strategy. Below is a summary of each force.

Bargaining power of suppliers

Every product or device requires inputs, which are provided by suppliers. Examples of inputs include the physical components required to assemble a product, intellectual property that must be licensed to create a product, and employees required to provide a service. Powerful suppliers reduce the profitability of an industry because they can charge higher prices, which reduces the margins of firms that rely on them. Supplier power in an industry is high when there are few suppliers relative to buyers, buyers have few substitutes for the suppliers' products, and switching between suppliers is difficult or costly [2].

A classic example of an industry with powerful suppliers is the airline industry. Hundreds of airlines in the world rely on the airplanes of just two dominant suppliers, Boeing and Airbus, to operate. This provides considerable negotiating power to Boeing and Airbus while weakening the position of individual airlines. In anesthesia, when a novel drug such as dexmedetomidine or sugammadex is first FDA-approved and placed on the market, it typically has only one drug manufacturer/supplier, and high costs are thus expected.

Bargaining power of buyers

Buyers are the individuals or organizations that make purchasing decisions. Powerful buyers wield their strength to obtain favorable prices and terms from producers, which reduces producer profitability. Buyers are powerful when there are few buyers relative to producers, substitute products are readily available, and switching between suppliers is easy and inexpensive [2].

In healthcare, the buyer is often not the end user of a product. For example, a novel airway device will be used by clinicians who administer anesthesia. However, the buyer of the device is likely the purchasing agent of a multihospital health system or a group purchasing organization (GPO). Many hospitals and provider organizations have consolidated or joined GPOs to increase their buying power. It is important to identify and differentiate between buyers and users both when developing products and performing the Five Forces analysis.

Threat of new entrants

New entrants are a significant threat to existing firms in an industry if they can readily enter the marketplace and gain market share, which reduces the profitability of existing firms. The threat of new entrants is generally low in industries with high startup costs, significant economies of scale, highly differentiated products (particularly those with intellectual property protection), well-established distribution channels, and high switching costs [2].

The anesthesia machine industry is an example of an industry with a low threat of new entrants. Anesthesia machines are complex, requiring sophisticated design, engineering, and manufacturing capabilities. Development of a novel machine would be expected to take several years at a significant cost. Thus established anesthesia machine manufacturers currently face a low threat of new entrants.

Threat of substitute products or services

A substitute replaces another product while accomplishing the same goal. Substitutes decrease a product's value to buyers and increase buyers' bargaining power. Industries with few substitutes are generally more profitable than those with many substitutes because customers have limited negotiating power. The threat of substitutes is high in industries with relatively undifferentiated products, low switching costs, and minimal brand loyalty [2].

The intravenous (IV) catheter is an example that faces a high threat of substitutes. Customers have many options to choose from, design differences are often trivial, switching costs are negligible, and there is minimal brand loyalty. Thus buyers can easily switch between different IV catheter types and brands, which decreases producer pricing power. This is known as a commodity product, which is the same across producers.

Rivalry among existing firms

How intensely do firms in an industry compete against one another? In some industries, competition is fierce—firms engage in cutthroat competition with their rivals to gain customers and market share. This drives down prices (through discounts, rebates, and price cuts) and increases operating costs (through increased advertising and marketing costs), which can threaten emerging businesses and weaken the entire industry. Conversely, in other industries competition between rivals is weak—firms are generally content with their market position and do not aggressively attack their rivals. The intensity of rivalry is generally high in industries when there are many nondominant firms, high fixed costs, low switching costs for customers, minimal product differentiation, and slow industry growth [2].

Rivalry among healthcare providers is rarely strong. In the hospital industry, rival health systems rarely engage in aggressive competitive actions such as price cuts or negative advertising campaigns, though these are common in many other industries such as consumer goods and automobiles. Instead of directly attacking rivals or reducing prices, hospitals generally compete for customers by promoting quality, advanced technologies, or convenience for patients.

Porter's Five Forces summary

A great idea in an unprofitable industry will likely be a business failure, while a mediocre idea in an attractive industry could become a major success. Understanding Porter's Five Forces will help you understand the dynamics of your industry and avoid investing time and capital in a venture that is unlikely to succeed. The insights gained from analyzing each of the five forces will help you anticipate challenges and shape your business strategy to respond accordingly.

Identifying your competitors

After understanding your industry, the next step is to study the current and emerging competitive landscape. The goal of competitor analysis is to determine the strengths, weaknesses, objectives, and threats of competing firms. This knowledge is used to prepare a defensible business strategy that will optimally position your company in the market, either by choosing a position that will exploit competitors' weaknesses or by selecting market segments that will pose little threat to your competitors, thereby reducing the likelihood of strong retaliatory actions [3].

A competitor is any company that satisfies the same customer needs as your business. Most businesses have several direct competitors and even more indirect competitors. For example, if your company produces a new video laryngoscope, your direct competitors are the producers of other video laryngoscopes. The list of indirect competitors is much larger, including all substitute products that allow a clinician to oxygenate and ventilate an unconscious or distressed patient. Such products include standard laryngoscopes, laryngeal mask airways, bronchoscopes, cricothyrotomy surgical kits, and many others.

Attribute	Competitor A	Competitor B	Competitor C
Headquarters			
Company size			
Product Name			
Differentiating feature #1			
Differentiating feature #2			
Differentiating feature #3			
Target customer			
Sales strategy			
Marketing/Advertising			
Pricing strategy			
Distribution Strategy			
Brand reputation			

Note: Important company and/or product attributes are placed in the first column. Competitor names are listed in each column. The table is completed as each competitor is researched.

Figure 6.1 Competitor profiling template.

As you identify your competitors, create a system to compile and organize key information, including names of competing products, differentiating product factors, price, marketing, sales and distribution strategy, and brand awareness and loyalty. Much of this information is available on company websites; additional information can be obtained by interviewing distributors and customers. Detailed company information about publicly traded companies is published in each company's 10-K (filed annually with the Securities and Exchange Commission) and annual report to shareholders. These documents contain a wealth of information, including a description of the company's business operations, financial performance, industry and market analysis, risks, and outlook for the future.

An example of a table that can aid in your competitor's data collection and analysis is shown in Fig. 6.1.

Competitor analysis (below)

After you have identified your competitors, the next step is to analyze their strengths and weaknesses. A commonly used tool for evaluating competitors is a SWOT (strengths, weaknesses, opportunities, and threats) analysis. A SWOT analysis helps you evaluate factors internal to the firm (strengths and weakness) and external to the firm (opportunities and threats) that impact its competitive position [4]. Examples of internal factors include personnel, intellectual property, financial resources, existing customer relationships, and distribution networks. Examples of external factors include regulatory changes, shifting customer preferences, and technological advances. A SWOT analysis is typically presented in a 2×2 matrix as shown in Fig. 6.2.

After completing SWOT analyses of your competitors, a relative strength table can be used to help compare and rank your competitors. To create a relative strength table, first list factors that dictate success in your industry (i.e., new product development, brand reputation, technological expertise, pricing, marketing, distribution). These will be the rows of the table. The names of your company and your competitors are assigned the columns. Each company is then given a ranking for each success factor. Finally, the values from each column are added to provide a final score for each company—those with the highest scores will likely be your strongest competitors. An example of a relative strength table is provided in Fig. 6.3.

A closer examination of the strength table provides valuable insights into opportunities within an industry that can guide your competitive strategy. Are there domains in which the competition is collectively weak? If so, could your business develop capabilities within these areas and then exploit these strengths to gain a competitive advantage? Rigorous analysis of your competitors will help you determine how to position yourself within the market, while also allowing you to anticipate how your rivals will respond to your actions. This knowledge will provide the foundation upon which you can build a successful business strategy.

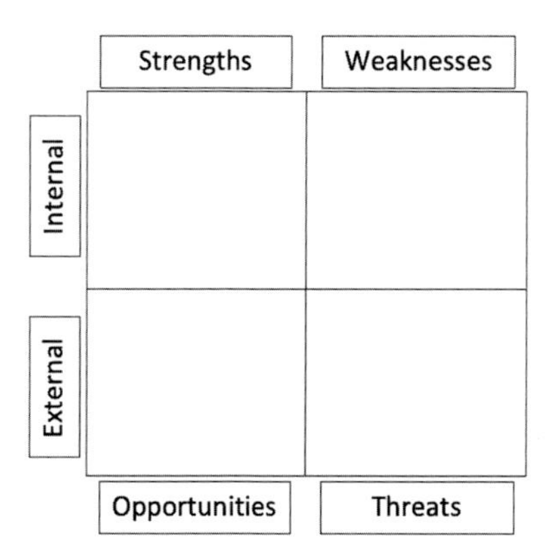

Figure 6.2 SWOT diagram.

Attribute	Competitor A	Competitor B	Competitor C
Brand reputation	3	2	1
Pricing	1	2	3
Product features	2	3	3
Portability	1	3	3
Customer service	3	3	1
Warranty	3	2	1
Innovation/new product development	1	2	3
Total	14	17	15

Note: 1 = unfavorable, 2 = average, 3 = favorable

Note: Attributes chosen will vary based on the nature of the product/industry.

Figure 6.3 Example relative strength table.

Conclusion

For your new product or business to be successful in the marketplace, it is critical to have a thorough understanding of both the industry you are entering and your competitors. Porter's Five Forces provides an overview of the attractiveness of the industry and helps the entrepreneur determine if the industry is worth entering. If the industry is attractive, then attention should be turned to individual competitors, as knowing their strengths and weaknesses will inform your business strategy.

References

[1] Porter ME. How competitive forces shape strategy. Harv Bus Rev 1979;57 (2):137–45.

[2] Porter ME. The structural analysis of industries. Competitive strategy: techniques for analyzing industries and competitors. New York: Free Press; 1980. p. 3–33.

[3] Czepiel JA, Kerin RA. Competitor analysis. In: Shankar V, Carpenter SG, editors. Handbook of marketing strategy. Edward Elgar Publishing Limited; 2012. p. 41–57.

[4] Benzaghta MA, Elwalda A, Mousa M, Erkan I, Rahman M. SWOT analysis applications: an integrative literature review. J Glob Bus Insights 2021; 6(1):55–73.

7

Product requirements for emerging anesthesia technologies

Umme Rumana[1] and Ettore Crimi[2]

[1]Department of Anesthesiology, University of Central Florida College of Medicine, Orlando, FL, United States [2]Department of Anesthesiology, Atrium Health Wake Forest Baptist, Wake Forest University Health Sciences, Orlando, NC, United States

Chapter outline

Abstract

The domain of product requirements focuses on discovering, communicating, and validating stakeholder needs or concerns to provide novel, faster, and more convenient solutions that satisfy business needs. Knowing the exact requirements of the right people to accurately define the solution is vital to successful product development and use. Effective methods to discover requirements include interviews, questionnaires, surveys, brainstorming, mind mapping, document analysis, observations, group workshops, expert opinions, and prototyping, among others. Missing requirements are minimized using diagrams, tables, decision trees, wireframes, models, business cases, use cases, and user stories where needed. Additionally, requirements are analyzed, refined, prioritized, documented, and validated

Innovation in Anesthesiology. DOI: https://doi.org/10.1016/B978-0-12-818381-6.00026-7

for completeness. Adding or modifying a product or service introduces a cascade of desired or undesired changes in tasks upstream and downstream; therefore knowing the actual problem, stakeholders, project goals, scope, context, risks, cost, feasibility, and change approval are all critical to successful technology development and adoption. As a result, it is crucial to match user requirements to user demands when commercializing a product. When developing a new product, most businesses streamline the process by employing project managers, business analysts, and domain-specific subject matter experts to ensure that no requirements are missed and that the project is completed successfully.

Keywords: Product requirements; business requirements; stakeholder requirements; solution requirements; functional requirements; nonfunctional requirements; transition requirements; user requirements

Key points

- Product requirements discovery and analysis is iterative work to build a shared understanding of thoughts and ideas among different stakeholders to solve a problem
- Several techniques and models assist with breaking down a problem to support the discovery and analysis of product requirements. Mapping out what consumers want and need is crucial for any product's eventual commercial success.
- High-quality product requirements enable product delivery on time, within scope, within budget, and users' expectations

Glossary terms and definitions

Product requirements Describe how the system must function to satisfy the needs of the business and the users.

Business requirements statements of goals, objectives, and outcomes for initiating a change.

Stakeholder requirements the needs of stakeholders that must be met to achieve the business requirements.

Solution requirements the capabilities and qualities of a solution that meets the stakeholder requirements. The solution requirements are subcategorized:

Functional requirements The scope of the product's capability is specified by its functional requirements. The product function is a sum of the inputs, the system's operation, and the outputs. Actions can be used to infer the computational, data-manipulative, business-process-related, user-interface-related, and other functions of systems. Functional requirements describe the performances and features for the product to be effective.

They set standards for:

Administrative functions How the organization's operational procedures will be managed

Audit tracking Document the chronological sequence of activities affecting a specific action, method, process, or device

Authorization levels Identify the user's level of access to the system

Business rules Specify how a business is run, what terms are used, and what limitations exist within the business

Legal, regulatory, and certification requirements Rules, regulations, policies, and procedures mandated by a government agency or accreditation council

Data management How information will be gathered, organized, protected, and used by an organization for business analysis

External interfaces The interface used to access the Internet or a wide-area network

Interoperability How the product or system will work with other systems

Nonfunctional requirements define how a solution is meant to be implemented or the conditions under which it will run the characteristics a solution must possess, and any prerequisites it must meet. It specifies the environmental conditions or qualities expected for the product to be user-friendly. They establish criteria for

Accessibility How simple does the answer need to be for users with disabilities to implement

Design The solution's intended visual aspects

Documentation Type and quantity of required or desired written records

Information architecture How the solution's data must be structured or organized

Maintainability How simple or complex it is to make changes to a system

Performance Speed and accuracy with which the solution carries out its intended function in response to user input

Reliability Consistent availability and effectiveness of the solution

Scalability The ability of a system to handle increasing workloads by adding resources

Security The standard of safety expected for the system and its data

Usability and user expectations The required level of simplicity for the solution

Transition requirements describe the terms and conditions of the solution needed to facilitate the transition from the current state to the future state.

User requirements Standards set by the end-users or target audience. To create a product that caters to the needs of the target audience, one must first have intimate knowledge of those needs. Incorporating user feedback may lead to increased user satisfaction, lower maintenance and training costs, higher productivity per unit of effort, and better final products. This method of solution design is known as "reverse innovation" or "agile product development," and it relies on iterative product development informed by user feedback.

Introduction

Requirements are ways to communicate with solution providers about the problem or need; innovations are ways to discover attractive solutions that add value to the end-users and product owners. Requirements management accounts for a significant portion of product success. According to a 2014 research by the Product Management Institute: 37% of businesses cited inaccurate collection of requirements as a major reason for project failure. Aside from a shift in organizational goals, ineffective requirements management is the most common reason for a project's failure. Gaps in communication, misinformation, or assumptions due to a lack of written requirements can result in conflicts, delays, and failure to deliver the desired product promptly.

Before digging out product requirements, the following preliminaries are crucial for the product requirements life cycle (Fig. 7.1) to avoid costly errors and gridlocks: (1) *Knowing the problem*: Is it a real problem or need? How did it impact productivity and efficiency? What are the limitations and

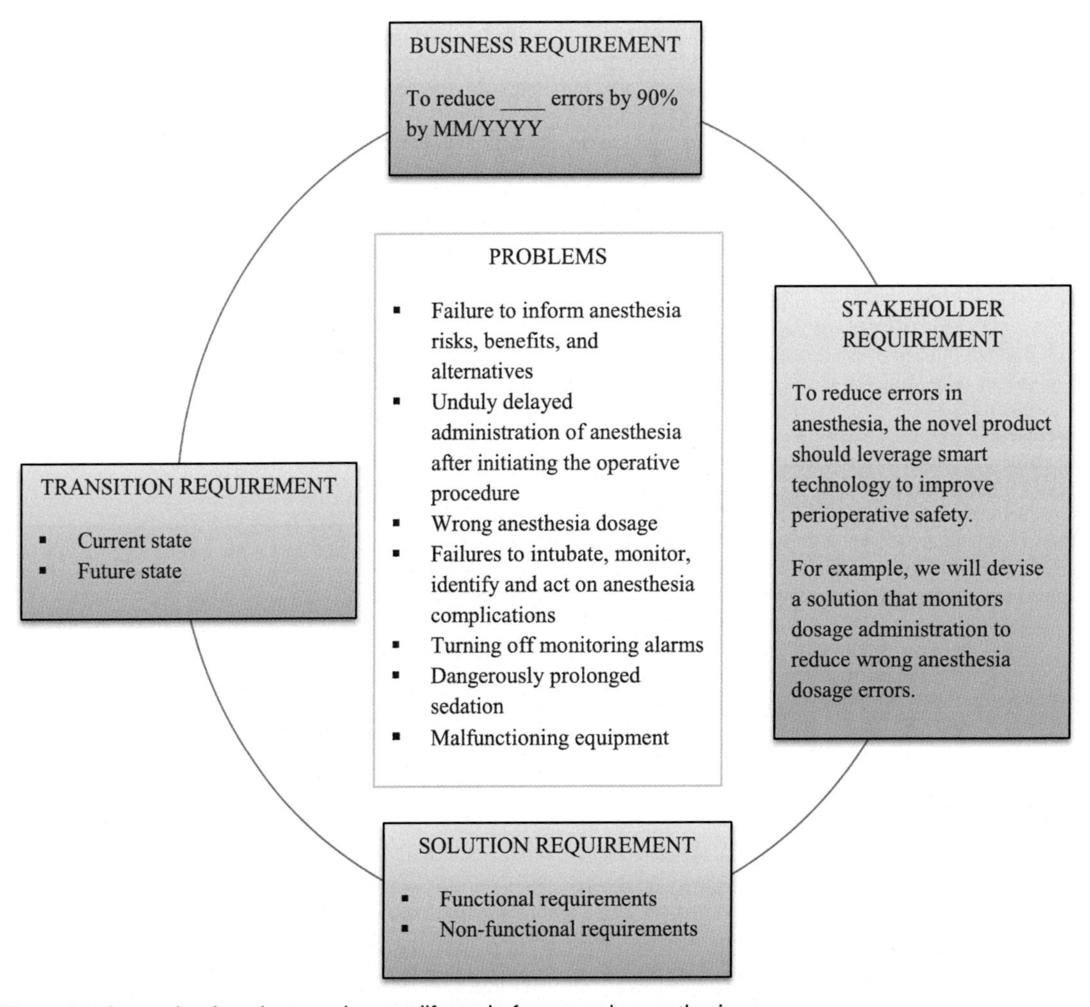

Figure 7.1 A sample of product requirement life cycle for errors in anesthesia.

drawbacks of existing solutions? (2) *Defining the purpose*: What problems will the product solve? Is there more than one solution? Which solution is considered more specific and valuable? What are the outcomes? Are the outcomes measurable? Why are the investors financing the product, and what benefits do they anticipate? (3) *Defining the product scope*: What functionalities and workflows will the product leverage, automate, improve, refine, replace, or omit? (4) *Identifying the stakeholders*: Who can affect the product or whom does the product affect? (5) *Knowing the constraints*: Is the solution within or out of scope concerning time, budget, and expectations? Is the product a worthwhile investment of the resources? What are the risks, costs, facts, assumptions, and alternatives?

Defining problems or needs is essential to determining scope and objectives, drawing high-yield requirements, and devising desirable solutions. Well-defined problem statements pave the path to delivering successful solutions. Furthermore, it is essential to determine who else has a stake or an interest in the product's success, such as developers, designers, testers, financers, and analysts. A lack of input from stakeholders, end-users, and decision-makers directly impacted by the proposed changes may force unsolicited solutions and failed embracement.

Discovering requirements

Discovering requirements is the first step in gathering input from those who will be directly affected by the solution, change, or improvement. In many projects, the time spent on gathering and analyzing requirements is substantial. Requirements Elicitation is the cycle of planning and gathering requirements to have enough data to begin developing a solution. It is necessary to elicit and validate requirements to define both known and unknown needs because they are not readily available or structured for collection upfront. Four groups are created from the requirements during evocation, each representing a different area of expertise held by a different group of stakeholders: (1) complete awareness of existing norms and standards; (2) partial assurance of existing norms and standards; (3) uncertainty about existing norms and standards but awareness of shortcomings; and (4) total ignorance of existing norms and standards.

There are three steps involved in discovering requirements: planning, elicitation, and iteration.

- **Requirements planning**: involves (1) assembling appropriate discussion materials and resources to define a problem or solution or answer questions; (2) identifying all participants involved in decision-making, their roles, availability, and medium of collaboration; (3) ensuring no stakeholders are neglected or overlooked during the requirements induction stage.
- **Requirements elicitation**: setting the stage by defining objectives and stating the value gained if goals are met. Various productivity tools, activities, and technology are available to boost creativity, analytical thinking, and problem-solving. Commonly used software applications and activities to conceptualize product requirements include:
  - *Interviews:* a process of collecting information one-on-one or in groups using structured and unstructured questions. They are the most commonly employed technique in data collection for a wide range of unknowns or problems.
  - *Surveys or questionnaires are a method of probing open-ended, closed-ended,* and contextual and combinatory questions to provoke requirements. While open-ended questions explore uninhibited ideas without targeted solutions, structured questionnaires and surveys coalesce preferential information, limiting creativity and outside-the-box thinking.
  - *Document analysis:* a process of retrospectively analyzing requirements gaps in the work area being studied, problem reports, existing product manuals, organizational charts, industry guidelines, and marketing studies. Some limitations of retrospective analysis include the lack of accessibility of charts, the degree of validity of problem reports, and the timeliness of industry guidelines.
  - *Process analysis:* a method of identifying root causes, delays, errors, and areas of improvement and limiting waste.
  - *Observation:* a process of inspecting and learning the problem areas by showing stakeholders in their work environment to elucidate tasks that are difficult to describe. Although attentively watching a person or tasks may provide realistic insights into problems and requirements, the awareness of being observed changes the performer's behavior, making the findings variable across observers.
  - *Brainstorming:* a creative way to generate and share several possibilities with free associations to solve a problem.

Group activities from different stakeholders rapidly produce many ideas; a noncritical and nonjudgmental environment is highly encouraged to minimize hesitancy in sharing ideas and maximize stakeholder input. Similarly, *workshops, focus groups, and collaborative games* are other ways of gathering innovative ideas from a large number of stakeholders in a short period. Team meetings with different stakeholders facilitate the reconciliation of overlapping requirements. Feedback from subject matter experts refines requirements. The requirements are more likely to be implemented when all stakeholders and solution teams assemble to resolve differences and reach concordance.

  - *Mind mapping:* an innovative way to create a cluster of concepts and solutions and explore problem-centered relationships. However, due to the randomness in individual thought patterns, shared understanding is challenging to achieve.

- **Requirements iteration:** a continuous process of reassessing needs; reordering high-priority needs; detecting missing functionality; managing change based on fluctuations in requirements to ensure consistency, accuracy, and timeliness; gaining consensus on changes; tracing relationships between requirements; and conforming solutions to requirements, including scope, risk, time, and cost.

Analyzing requirements

Analyzing requirements: collected requirements are analyzed to explore details necessary for product development. Analysis generates additional questions to clarify ambiguities, solidify vagueness, and add more information about a particular topic. Evaluating requirements is necessary to filter multiple solution options and select requirements that meet outcomes and constraints acceptable to key stakeholders. Assessment aids in understanding whether the business can accommodate technology change and maintain a sustainable competitive advantage. Root cause analysis using the Fishbone diagram (see Fig. 7.2, root cause analysis in anesthesia dosage errors) and Five Whys helps identify the problem source. Business rules analysis unveils day-to-day operations requirements that influence organizational policies. Data dictionaries and data flow diagrams analyze standard terminology requirements that promote shared understanding. Functional product requirements are analyzed by decomposing

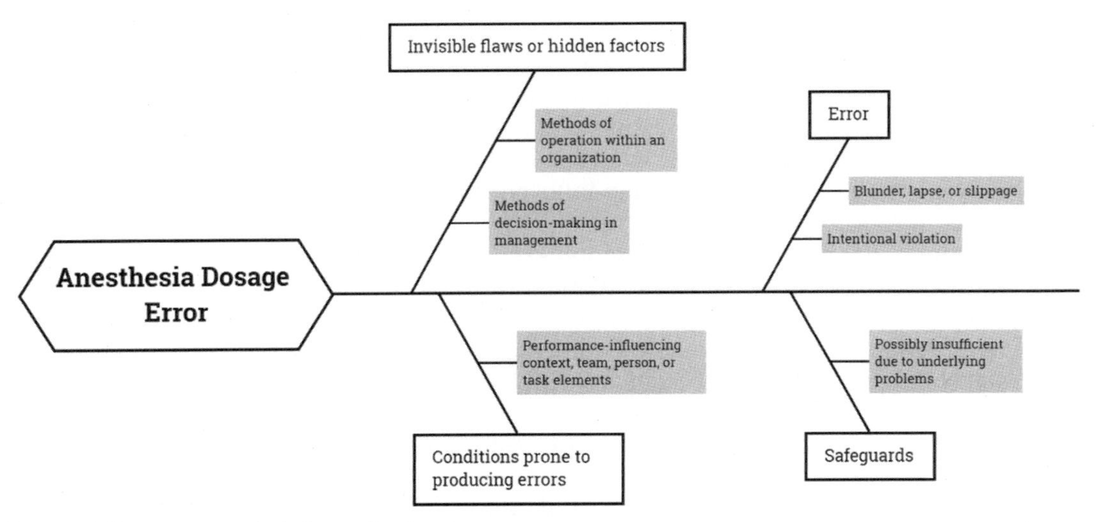

Figure 7.2 Root cause analysis for medication dosage errors.

complex problem areas into functions, sub-functions, processes, and activities using various diagrams—use case, flow, tree, state transition, sequence, components, and others. Nonfunctional product requirements define user expectations; therefore analysis is highly subject to user preferences. Table 7.1 lists a few examples of functional and nonfunctional requirements for anesthesia dosage errors.

Requirements analysis consists of modeling, verification, and validation.

- **Modeling requirements:** a visual representation of goals, features, feasibility, and dependencies between requirements, types of stakeholders, users, and others, to convey complex data, analyze and refine requirements, and find missing information. For example, *process modeling* describes workflows, tasks, or problem areas using a use case diagram, user stories, and process flow; *scope modeling* defines product boundaries using objectives, context diagrams, feature models, organizational charts, and dependencies.
- **Verifying requirements**: a process of proofreading requirements for quality and errors. Verification is done through domain experts, quality assurance, or testing agencies to ensure correctness, consistency, completeness, and clarity of definitions. Standard verification methodologies include item tracking, peer reviews, metrics, and key performance indicators to measure the performance of solutions.

Table 7.1 Functional and nonfunctional requirements using anesthesia dosage error as the problem.

Business requirement	Reduce anesthesia dosage errors by 90% by MM/YYYY.	
Stakeholder requirement	The novel product should calculate the dosage for drug administration safety.	
Solution requirement	Define the function, data, quality, or constraints of the product. The solution requirements are classified into functional and nonfunctional requirements.	
	Functional requirements	**Nonfunctional requirements**
	Functions requirement: calculate dosage amount *Informational requirements*: check the following: medication, units of measurement, patient's age, glomerular filtration rate, and allergies; calculating dilution proportion using a formula or algorithm derivable from the data elements	*Performance requirements*: Is the product fast enough to meet the demands of dosage frequency? Could it calculate drug dosages for multiple drugs? Could it store the information for data mining, transfer, and retrieval? *Constraints*: margin of error, user authentication *Visual design*: ease of access, understandability, ease of navigation, color, font, icons, illustrations, esthetics *User experience*: usability, reliability, compatibility, interoperability, maintainability, extensibility

- **Validating requirements:** a process to ensure requirements solve the problem, deliver the intended benefits, and are within the scope of the expected solution. Techniques for validation include document analysis, financial analysis, item tracking, metrics and key performance indicators, peer reviews, and risk analysis and management.

Business and stakeholder requirements define design options and depend highly on the constraints. Finding the right combination of requirements to design a solution is the key. For example, many number combinations equate to twelve, but not all combinations are the same. Similarly, not all requirements follow a universal format; hence, strategic trade-offs may be required among design alternatives.

Knowledge of change strategy, business requirements architecture, and analyzing potential business value is vital to building the most favorable solution. Proper techniques to develop design options after gathering all requirements include interviews, surveys, document analysis, brainstorming, research of suitable vendors that exceed expectations, conducting benchmarking studies to compare organizational practices against the state-of-the-art, market analysis, root cause analysis, and SWOT analysis (strengths, weakness, opportunities, and threats).

Examples of product development in anesthesiology

A novel device to prevent medication dosage and dispensing errors in anesthesiology

Drug errors in anesthesia are a complex problem that necessitates a multivariate analysis to account for all potential contributing factors. Dr. Xing Wu of the Anesthesiology Department at the Hangzhou Children's Hospital in China sought to devise a system to reduce the likelihood of anesthesiologists picking up the wrong syringe by accident. Several significant processes in a typical clinical scenario involving the identification of medication and equipment, followed by dosing and dispensing, are addressed by the device.

A syringe is loaded into a device, and the user speaks its name and dose into the device, which it stores. The electronic component alerts the user by broadcasting the drug's name when the syringe is removed, prompting a manual double-check by the user. The recording feature of the electronic component allows the drug dosage to be recorded and rebroadcast to users at a later time to remind them of the correct dosage (Fig. 7.3). As a result of this automated redundancy, fewer mistakes and better patient outcomes are expected.

Flutes and acoustical energy for sputum mobilization

Pulmonologists, critical care specialists, and anesthesiologists have sought techniques, devices, medications, and maneuvers to decrease the viscosity of mucus so it can reach larger airways to be eliminated. In 2009 a handheld device was approved by the FDA consisting of a mouthpiece on a hardened, square, plastic flute-like long tube attached to a fluttering Mylar reed. The distal end is flared at the larger base to increase air mass, providing acoustic impedance. When air is exhaled into the "The Lung Flute," the reed will oscillate at about 16–25 Hz, approximately the resonance frequency of pulmonary secretions, lowering their viscosity. The mechanical vibrations thin and mobilize mucus, improving expectoration. The

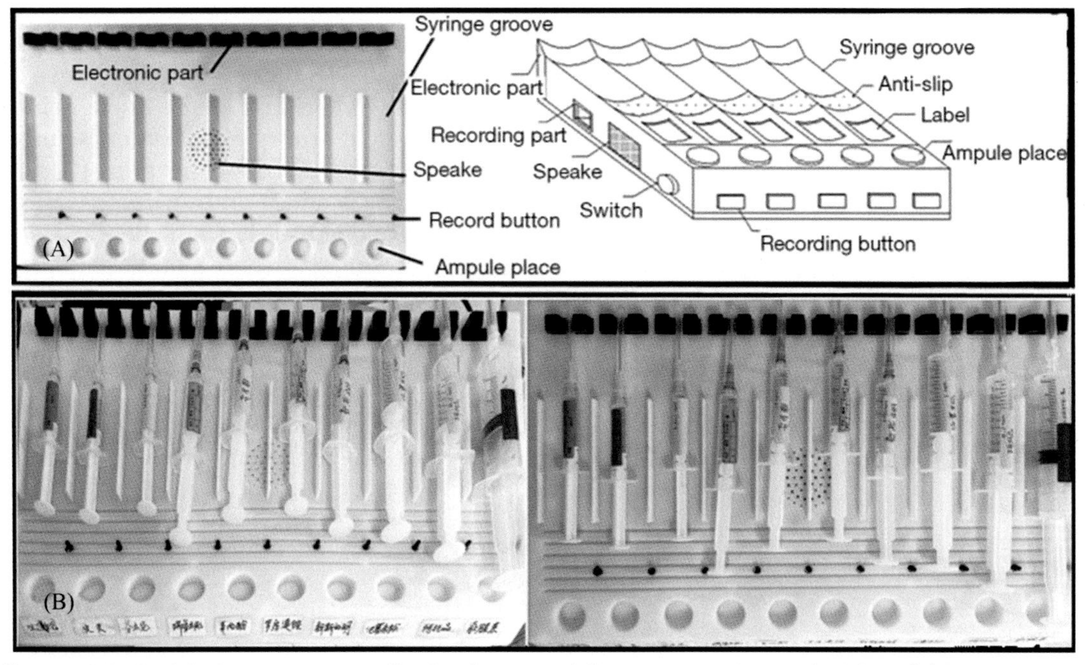

Figure 7.3 A novel device to prevent medication dosage and dispense errors in anesthesiology: (A) Device structure and (B) device prototype. Adapted from Wu X, et al. A novel device to prevent errors in medication dosing and dispensing. Transl Pediatr 2020;9(2):133–6. doi:10.21037/tp.2020.02.05 (CC BY-NC-ND 4.0).

FDA approved the device as a positive expiratory pressure (PEP) device, not as a mechanical sputum inducer (2020). The PEP prevents alveolar collapse, acting like an alveolar "stent." The "Flute" was deemed as effective as its predicate device, Acapella, which produces oscillations using a counterweighted lever and magnet, and also deemed an airway clearance device using PEP Therapy (2010).

However, using the criteria above, while the longer "Lung Flute" was efficacious, patients found the length impractical and unwieldy, making it difficult to use correctly. Some hospitals that did incorporate The Lung Flute or Acapella into their pulmonary rehabilitation program eventually abandoned it, for lack of reimbursement by insurance companies, despite its FDA approval.

Conclusion

Products are engineered by developers, aestheticized by designers, publicized by marketers, sold by sales representatives, and capitalized by inventors. During the development phase, interactions with all stakeholders occur to gather, analyze, prioritize, and validate requirements concerning product development. Viable technology development requires resource-efficient innovation processes and different strategies to design solutions specific to stakeholder requirements. Identifying unmet patient safety needs and technology gaps is critical to user-centered product development. Products developed solely from the innovators' point of view have high failure rates. A shared understanding of product requirements and knowledge exchange between innovators and users is essential for successfully implementing and adopting innovative technologies. Identifying and tailoring requirements to user-specific needs, rather than forcing features and specifications upon them, involves thorough research and understanding of the user environment and preferences. Therefore mapping user requirements to needs plays a critical role in commercialization.

8

Partner with a successful expert

Umme Rumana[1] and Ettore Crimi[2]

[1]Department of Anesthesiology, University of Central Florida College of Medicine, Winston-Salem, FL, United States [2]Department of Anesthesiology, Atrium Health Wake Forest Baptist, Wake Forest University Health Sciences, Orlando, NC, United States

Chapter outline

Abstract

To succeed in today's increasingly complex business environments, it is crucial to cultivate and sustain partnerships that facilitate information exchange and entry into key distribution channels. Creating meaningful alliances helps businesses develop into something bigger than the sum of their parts. Many companies improve their core operations by forming partnerships with experts in related fields such as sales training, marketing, and brand development. Selecting a compatible partner is a challenging endeavor. Look for a business partner who is trustworthy, who is compatible, who has a history of forming strong bonds, who can encourage you to explore uncharted territory and take calculated risks, who can serve as a mediator when conflicts arise, and whom you can count on in times of trouble. Despite best intentions, miscommunication and distrust are the leading causes of a partnership's failure. The success of a partnership depends on the mutual benefit of all involved.

Keywords: Strategic partnership; alliance; contractual partnership; joint ventures; equity partnership; supply chain partnership; partner ecosystem; partnership agreement

Innovation in Anesthesiology. DOI: https://doi.org/10.1016/B978-0-12-818381-6.00011-5

Key points

- Successful businesses continually look for ways to expand their customer base and earnings potential due to economies of scale. These kinds of pressures possibly could lead to the formation of partnerships between competitors that were previously unimaginable
- Whether the scope of the partnership is local, regional, or national, it can be difficult to form alliances with organizations that have vastly different histories, corporate philosophies, and priorities
- Partners should work toward a joint operational mission after the union while appreciating the individual strengths of both parties

Glossary terms and definitions

Strategic partnership An agreement between two or more businesses to pool their resources, knowledge, or information to benefit both parties is a strategic partnership. A corporation frequently looks for a strategic partner to raise the likelihood of success and income, cover a gap in one of its areas of strength, or both.

Alliance is often called a strategic partnership, a relationship between at least two independent organizations seeking to achieve a competitive capability that each cannot achieve independently.

Contractual partnership is a legally binding arrangement whereby organizations that offer complementary services work together to develop and market a new product. It is the most common form of partnership since it is easy to set up and modify, making it suitable for the ever-changing business world. The parties to a contractual alliance forgo the creation of a new joint venture or exchange shares between their respective companies.

Joint venture When two or more businesses work together to accomplish a common goal, they engage in a joint venture. This may be the start of a brand-new venture or the continuation of existing business practices. Each member of a joint venture is solely responsible for his or her share of the venture's gains (or losses) and expenditures (or "shared costs"). Temporary partnerships are formed for specific projects and dissolve once those projects are finished.

Equity partnership A partner's equity in a business partnership represents a person's stake in the business or the proportion of the business they own. On a partnership's financial statement, equity represents the sum of all partners' investments plus retained earnings. An equity partner invests a large sum of money upfront to gain partial business ownership.

Supply chain partnership Collaboration within the supply chain is coordinating between different groups to ensure the most efficient distribution of goods.

Partner ecosystem A group of synergistic partners, plug-in suppliers, and independent vendors that coordinate to serve users with comparable needs about an organization or sector's core technology and complementary product lines. Companies within an ecosystem work together, both cooperatively and competitively, to support new products, meet customer needs, and ultimately incorporate future innovations. Partners provide interconnected sets of services through which users can fulfill various cross-sectional needs in one integrated experience.

Partnership agreement A partnership agreement is a legally binding contract that specifies the rights and responsibilities of each business partner and the operation of the business as a whole.

The partnership

Information and communication technology and informatics innovations such as artificial intelligence (AI), machine learning, cloud computing, genomics (Genome), and others are gaining increased traction in the healthcare sector. According to the Future Health Index 2021 survey, 41% of healthcare executives in 14 countries believe that strategic partnerships and collaborations are crucial to successfully implementing digital health technologies in their hospitals or healthcare facilities. Healthcare organizations seek the development of novel, often technology-enabled solutions to address the expanding customer demands, shifting healthcare standards, increasing disease prevalence, and heightened focus on cost-effective and high-quality care. When healthcare providers and other interested parties work together, they may be able to adapt to patients' evolving needs and improve the quality of care they provide.

Forging healthy partnerships in the era of soaring partner ecosystems that facilitate the sharing of resources and allow access to large distribution channels is the key to rapid success. Yet, according to data compiled in 2007 by the Association of Strategic Alliance Professionals, more than half of all alliances fail within the first three to five years of operation. If the success rate is so poor, why should entities consider forming new strategic alliances? Why not build or acquire resources warranted for success? The answers to building strategic alliances are varied: (1) *partnering aids in rapid growth*: as technology progresses, product life cycles shorten, emphasizing the need for building new products rapidly. To combat this, businesses are increasingly opting to form partnerships, which facilitate the distribution of workloads and the timely production and supply of new products, (2) *partnering conspires to global success*: Because of the increasing demands placed on single businesses as a result of globalization, more and more organizations are forming partnerships across national boundaries to meet the needs of their expanding customer bases, (3) *partnering enhances profitability*: through broad access to a range of industries, geographies, better networks, and buyers more profits are made, and (4) *partnering curbs the cost*: with the fast-changing global economy, overwhelming speed and complexities of new technologies, and competition for innovative assets, scalability remains an imperious challenge, amalgamating resources offer affordability.

Selecting a partner

The purpose of a partnership is to expand commercial value. Securing new clients is a laborious, time-consuming, and expensive process. It often takes a small or startup company years of hard work and significant investments in marketing activities to establish the same level of market awareness as a well-established company brand. Through strategic alliances with established businesses, small businesses can tap into their extensive channel of distributors, dealers, resellers, value-added retailers, and others. Promoting a new brand by associating it with an established one increases its visibility and credibility. It conveys a sense of reliability, dependability, and consistency.

Similarly, larger companies form partnerships with smaller ones to fill their technology gaps, increase their market share, and produce and market a more robust product. It is in everyone's best interest for businesses to keep thriving as sources of innovation and economic independence.

According to a survey conducted by Breezy, a strategic partner discovery firm, finding the best possible partner takes up about a third of the time needed to build a partnership. When looking for a partner, it is essential to ask questions such as: Where are the gaps? Where precisely are the opportunities? How many alliances are needed? How much money and time will be set aside for them? Structures of alliances? Long-term partnerships or temporary alliances?

After gathering partner requirements, create a list of companies and individuals who could be valuable contacts (research partners, conference speakers, suppliers, industry-specific organizations, associations, trade shows) and board members who are experts in their field, chief executive officers, chief technology officers, product managers, vice presidents of sales and marketing, and so on. More can be learned about the potential future partners' by researching LinkedIn groups, user forums, blogs, conferences, social media platforms, and market report feedback.

Selecting a partner who shares the passion, is dependable, is a good fit, has experience building strong relationships, has a proven financial sense, is creative, is open to new ideas, is willing to take risks, is a good mediator and problem solver, and can bounce back from adversity are essential. Along with carefully curated alliances, businesses may choose to invest in personal connections, transparent contracts, and vetted partners.

After selecting potential partners, a partnership agreement or contract is set between one or more businesses that defines initial and future partner contributions. The document outlines business decision-making, partnership percentages, business management, and more. Contracts should be read and negotiated carefully to protect oneself and others (Table 8.1). Successful commercialization requires a thorough familiarity with the various business alliances (Fig. 8.1), partnerships (Fig. 8.2), and partners (Table 8.2).

Potential pitfalls to a successful partnership

Because each partner views themselves as the leader, partnerships tend to escalate into power struggles. That is one of the many causes of failed business collaborations. Companies join forces hoping to accomplish great things. There are several avoidable causes for their demise. Likely stumbling blocks in healthy partnerships include (1) lack of forethought in negotiating deal terms, despite having a clear vision and strategy in place, (2) going straight into contract details before fully fleshing out the long-term goals and strategic direction, (3) not being clear on the business plan while defining the operational and governance plan, (4) making a mistake in prioritizing deal-breaking factors, (5) both a lack of leadership dedication and a lack of reliable accountability, (6) failure to monitor progress or measure the partnership success periodically, and (7) not having a plan for when the partnership is no longer working out or no longer bringing in the desired results.

Additionally, many obstacles must be overcome when working together, and one of them is sharing leadership or difficulties in delegating authority. Partnerships are fragile. Relationships rely on

Table 8.1 Listed are some contracts terms as well as specific contract recommendations.

Commissions	Explain how you'll make money and what you'll need from the large company if they don't pay a commission or royalty fee.
Commitment	The optimal time frame is three to five years. Try negotiating a three-year term with automatic renewal after the first year and a month's notice before the contract is no longer valid. Obviously, a year is not enough time.
Conflict resolution	Define who is responsible for resolving conflicts? Describe in detail how and when a problem is escalated. Point people on both sides will take care of business-related matters, relationship management, pricing, and technical/developmental difficulties.
Discounts on a sliding scale	In the contract, volume or percentage price breaks may be stated. Ask the team if these discounts are reasonable based on the sales projections given during negotiations.
Exclusivity	You are limited in your ability to explore other options in the market if you sign an exclusivity agreement. Don't cave. If you can't agree, attempt a six-month exclusivity term. A better technique is to limit the exclusive terms to a selected group of merchants for a limited time.
Exit plan	Corporate mergers and acquisitions pose a risk to interpersonal connections. Unfairly rejecting new ownership should be expressly forbidden in contracts. Make sure you have safe exits planned. During the 90-day notification period, the acquiring company will not make any acquisitions. Explain how you intend to keep the relationship productive for both parties.
Intellectual property (IP)	Your most valuable asset is IP, and the top three reasons partnerships fail all have to do with IP issues. Define ownership. If a product is created together, create an application programming interface (API) to separate code and IP.
Liability and repayment	The weaker party should not agree to an overly onerous indemnification agreement. Finding a way to reduce the financial impact on your company by negotiating liability insurance is a sure bet.
Marketing	The inclusion of the company's name with the product name of the partner is key to successful marketing communication. This can include: Joint press releases, marketing at trade shows and events, access to customer lists, and other key communications pieces. It can also mean an OEM-free white-label agreement.
Planned events and benchmarks in the development of the product	Plans must be kept, and promptness is required. When planning a launch, it's important to set realistic deadlines before development begins. However, if possible, build in some flexibility to account for delays that may occur due to events beyond your control on either end.

(*Continued*)

Table 8.1 (Continued)

Prepayment	Spending money on development detracts from product goals and future ambitions. Request an advance on the first fees before signing the contract.
Source code	If you go bankrupt or can't fulfill your contract, put the source code in with a third party that holds two parties' money and property until all contract terms are met.
Support	It's best to avoid Tier 1 support. Due to the disparity in company sizes, you may be inundated with calls. Instead, provide advanced developer support.
Training	Training improves partnerships. Smaller partners may not be able to afford sales and tech support training. To use the larger company's training channels and resources, negotiate for training materials and programs.

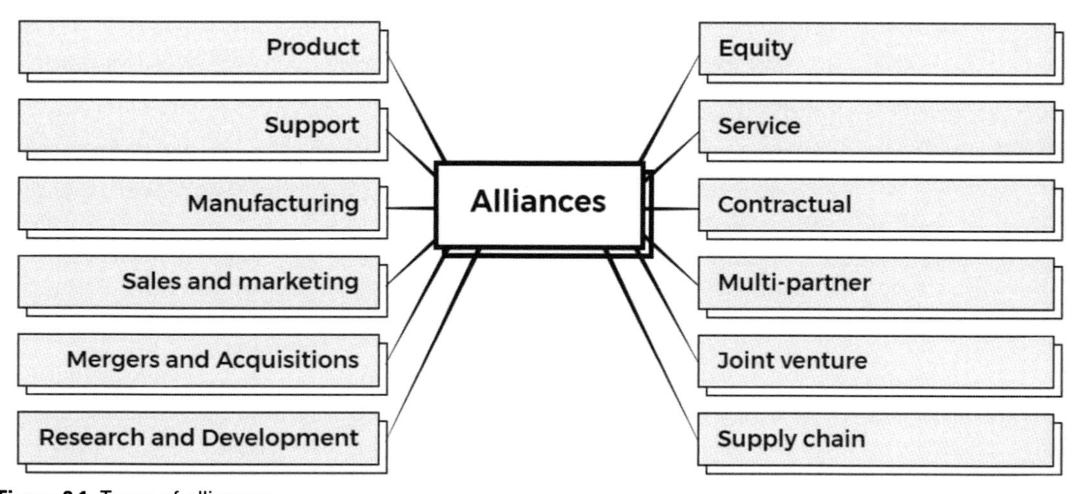

Figure 8.1 Types of alliances.

people and are therefore vulnerable. Ineffective two-way dialog and distrust between partners is the leading cause of partnership breakdown. Unrealistic expectations, lack of hierarchy, open-ended contracts as opportunity and threat, breach of the confidential

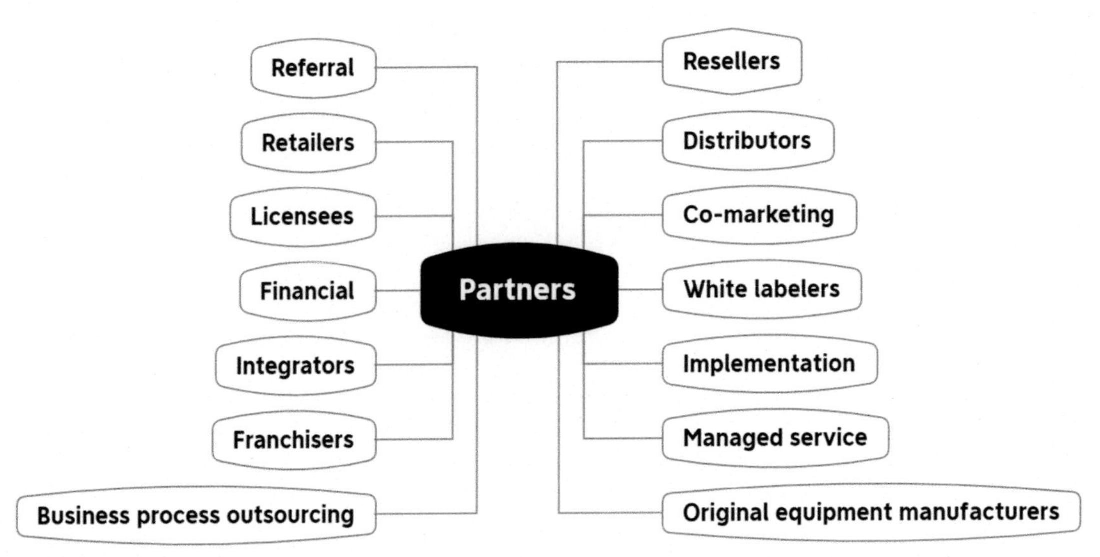

Figure 8.2 Partner types.

Table 8.2 Types of business partnerships.

Sole proprietorship	Sole proprietors offer services without being employees or owners. A sole proprietorship doesn't cover personal losses. Only bring on a partner if he has the necessary funds or contacts.
General partnership	If you agree to work together on a business, you have a General Partnership. This is easy to get, like Sole Proprietorship. In this type of partnership, profits and losses are split evenly.
Limited partnership	Limited Partnerships work like General Partnerships, with managing partners maintaining responsibility and liabilities. This structure is used for larger companies with silent partners. Limited partners are liable only for capital invested and have no say in day-to-day operations.
Limited liability company (LLC)	LLC are taxed at the personal level, not the corporate rate. This company type is tax-friendly. If a lawsuit is filed against a LLC, the owners and employees are not personally liable. Having at least two members has legal and tax advantages.
Corporation	A corporation is a separate tax entity and needs a separate return. Corporations have evolved over time, and case law is vast. Most small businesses lack the resources to comply with these laws and opt for the LLC format. It's easier to go public and sell than LLCs.
S Corporation	S Corporation combines Corporation and Partnership. S Corporations have all the benefits of Corporations but file taxes like Partnerships. This structure requires extensive paperwork.

shareholder agreement, fear of the alliance being disbanded, disputes, accusations, lack of transparency, betrayal, and parties not always functioning in the desired manner cause 60%–65% of strategic partnerships to fail to achieve their objectives. When there are too many rules to break, creativity and innovation suffer, while when there is too much leeway, selfish motives can upset the partnership dynamic. Partnership success calls for ongoing observation, communication, and adjustment of expectations.

Strategies for a successful partnership

For a partnership to be successful, all parties involved must benefit something from the business agreement. Look for a solution that benefits both parties. To have fruitful partnerships and reduce the likelihood of failure, it is essential to employ the following strategies: cultivate long-lasting trust, unwavering mutual respect, united belief, shared humility, nurturing generosity, and compassionate empathy; communicate openly and honestly; be customer-centric; celebrate individuality; be results-oriented; be professional; work together effectively; focus on shared goals; be committed to achieving those goals; expect to make blunders and be responsible for taking accountability; resolve conflicts and make learning from conflict a priority; maintain composure in the face of adversity; convert disagreement into creativity.

Examples of partnership in anesthesiology

Emergency partnerships

During the COVID-19 pandemic, multiple partnerships became necessary to overcome resource limitations and maintain patient care. With the imposition of quarantine, previously underutilized technologies, such as telemedicine critical care, rose to the forefront. Engineers from NASA's Jet Propulsion Laboratory collaborated with the American company STARK Industries to create a simple, low-cost ventilator called Ventilator Intervention Technology Accessible Locally in response to a worldwide shortage of ventilators during the COVID-19 pandemic. The FDA approved their use in an emergency, and STARK Industries was permitted to mass-produce ventilators worldwide.

Public–private partnerships

The International Anesthesia Research Society and the Food and Drug Administration of the United States cohosted a meeting in 2009 to establish public–private partnerships to find novel solutions for developing safe anesthetics, sedatives, and analgesic drugs. Partners from academia, industry, and the National Institutes of Health have collaborated to develop novel drugs. Research projects related to this goal include the SAFEKIDS Initiative (to ensure the safety of key inhaled and intravenous drugs for children); the SmartTots Program (to study anesthetic neurotoxicity in developing brain); ACTION (the Analgesic Clinical Trial Innovations, Opportunities, and Networks); The INOVATE-Pain (Interdisciplinary Network on Virtual and Augmented Technologies for Pain management). Equally, several public–private strategic partnerships have developed in the area of medical AI. For instance, Microsoft, Omron, and SoftBank have collaborated with universities to create AI technology that uses natural language processing tools, patient biometrics, and voice recognition to identify psychiatric derailments in intensive care unit patients. Data scientists, developers, and engineers from the Massachusetts Institute of Technology have joined forces with doctors from the Japanese Society of Intensive Care Medicine to advance intensive care medicine with AI.

Academic partnerships

Through collaborative research and teaching, universities in countries with scarce resources have closed the training gap. For instance, to better equip Ethiopian anesthesia residents with the knowledge, skills, and confidence to perform emergency cricothyroidotomy, the University of Michigan's department of anesthesiology has developed a low-cost, high-fidelity 3D printed airway simulator and Cricothyroidotomy Skills Maintenance Program. Another case in point is the expansion of surgical and anesthetic services in 16 low-resource countries through the support of 24 peer-to-peer training initiatives funded by a health partnership scheme based in the United Kingdom. Similarly, Kybele, a multinational humanitarian organization, has established stable global health partnerships in Croatia, Serbia, and Ghana to improve maternal and newborn outcomes during labor and delivery. Kybele used a partnership model that linked anesthesiologists in the United States and Europe with doctors in developing nations to teach them how to provide safe obstetric anesthesia. Kybele's guiding principles for the collaboration were champion engagement, stakeholder support, partner involvement, and data-driven discovery.

9

Conceptualizing idea to solution

Chunyuan Qiu
Anesthesiology & Perioperative Medicine, Kaiser Permanente Southern California, Pasadena, CA, United States

Chapter outline

Abstract

Anesthesiologists undergo a long, vigorous training process that is a double-edged sword. While the pilot-like learning framework greatly benefits patient safety, curiosity and imagination are handcuffed. Anesthesiologists who can free themselves from these constraints and venture out of their comfort zones are in a better position for innovation and creativity. Innovation and creativity begin with ideas, which are converted into concepts, which are transformed into solutions. While many ideas will not come to fruition, collective efforts eventually often lead to a positive result. When they do, the process of converting ideas into solutions is the necessary passage to success. This chapter provides an overview of how to overcome the specific barriers unique to anesthesia professionals and convert creative and innovative ideas into concepts.

Keywords: Idea; concept; solution; innovation; creativity; anesthesia

> *If you can dream it, you can do it. Always remember that this whole thing was started with a dream and a mouse.*
>
> **Walt Disney.**

Anesthesiology is at the forefront of healthcare reform today. Despite providing excellent cutting-edge care to surgical patients through preoperative optimization, intraoperative management,

Innovation in Anesthesiology. DOI: https://doi.org/10.1016/B978-0-12-818381-6.00020-6

and postoperative enhanced recovery processes, anesthesiology is now facing challenges in redefining its services, cost, and value. Historically, academic institutions have led the way during uncertain times, but this traditional model may not be enough at present. Generally, research funding for anesthesiology has been decreasing year after year. Coupled with the uneven distribution of research funds, which are highly concentrated in the top ten academic anesthesiology departments [1], the advancement of anesthesiology has been lagging and the scope of its competitive advantage has been narrowing. Consequently, anesthesiology is at elevated risk of being stalled in the new era of this emerging healthcare industry. By unleashing innate creative ability and transforming ideas into a practical reality, today's anesthesiologists can reverse the trend and build a brighter and better future for generations to come.

Anesthesiology attracts very talented and hardworking people, but their creative talents and innovative capacity are often underappreciated or overpowered by daily routines and busy clinical schedules. Anesthesiologists are preoccupied with rules, regulations, evidence-based practices, routines, and standards. They inherited a double-edged sword whereby safety protocols and checklists have generated a remarkable safety record over the last few decades, but have also positioned anesthesiologists in a direct collision course with the fundamentals of creativity. *When in doubt, go back to the basics.* This basic belief system has warded off innovation for many. Anesthesiology, like many medical specialties, is an "evidence-based practice" that is tried and true but often outdated, incomplete, and slow in adopting anything new. Rather than attempting to generate new evidence to replace the old, the "pilot-like" training in anesthesia effectively has deprived anesthesia providers of their born capability of curiosity and imagination. Anesthesia professionals must decouple themselves from this hardwired framework before they can embark on the journey of a creative life. In this journey, their motto should change to *When in doubt, adventure out.*

The declaration that waste exists in anesthesiology may immediately evoke concepts of poor operating room productivity or efficiency, medication costs, and wasted supplies. However, the biggest waste in anesthesiology may be the underutilization of intellectual power within each anesthesia professional—from anesthesiologists to certified registered nurse anesthetists, anesthesia assistants, and anesthesia technologists. Currently, converging technologies such as robotics, machine learning and artificial intelligence, 5G communication, and big data are accelerating both in velocity and scope. When considering that medical, and specifically anesthesia, practices are still largely based on antiquated methods and a slower pace of innovation than some other industries, the gap between the technological capabilities of other industries and the current standard of anesthesia practice is embarrassingly large. Anesthesia professionals cannot afford to waste their creative talents or delay the use of their full intellectual power. It is the time to execute one's genius.

Idea, concept, invention, and innovation

There are subtle but fundamental differences between ideas and concepts, inventions and innovations. An idea is the primitive form of novel thinking that often presents as nonverbal mental images. These images often reside in between one's subconscious and conscious worlds. The concept is the cultivated and granulated actualized appearance of this novel thinking, often descending from one or a cluster of related ideas. This concept is the edited form of ideas that can be described by words, figures, or mathematics.

Invention, also often referred to as creativity, is comprised of novel and original ideas that solve or address actual or potential challenges. The invention is characterized by two distinguished requirements: novelty and usefulness. Innovation is an implementation method and process by which ideas are brought to life. Invention and innovation are equally vital for bringing out the intrinsic value of an idea; however, they contribute to the final success differently. For example, Steve Wozniak and Steve Jobs are two technology giants in the history of Apple and technology in general. The two Steves are very different but very complementary. Steve Wozniak was a genius responsible for the creation of Apple I and II, while Steve Jobs was a genius responsible for the innovative sales and marketing that eventually brought Wozniak's invention to life. Steve Jobs famously said, "Some people say, give the customers what they want, but that is not my approach. Our job is to figure out what they are going to want before they do." Invention and innovation must go hand in hand for ideas to be successfully realized, whether they are carried out by one individual or a group of people.

Anesthesiology needs both inventors and innovators, people who can start transforming ideas into concepts and then finding solutions on a clear path. As a person experienced in both clinical and technological innovation, as exemplified by the creation of the perioperative surgical home [2–4] practice and by many patented new ideas [5], I will guide you, my colleagues, on how to sow, grow and harvest a new idea; how to conceptualize ideas into productive solutions. It is easier than you think, and they are within our reach.

Ideas are everywhere

Ideas are novel mental images or expressions arising from connections of existing information and knowledge that have been warehoused in the intangible recesses. They are simply new neural crossings among the existing neurons where the information or knowledge was harbored. By connecting previously unlinked neurons or mental dots, a new idea is born.

New ideas are miracles regardless of whether a similar idea has already existed in another person's mind or been actualized in the world. The new idea and its underlying neural connection are more valuable to you, the beholder, as Descartes said, "I think therefore I am." Anesthesiologists already possess a large built-in capacity for great ideas, based upon the statistical probability of novel connections given so many years of training, exposure, and information absorbed. Unfortunately, this advantage is underexploited by many anesthesiologists.

If ideas are everywhere, then why can we not be more creative than we are right now? It turns out that past training and current misunderstandings about ourselves are inflictions. Our anesthesia training has, in some respects, preprogrammed our brains to follow instead of create, which is a significant barrier to overcome. We may mistakenly assume our lack of creativity is due to a lack of knowledge; rather, the problem is that how we are trained to think and practice has been inhibiting our curiosity and clouding our imagination. We need to detach ourselves from our programmed past and be confident about our innovative and inventive capabilities. Einstein once said, "Imagination is more important than knowledge, for knowledge is limited, whereas imagination embraces the entire world, stimulating progress, and giving birth to evolution."

Generating new ideas

Let's hone it in! Unique ideas may emerge for anesthesia professionals in two ways:

One manner is by zooming in on proven logic and theories of accumulated knowledge, and then examining the gaps and flaws between what we know and what we are doing. The obvious disconnects occurring frequently in clinical practice are begging for different answers or better ideas. Many incremental progressions in the field of anesthesiology are due to fresh ideas that come out of the gap in our practice.

The second route of idea emergence occurs when we zoom out from the intrinsic logic and theories and start questioning. All theories and practices contain inherent limitations, conditions, and assumptions of application. The truth is that a lot of the theories, practices, and beliefs in anesthesiology are old and older than our ages. Their validity in current "omics" and precision medicine needs reevaluation, which, to a large degree, has not been done. For example, the safety of anesthesia is remarkable when mortality is the main criterion; however, from the angles of surgical site infection, cancer recurrence, brain health, and functional recovery [2–4], the record of anesthesiology is less pristine. Disruptive ideas and revolutions in anesthesiology often originate from here. By simply following these "zoom in and zoom out" methods, the streams of ideas will come out of one's subconscious world, especially when one locks onto terms such as "perfect," "golden standard," "routine" or "ambiguities," "impossible," "paradox," "gaps," and "dilemmas."

A person may generate 80,000 to 100,000 ideas per day. The novel ideas may come from different bodies of existing knowledge or an internal alternative reality such as hobbies, interests, and arts. The human brain has over 100 billion neurons and countless synapses encompassing many scientific principles of which anesthesiology is only a small part. Often stored and used in silos, these principles can be and should be connected through purpose-driven rewiring and restructuring. Currently, there are large gaps between advanced technology and the healthcare industry. There is also uneven penetration of these technologies within each of the healthcare specialties. This concerning trend is manifested by the fact that the healthcare industry appears to be lagging behind current global scientific and technological standards. When this is coupled with ineffective implementation, there is an enormous creative and innovative opportunity for anesthesia professionals. By "borrowing" and "imagining" the new technologies in the field of anesthesiology, one can generate countless new ideas. For example, robotic technology in surgery was first created at the Stanford Research Institute in the 1980s under a US Army contract. It was not fully leveraged in clinical practice until the creation of Intuitive Surgical, Inc. in 1995, a 15-year gap [6], and it still has yet to be introduced to clinical anesthesiology. Anesthesia professionals should not underestimate the power of creative borrowing, as it is a proven source for disruptive ideas and new practices as well as a potential for new billion-dollar companies. After all, Google started with the idea of combining a digitalized library with the World Wide Web, and the smartphone was created by marrying the cellular phone with a personal computer.

For serious inventors, setting up an idea quota is a useful exercise, which should be initiated early. As mentioned before, there are thousands of ideas that go through one's mind daily. While a majority of them represent nothing more than a simple flash of subconscious activities, some of them do surface to our explicit world. It is not difficult to capture one or more new ideas a day, or 10 or more ideas a week, or hundreds in a month. However, few of us do. The reason is clear. Only a few of us trust the importance and power of our ideas. This idea of quota exercise can change that. It will help us to realize the genius within ourselves and reclaim lost curiosity and imagination. It will drive us to the limit and edge of the ordinary, where the extraordinary is bound to happen. When ideas strike, we must be ready to capture them.

Getting it on paper

Once the ideas bubble out of our subconscious universe, they warrant a fair chance of being acknowledged and incubated. We need to treat our ideas with respect and pride. If it is our idea, then it deserves to be recorded in a written history of ours. Ideas are only temporary new connections or synapses between neurons, flashes of mental images that have to be transformed from inexplicit images to their explicit copies in words and figures. Simply writing them down is the least we can do for our ideas. Granted, a vast majority of these ideas are worth "nothing" in the traditional sense and may be "rejected" by even yourself in the next hours or days; however, the value of such ideas in underlining new neural connections cannot be underestimated. When they are simply written down, those seemingly useless ideas become a precious fuel that can power a long journey. They are the new fertile ground with all the necessary ingredients for growing secondary ideas, which may be even more productive. Ideas beget ideas, so do not fail to recognize the value of such a simple act of documentation. Famous for his abundant [failing] ideas, Thomas Edison was well known for keeping several notebooks at arm's length, jotting down ideas and sketches whenever his ideas would strike him.

Historians were perplexed when attempting to figure out the inner connections between his ideas, but this habit served him well in materializing his inventions and protecting him from infringement by others in his lifetime. Many of Edison's ideas failed to produce meaningful products, but some of his inventions markedly changed our world. Edison once famously argued about his failed ideas and inventions, "I have not failed; I have just found 10,000 ways that won't work."

"If it was not documented, it did not happen," a medical-legal motto with which anesthesiologists are familiar, also holds for our ideas. Documenting the ideas are important milestone regardless of their level of "usefulness," "genius," "stupidity," or "craziness." Documentation has at least three purposes here. First, it is a birth certificate for our ideas, a proof of their existence that is beyond any reasonable doubt. Even though they are mainly for our use, they are still sacred to us. Secondly, capturing the ideas in words and figures can help us find their meanings later. As Francis Bacon said, "Writing [makes] an exact man." Frequently, our newly surfaced ideas are abstract in their presentation and may not even have an obvious connection to a problem of which you are aware at that moment. By scripting them down and visiting them later, after a period of incubation, the inner links between the idea and its real meaning can become clear. Third, time-stamped documentation can serve you well in any unforeseen legal battles if your invention conflicts with others. Jot down ideas any way you want: a pen and piece of paper, notebook, smartphone, voice recorder, or mixed methods. Generally speaking, a system and method that allows you to quickly and conveniently record ideas at any time of the day, even in the middle of the night, are preferable. Once the idea grows in scope, you may want to group multiple idea components and categorize them for easier access later. Categorizing ideas may seem trivial or even unnecessary at the beginning when you only have a few ideas with which to work, but it will serve you well in the long run.

Concepting idea

Concepting an idea into a solution is to validate the idea in a goal-directed way, in which both the value and novelty need to be determined. Far different from clinical practice or a board exam, there are no wrong ideas or concepts in the creative world but simply many alternatives, which may be better or simpler than the original one. As a result, we need to look for the siblings and cousins of the original idea, and then extend it to the other relatives in the same family tree or beyond. All ideas in their raw form are only a temporary new link between a problem and a solution that may or may not be true, necessary, or important. Only granulated ideas that can pair a problem with a unique solution, both correctly and proportionally, possess the potential to become a concept. While the purpose of generating ideas is to have as many new ideas as possible, the process of conception is all about idea reduction and selection. Conception involves identifying the best idea from an abundance of them. The most important step in concepting ideas into a solution is to study the relationship between your idea and the identified problem.

Though a detailed discussion about the relationship between problems and solutions is beyond the scope of this chapter, evaluating the intimate relationships between the two in terms of "correlation," "association," and "causation" will assist you in the selection process and help you to convert your ideas into solutions. The more you know the problem, the deeper you understand its nature and the better you can assess the gap between your idea and the problem you are trying to solve.

While ideas can come from anywhere, even in random or spontaneous spurts, concepting ideas into solutions is a purpose-driven process that requires dedication, focus, and strategy. Not all ideas can or should go through the concepting process. We should allow our brains to go wild, uninhabited, and unrestricted so that the original idea can flow unobstructed with the additional help of more imagination—almost like a meditative state, where we purposely allow ourselves into a promising environment, where the act of concepting can invite us to a perfect world without affixed restrictions, conditions, or assumptions. Here, matching our idea with the problem can be done without the burden of technical feasibility, practicality, or even cost. Those will be confronted later, but not right now. Often, we will be happily surprised that if we can dream it, then it will be made for us. Recall Walt Disney's quote at the beginning of this chapter, "If you can dream it, you can do it. Always remember that this whole thing was started with a dream and a mouse." However, if our ideas cannot measure up to the magnitude of an identified problem even in this perfect dream world, then there is little reason to move on beyond this point. Note that this is different from "giving up." For example, the problems one is facing may not be universal, or one's solution may prove even worse than the problem itself. It is also entirely possible that one's original idea may cease to excite him or her after this rigorous process. Normally, only a few ideas can pass through this stage of inventorship.

Here is a real story of one idea's evolution into a concept and progression into a remarkable reality. Larger in death than himself in life, John F. Kennedy (JFK), the 35th President of the United States of America, was a legendary President with great visions. Using his unparalleled talent in articulating and communicating such visions, JFK presented his initial idea of landing the first human on the moon to Congress on May 25, 1961, "before this decade is out, of landing a man on the Moon and returning him safely to the earth." He then illustrated his detailed concept in his "we choose to go to the moon" speech at Rice University on September 12, 1962, in which he concepted his earlier idea into a solution. "If I were to say, my fellow citizens, that we shall send to the moon, 240,000 miles away from the control station in Houston, a giant rocket more than 300 feet tall, the length of this football field, made of new metal alloys, some of which have not yet been invented, capable of standing heat and stresses several times more than have ever been experienced, fitted together with a precision better than the finest watch, carrying all the equipment needed for propulsion, guidance, control, communications, food and survival, on an untried mission, to an unknown celestial body, and then return it safely to earth, re-entering the atmosphere at speeds of over 25,000 miles per hour, causing heat about half that of the temperature of the sun—almost as hot as it is here today—and do all this, and do it right, and do it first before this decade is out—then we must be bold." This speech, from idea to detail concept, was an impossible mission judged by many including many leading scientists in that era. The impossible became reality on July 21, 1969, 6 years after his tragic death, when Neil Armstrong and his fellow Apollo crewmember, Buzz Aldrin landed on the moon, then safely returned to earth on July 24, 1969, after 8 days in space.

In the triumph or failure of the ideas and concepts, it is not the eventual success that defines it but rather the many failures along the way that, ultimately, make success possible. Start the exciting journey of creativity and innovation! There is no better

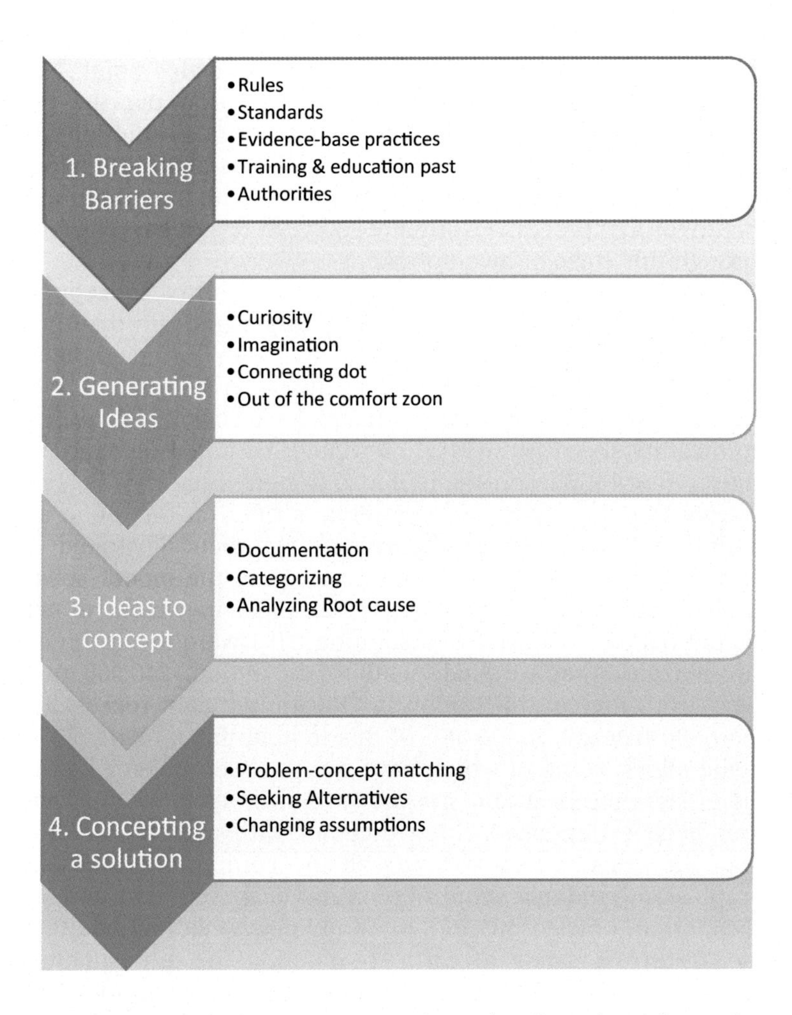

Figure 9.1 Four steps for concepting idea into a solution.

time than now. Fig. 9.1 summarizes the four interdependent steps for generating and harvesting ideas and concepting them into solutions.

References

[1] Chandrakantan A, Adler AC, Stayer S, Roth S. National Institutes of Health–funded anesthesiology research and anesthesiology physician-scientists. Anesth Analg 2019;129(6):1761–6.

[2] Qiu C, Lin JC, Shi JM, Chow T, Desai VN, Nguyen VT, et al. Labor epidural analgesia and risk of autism spectrum disorders in offspring. JAMA Pediatr 2020; Published online.

[3] Qiu C, Cannesson M, Morkos A, et al. Practice and outcomes of the perioperative surgical home in a California integrated delivery system. Anesth Analg 2016;123(3):597–606.
[4] Qiu C, Rinehart J, Nguyen VT, et al. An ambulatory surgery perioperative surgical home in Kaiser Permanente settings. Anesth Analg 2017;124 (3):768–74.
[5] JUSTIA. Patents by inventor Chunyuan Qiu, <https://patents.justia.com/inventor/chunyuan-qiu> [updated 08.09.20, accessed 09.10.20].
[6] Saddawi-Konefka D, Schumacher DJ, Baker KH, Charnin JE, Gollwitzer PM. Changing physician behavior with implementation intentions. Academic Med 2016;91(9):1211–16.

10

Prior art searching

Douglas Portnow

Douglas Portnow, Schwegman Lundberg & Woessner, P.A., Minneapolis, MN, United States

Chapter outline

Abstract

Prior art searching is an essential tool for determining which aspects of an invention are patentable as well as for investigating whether a product infringes any third-party patents. While not mandatory, conducting prior art searching will help reduce your risk of liability and strengthen your intellectual property position when dealing with patent examiners, investors, strategic partners, or other parties interested in intellectual property.

Keywords: Patent; prior art; searching; novelty; obviousness; freedom to operate; product clearance; validity; invalidity

Innovation in Anesthesiology. DOI: https://doi.org/10.1016/B978-0-12-818381-6.00029-2

Introduction

Medical device companies rely heavily on intellectual property (IP) since the industry is highly litigious and a fundamental part of a company's IP strategy should include prior art searching.

Prior art searching is a powerful tool that can help an organization analyze its IP position. Two main IP issues are typically addressed during the product development and fundraising or exit phases of a company's evolution. The first and perhaps most important is freedom to operate (FTO). FTO is also sometimes referred to as product clearance and involves searching and identifying relevant patents owned by third parties that block you from making, using, selling, or offering to sell your product or service. Without FTO, you are at risk of being sued for patent infringement which can result in royalty payments to the patent owner or an injunction that shuts down business operations.

Prior art searching is also helpful with patentability studies. Patentability refers to analyzing an invention to determine what aspects of the invention are patentable and protectable. This analysis may be applied before filing a patent application, or during the patent application process, or it may be applied retroactively to a granted patent to determine if the granted patent is valid or invalid. In either case, the claims of a patent or patent application are reviewed and compared to the prior art. The claims are what define the boundaries of the invention protected by the patent. Anything that is claimed that has a useful purpose, is novel and nonobvious over the prior art is patentable, and therefore results in a valid patent claim. If all the elements of a claim are disclosed by a single prior art reference, then the claimed invention is not novel and not patentable since it is not new. An invention may be obvious over the prior art if several prior art references can be combined to teach the claimed invention, or if the claimed invention is an obvious variation over the prior art.

To obtain a patent, an invention as defined by the claims must be novel and nonobvious over the prior art. Therefore conducting a prior art search provides a better understanding of an invention's novelty or nonobviousness compared to the prior art. In the situation where the invention is your own or your company's, knowing the prior art helps you to assess what sort of claims may be possible in your patent applications and therefore determines the scope of patent protection that can be obtained. In the case of a third-party patent or patent application, knowing the prior art helps you to understand what claims the third party can either obtain or if already issued in a patent, knowing the prior art will help you understand whether those claims are or may be valid or whether those claims may be invalid. If invalid, a third party's patent will not be a concern from an FTO perspective.

Prior art searching helps the analyst look into the proverbial crystal ball and evaluate patentability and freedom to operate positions which are essential in crafting an IP story about the strength of a company's intellectual property position. This is often necessary during fundraising, or an exit strategy such as a merger, acquisition, or an initial public offering (IPO) since investors, acquirers, or underwriters evaluate intellectual property during a business transaction to minimize their risk.

What is prior art?

Understanding what counts as prior art is critical so that search efforts can be properly performed, and only identify relevant prior art. Prior art is considered any publicly available information anywhere in the world, that describes aspects of your invention and predates the priority date of your patent application or the patent being evaluated. If the prior art discloses your exact invention as claimed, then the prior art is relevant to the novelty of your invention and if the prior art discloses a similar invention to yours, then it may be relevant to whether your invention is obvious.

To obtain a U.S. patent, your claimed invention must be novel and nonobvious over the prior art. Prior art that has claims that cover your invention is relevant from an FTO perspective. Prior art is most often in the form of a printed publication that describes the invention in writing, illustrations, or a public display of the invention such as demonstrating the invention at a conference. However, prior art is not limited to these constraints, as it can be anything that teaches your invention before your priority date (the date when you first filed a patent application disclosing your invention). Some common examples of prior art include printed patents and published patent applications, scientific literature, marketing brochures, videos, demonstrations at a trade show, presentations at a conference, sale of the invention, an offer to sell the invention, nonexperimental use of the invention, internet websites, school theses, etc. Thus prior art is construed broadly to include a wide array of publications and events.

One important point is that the prior art must be an enabling disclosure to count as legitimate prior art. Enabling disclosure means that the prior art must provide enough disclosure that a person of skill in the art can understand the invention and reduce it to practice based on the prior art disclosure without undue experimentation. Thus in the situation where a journal article talks only about a new endotracheal tube that is easily inserted and accurately placed would probably not be prior art against an invention that is related to an ultrasound-guided endotracheal tube since the journal article does not provide enabling information about the ultrasound guidance. Information that is maintained as a secret or shared under a nondisclosure agreement is generally not considered prior art since it is not publicly known.

Priority date

The priority date is a key part of any prior art search criteria. The priority date is the date when your invention was first filed in a patent application. Therefore, anything that counts as prior art must have been publicly available before your invention's priority date. For freedom to operate searches, the priority date is less important and any patent that is granted before you practice your invention is potentially relevant.

Types of prior art searches

Patentability

There are several types of prior art searches. As previously discussed, the most common prior art search is for patentability analysis. That is, to determine what aspects of your invention are novel and nonobvious over the prior art and therefore what aspects of your invention are patentable. A few simple examples illustrate this concept.

Example 1

Your invention is a new endotracheal tube having a unique drug coating that prevents pathogens from building up on the tube and passing into the bronchial passageway. The drug is newly developed and nothing like it previously existed. The prior art shows endotracheal tubes with drug coatings that reduce pathogens but nothing in the prior art teaches your unique drug coating. Therefore the drug alone and in combination with the endotracheal tube are both probably patentable because they are novel over the prior art and probably not obvious variations over the prior art.

Example 2

Your invention is a new endotracheal tube that uses a standard known antibiotic coating to prevent pathogens from building up on the tube and passing into the bronchial passageway. A prior art search identified several articles in the scientific literature that each report uses the same antibiotic coatings on an endotracheal tube for the same purpose as you have proposed. Because the prior art teaches the same invention as yours, the invention is not novel and therefore not patentable.

Example 3

Your invention is a new endotracheal tube that uses a standard known antibiotic coating to prevent pathogens from building up on the tube and passing into the bronchial passageway. A prior art search reveals a single reference that teaches the same endotracheal tube as your design and proposes the use of antibiotic coatings but does not specifically disclose the antibiotic that you want to use. The scientific literature is full of articles reporting the use of your proposed antibiotic to kill pathogens on medical devices. Therefore because no single reference teaches all of the features of your invention, your invention is novel. However, because the combination of the two references teaches all the features of your invention, it is probably obvious and therefore not patentable.

Freedom to operate

The other type of prior art searching is FTO (freedom to operate) searching. While this type of searching is more typically referred to as an FTO or product clearance search rather than a prior art search, it nevertheless is still considered prior art searching. For purposes of this chapter, FTO searches are included in the general category of prior art searching. While a patentability search focuses on the teachings of any prior art, an FTO search is primarily concerned with granted patents since only granted patents can be enforced against infringers. But nonpatent literature prior art can also be used during an FTO analysis. Pending patent applications that have been published may also be considered in an FTO search, although this is more challenging since granted claims have not yet been issued and require speculation about what claims will be issued in the patent. FTO searching will primarily focus on the granted claims more so than the teachings of the patent or published patent application. A few simple examples of FTO analysis will help illustrate the concept.

Example 4

Assume that your invention is using a unique drug, CRM114 coated on an endotracheal tube to kill pathogens and prevent them from entering the bronchial passageway. CRM114 relies on a newly developed mechanism for killing pathogens that previously did not exist.

Assume you conduct an FTO search and find a granted patent with the following claim:

1. An endotracheal tube, comprising:

an elongate tube having a proximal end and a distal end, wherein the distal end is configured to be disposed adjacent to a patient's lungs, and wherein the proximal end is configured to be disposed adjacent to the mouth of the patient;

an expandable cuff adjacent to the distal end of the elongated tube, the expandable cuff having an expanded configuration and a collapsed configuration, wherein the expanded configuration is configured to engage and seal the elongated tube against tissue in a larynx of the patient, and wherein the collapsed configuration is sized for passage through the larynx; and

a drug LV426 coated on the endotracheal tube, wherein the drug is configured to eliminate or reduce pathogens on the endotracheal tube.

Your device has the elongated tube and the expandable cuff but uses drug CRM114 and does not have the drug LV426, therefore your device does not literally infringe the patented invention.

Example 5

Assume your invention is the same as previously described in Example 4 above, except that the drug CRM114 is an antibiotic. Your FTO search identifies a reference that describes an endotracheal tube with an antibiotic coating including drug CRM114 but the claim does not expressly recite your drug. The patented claim recites:

1. An endotracheal tube, comprising:

an elongate tube having a proximal end and a distal end, wherein the distal end is configured to be disposed adjacent to a patient's lungs, and wherein the proximal end is configured to be disposed adjacent to the mouth of the patient;

an expandable cuff adjacent to the distal end of the elongated tube, the expandable cuff having an expanded configuration and a collapsed configuration, wherein the expanded configuration is configured to engage and seal the elongated tube against tissue in a larynx of the patient, and wherein the collapsed configuration is sized for passage through the larynx; and an antibiotic drug coated on the endotracheal tube, wherein the drug is configured to eliminate or reduce pathogens on the endotracheal tube.

Because your device includes the elongated tube, the expandable cuff, and an antibiotic drug, your product is likely literally infringing the patented claim. Even though the claim does not expressly recite the drug CRM114, the claim does recite the more generic term "antibiotic" which encompasses drug CRM114, therefore the claim "reads" on your invention and further work is required on your product to reduce the risk of being sued for patent infringement.

Another common type of prior art search is referred to as a landscape search which means that the search intends to provide an overview of the technological space from both a patentabilty as well as freedom to operate perspective. It can also provide a good overview of competitors in your space. Thus a landscape search is a hybrid FTO-patentability search and sometimes may also be referred to as a state-of-the-art search.

The examples in this chapter are simplistic and are only used to demonstrate the basic principles of novelty, obviousness, literal infringement, or noninfringement. Also note that this chapter only addresses literal infringement of a patent claim and has not addressed other legal theories of infringement such as infringement under the doctrine of equivalents, or theories of indirect infringement such as inducing infringement or contributory infringement, which are beyond the scope of this chapter. For this reason, it is recommended as best practice to consult with patent counsel when performing such analyses.

Your patents and publications can be used against you as prior art

When conducting a prior art search you also need to review your own patents, publications, and activity as potential prior art. Anything that counts as prior art, even if it is your own (e.g., an old patent you filed, an old presentation or journal article, a trade show display, sale of the product, etc.) can be used against your future patent applications as far as patentability analysis is concerned. Almost any public information is subject to consideration. There are some exceptions, but in general, anything in the public domain can be used against your future patent filings. Also, in the United States, there is a one-year grace period from the time you publicly disclose something to allow you time to file a patent application. After that one year, novelty no longer exists. Most foreign jurisdictions do not provide a grace period, therefore if international patent rights are of interest, then you cannot rely on the one-year grace period.

Patents do not infringe patents

Contrary to a patentability analysis where your own publications can be used against you for patentability purposes, patents owned by you are not an issue with respect to freedom to operate since you are not going to sue yourself for infringing one of your patents. And even if the patent is jointly owned by several parties, you still have rights to the patent and still cannot be sued for infringing your patent.

Also, simply filing a patent application for an invention that you know infringes another patent is not an infringing activity. Patents do not infringe on other patents. It is only if you practice the invention (i.e., make, use, sell, or offer to sell the invention) that there would be infringing activity.

Who conducts the prior art search?

Anyone can perform a prior art search, but to determine who is best, you have to consider the cost of the searching, the quality of the searching, and the timeliness of the search, and then prioritize which of these criteria is most important. Searching is no different than any other project where you have to select a vendor or decide whether you are going to do it yourself.

Performing the search yourself could be as simple as running a Google search for your invention and reviewing the search results. Most people are usually comfortable performing this task. However, Google does not necessarily provide comprehensive prior art search results and you may want to search the patent or scientific literature which may require access to different databases that either require a paid account and may require some training and familiarity with the systems to conduct searches efficiently and accurately. If you do not have access to these databases or are not familiar with using them, you may not be able to obtain robust search results and therefore it may be prudent to use a search vendor.

The other thing to keep in mind and be realistic about is that entrepreneurs are busy and may not have the time or focus to conduct the searches. Therefore it may be quicker to simply use a search vendor to perform the search. Most vendors typically return search results in a couple of weeks and this may be more timely than waiting weeks for an entrepreneur to perform the exercise. Also, it is worthwhile mentioning that a professional search vendor performs this type of work full time, and therefore develops considerable skill and expertise performing this type of work which not only delivers efficiency and lower cost, but higher quality.

The cost of your time versus a vendor's time must also be considered. While there is no typical search, on average a prior art search can cost anywhere from about $500 to $20,000 depending on the complexity of the search, and whether any expediting charges are incurred

Therefore it may be prudent to start with some simple Internet searching that you can easily perform on your own. Based on those results, you can decide whether it makes sense to perform additional searching using a vendor and/or other databases.

Cost/timing of searching

If you prefer to perform the search yourself, the cost is limited to your time only and can be conducted at your own pace. However, many patent practitioners engage a professional search vendor to search. For most routine FTO or patentability searches, the searching can be completed in about two working weeks but this may be expedited if you are willing to pay an expediting fee to the vendor which can be a 50%–100% surcharge depending on how rapidly you want or need to accelerate the search.

The cost for a simple patentability search using a U.S.-based search vendor may range from around $1000–$2000. Smaller search vendors or offshore vendors with reduced labor rates can be less than $1000. For more complex searches, the cost can exceed $20,000. A simple Internet search for "patent search firms" or "freedom to operate search firms" will quickly identify many vendors and law firms that can perform this type of work. Additionally, the search vendor typically only provides a report listing the search results with citations to relevant portions of the document that the vendor feels are relevant to the search criteria. A search vendor typically does not provide a legal analysis of the search results and therefore an attorney is often required to review the results and provide an opinion which can add anywhere from $500 to $25,000 to the cost depending on the number of search results, the complexity of the invention and the type of summary report requested. For example, a search that identifies twenty relevant references will be less costly to review compared to a search that identifies 1000 references. Or, evaluating a simple mechanical device with a few components will likely be less costly than evaluating a complex machine with thousands of moving parts. Finally, the summary report can be a simple informal report with a few notes or comments provided by the searcher, or a more formal written legal opinion from an attorney which is more costly.

Define the scope of the search—casting a wide net versus too narrow

One key strategy as you conduct prior art searches is to carefully define the scope of the search. If your search is too broad, you will identify too many references, and if your search is too narrow, you won't capture a good cross-section of references. A few simple examples will illustrate this issue.

Example 6

You are interested in conducting a patentability search on an invention which is an endotracheal tube having an inflatable cuff and a drug coating on the endotracheal tube that reduces or eliminates pathogens from entering the bronchial passageway.

Using the U.S. Patent and Trademark Office's Full Text and Image Database and searching for the term "endotracheal tube" in the patent specification reveals over 7000 references. However, if the search is refined to include the terms "endotracheal tube" and "drug or therapeutic agent" then this is reduced to about 2500 hits. Further refinement of the search criteria will reduce the number of hits to a number that you are comfortable managing. There is no optimal number of hits in a search.

Another critical factor is your jurisdiction of interest. For patentability purposes, any public information from just about any country that predates your priority date is likely to count as prior art against the patentability of your invention. Therefore, best practice is to extend your search beyond the US. For most medical devices, if you search for prior art in the U.S. and Europe that's usually a good start. Adding the World Intellectual Property Organization which handles PCT (Patent Cooperation

Treaty International Patent Applications), will usually provide a well-rounded geographical search.

For FTO analyses, you only need to search jurisdictions where you think you may manufacture or sell your product. There's no point in performing an FTO analysis in a jurisdiction if you do not anticipate selling or manufacturing in that jurisdiction.

Duty of disclosure

Prior art searching is a bit of a double-edged sword. On one hand, the search results will increase your knowledge of what exists and will help you with FTO or patentability analysis. However, on the other hand, anyone working on patent applications before the U.S. Patent and Trademark Office (USPTO) has a duty of candor to disclose any relevant prior art that is material to the patentability of your invention. Therefore if you search, you may be obligated to disclose some or all of the search results to the USPTO. While this itself is not a huge issue, it does nevertheless require some additional work and cost to prepare and file the appropriate disclosure forms with the patent office. Patent Offices in some other countries have similar requirements too.

Prior art search and IP due diligence during fundraising or exit strategies

Not only is prior art searching critical for performing patentability and FTO analyses, but if your company is involved in fundraising or an exit strategy, such as a merger, acquisition, or initial public offering, then the investor, acquirer, or underwriter will undoubtedly conduct due diligence on your intellectual property portfolio. That due diligence will include an evaluation of the scope and validity of your patents as well as your freedom to operate position.

Conducting prior art searching will allow you to identify the most relevant art and explain to the party conducting the diligence how your inventions are distinguished from the prior art and why you do not infringe a third-party patent. Thus you will develop a robust position on your intellectual property that minimizes risk for the investor, acquirer, or underwriter and which helps advance business transactions.

Examples of vendors and databases

Many sources can be used for prior art searching. For example, Google Patents is easily accessible on the Internet and is free. So too is the U.S. Patent and Trademark Office's database. Conducting simple Google searches can provide a wealth of information. These are fairly easy to use and often a good place to start.

There are also several excellent databases that you can subscribe too but the cost can be prohibitive, perhaps in the several thousand dollars a month range. Some other commercial databases are offered by LexisNexis (TotalPatent), minesoft (Patbase), ProQuest (Dialog), Questel (Orbit), Thompson Reuters (Thompson Innovation), etc.

Simple Google searches are straightforward but do not always guarantee that you are searching for the best sources of prior art. For example, sometimes search results are more on point when you search a patent database such as the USPTO's EAST system (Examiner Automated Search Tool) which is available at the U.S. Patent and Trademark (USPTO) headquarters in Alexandria, VA, as well as any of the USPTO regional offices in Detroit, Denver, San Jose, and Dallas. Patent data is also searchable on the internet without requiring a visit to the USPTO offices (see www.uspto.gov). Other common databases include the World Intellectual Property Organization (WIPO) international patent application database (WIPO.int), Espacenet, and the European Patent Register.

Also, most law firms with an intellectual property department can conduct prior art searches for you and some vendors specialize in conducting searches. For example, some retired patent office examiners become individual searchers and there are many other search vendors in the United States as well as in foreign countries.

Conclusion

Conducting prior art searches allow you to evaluate the patentability of your invention and your freedom to commercialize your invention without infringing patents owned by third parties. While not required, prior art searching will help you reduce your risk of liability and strengthen your intellectual property position when dealing with the patent office, investors, acquirers, or underwriters.

11

Patent strategy—patent types and timing

Ettore Crimi[1] and Umme Rumana[2]

[1]Department of Anesthesiology, Atrium Health Wake Forest Baptist, Wake Forest University Health Sciences, Orlando, NC, United States [2]Department of Anesthesiology, University of Central Florida College of Medicine, Orlando, FL, United States

Chapter outline

Abstract

In the United States, patents are issued by the United States Patent and Trademark Office (USPTO). The purpose is to safeguard the work of innovators from using, modifying, or selling the invention, reward their pioneers, and bolster scientific progress. The USPTO issues three types of patents: utility, design, and plant. A utility patent is the most common kind of patent. In the United States, utility patents account for more than 90% of all patents. Obtaining a patent is an extensive process and involves a series of steps in chronological order issued by the USPTO. Preventable pitfalls in the patent application procedure that could speed up the patenting process include delaying filing a patent application, not keeping an inventor's logbook, failing to search for plagiarism, providing insufficient disclosure of claims, enablement, and specifications, filing an overly specific or overly broad application leaving room for imitators and failing to allocate sufficient funds. Patent protection is only applicable within

Innovation in Anesthesiology. DOI: https://doi.org/10.1016/B978-0-12-818381-6.00037-1

the borders of the issuing country; therefore inventors seeking international patent protection must file patent applications with each relevant foreign patent office, as a U.S. patent only protects an invention within U.S. territory and has no effect abroad.

Keywords: Patent; design patent; utility patent; plant patent; provisional patent application; pending patent; infringement; copyrights; trademarks; nondisclosure agreement; open innovation

Glossary terms and definitions

Patent a legal privilege provided by the patent office to the creator of a novel product, process, technology, or service with commercial value. A patent is a legal right granted by the relevant government agency in the nation where the patent application was first filed and is, therefore only valid within that jurisdiction. For a limited time, the patent holder has the sole legal right to produce, market, and otherwise sell the patented invention.

Utility patent a utility patent is granted for any novel process for a useful device, apparatus, instrument, tool, appliance, utensil, manufacturing method, or composition of an item, including intellectual products. The majority of novel concepts proposed by inventors fall into this class. The validity of the utility patent is 20 years from the filing date.

Design patent a patent in this category may be granted for any novel and readily reproducible improvement in the external appearance, configuration, ornamental design, or shape of a functional product. An upgrade made to an already-existing product either by using different materials or finishes or by taking a new approach to the product's design. For instance, a major renovation of a hospital's flooring or a total overhaul of a skyscraper. Commonly, computers are used to create graphic representations of new designs. Illustrations in design patents rarely include descriptive text, making them unsuitable for online keyword searching. Design patents are valid for 14 years after the date they were granted.

Plant patent new or improved methods of plant production, whether for fruits, vegetables, or wood by asexual reproduction or cuttings, typically horticulture rather than genetic manipulation, are protected by a plant patent. The duration of a plant patent is 20 years from the date of filing.

Infringement an act of violating the terms of an agreement, breach of contract, or unauthorized use of protected material under intellectual property laws.

Provisional patent application (PPA) a short-term patent that lasts for a year and allows the inventor to sell the idea to investors, test its commercial viability, or make improvements before filing a full patent application. The USPTO issues the provisional patent application to help protect a new invention from imitation for up to a year before a full patent application is filed.

Pending patent the phrase "patent pending" denotes that a provisional patent application has been filed with the United States Patent and Trademark Office. A patent application shows that the inventor is serious about securing legal rights to their creation. To discourage potential imitators, a pending patent acts as a formal notice of infringement for 12 months. If a regular, nonprovisional patent application is not filed within that year, the "pending patent" status will expire, and the invention will no longer be protected. The pending patent status opens the door for the inventor to make deals with interested businesses to sell or license inventions.

Copyrights a form of intellectual property right that protects the author or artist of any work that is fixed in a tangible medium of expression, whether it has been published or not. The United States Copyright Office grants the holder the limited, exclusive right to reproduce, modify, adapt, display, and perform an original piece of creative work. The artistic expression could be written, visual, or auditory. Copyright protection cannot be applied to ideas, methods, formulas, recipes, names, slogans, facts, research, or unrecorded speeches. The term of protection extends for the author's lifetime plus 70 more years (or for anonymous works, 95 years from first publication or 120 years from the date of creation—whichever is shorter).

Trademarks trademarks are defined by the United States Patent and Trademark Office as words, phrases, symbols, or designs, or any combination thereof, that identify and distinguish the source of the goods of one party from that of others. the ® symbol denotes a registered trademark, while the TM symbol denotes a trademark that has not been formally registered. A trademark's protections remain in effect indefinitely so long as the owner continues to use the mark in commerce.

Nondisclosure agreement (NDA) also known as a confidentiality agreement (CA), formalizes the existence of a confidential relationship between the signatories and prohibits the disclosure of any information gained through that relationship to any third party. An NDA is standard procedure for companies exploring the possibility of a joint venture. To protect the employer's proprietary information, many companies have their employees sign nondisclosure agreements.

Open innovation is a method used in businesses whereby ideas and solutions are solicited from various sources to spur original thought and development. Businesses need to consistently introduce new products and services into the market to stay current. By sharing the burden and the benefits of innovation with others, businesses can grow while increasing their competitiveness and flexibility, and minimizing the time and resources required to develop and launch new products.

Chapter outline

- Introduction
- The patent strategy
- Strategies for commercializing novel ideas without a patent
- Patenting pitfalls to avoid
- Examples of patents in anesthesiology

Key points

- Patents are exclusive property rights issued by the U.S. government to shield inventors' work for a limited period in exchange for a complete disclosure of the invention
- Before submitting a patent application, it is important to search existing patents using the USPTO Patent Search Database, Google Patent, or other resources recommended by the USPTO

- All patent applications are thoroughly examined, both technically and legally. A mere suggestion or idea does not suffice to obtain a patent

Introduction

The first patent was issued on July 31, 1790, by George Washington. The Patent Act was established in 1836, and the United States Patent and Trademark Office was founded on July 4, 1836. Twelve million patents have been granted since its establishment to (1) promote scientific progress, (2) protect an inventor's product, discoveries, and intellectual property, and (3) process patent applications (Article I, Section 8 of the United States Constitution). As of the end of 2022, there have been 11,661 anesthesiology-related patents issued in the United States.

Traditionally, the "first-to-invent" rule has been used to determine who gets a patent. However, since the passing of the America Invents Act (AIA) in 2011, the invention date has lost its precedential value in American law. The only issue that can be contested in the court is who filed the patent first, the inventor or someone else. The first-to-invent rule remains in effect for patent applications submitted before March 15, 2013. Patent applications submitted on or after March 16, 2013, are governed by "first-to-file" legislation.

USPTO nonprovisional patent fees range from $2000 to $3500 for a design patent and $15,000 to $45,000 for a utility patent, depending on the complexity of the invention. Filing, search, examination, issue, maintenance, attorney, and appeal fees are just some of the costs associated with obtaining a patent. Generally, utility patents take between 18 and 30 months to process, while design patents take 13 months.

The patent strategy

Sharing or publicly disclosing ideas before obtaining a patent exposes the innovation to the risk of being copied or stolen, necessitating the need to establish ownership in court. The strategies to reduce the likelihood of intellectual property theft are as follows: (1) **Patent search**: Inventors can either hire a lawyer to conduct a search or do it themselves. The ability to do one's own research on innovation helps immensely when it comes time to create a PPA and learn similarities. The USPTO outlines seven steps for searching U.S. Patents and Published Patent Applications (see Fig. 11.1). (2) Keep an **Inventor's Logbook** to consolidate your thoughts, research, and activities from the moment you have an idea. Keep a record that is accurate, chronological, and detailed. Sometimes the best defense is a well-documented offense. Learn the guidelines for keeping a logbook, such as using a bound notebook rather than a spiral one, not skipping lines or pages, and recording each step in sequential order along with the name of the person performing the step, the date, the nature of the work being performed, and any relevant drawings, specifications, tests, or prototypes, and having each entry dated and handwritten in ink and signed by an independent witness. Inventor's journal can be purchased from any bookstore, office supply shop, or online store. (3) File a **provisional patent application** and obtain a "**pending patent**" status before revealing the idea to anyone. PPA is much more affordable than a full patent. Having the invention officially recognized as "pending" will open

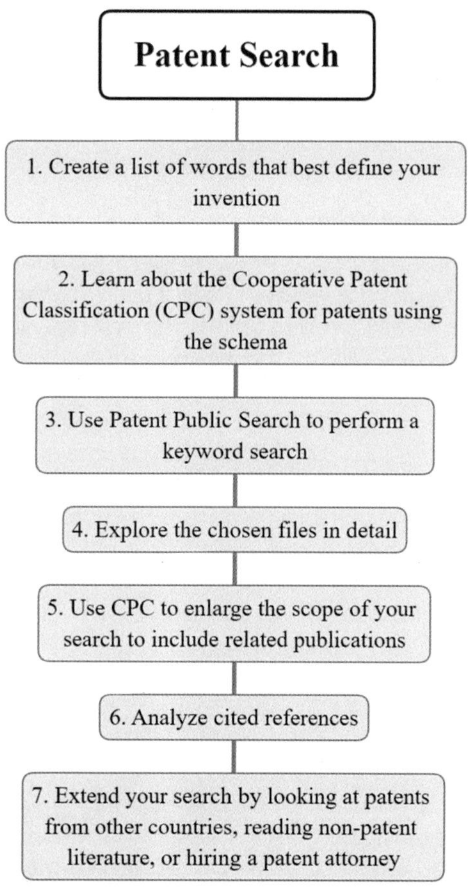

Figure 11.1 A seven-step plan for using the USPTO's free online resources to conduct preliminary searches of U.S. patents and published applications.

up commercial opportunities. (4) When the time comes to present ideas to investors, make potential parties sign a nondisclosure agreement (NDA) to protect proprietary information or trade secrets. To write a nondisclosure agreement, a patent attorney is a great resource. Companies may be hesitant to sign inventors' NDAs. In any case, seek legal advice regarding the NDA of the company to avoid accidentally disclosing private information or agreeing to a waiver of your right to secrecy. (5) Finally, file the official nonprovisional patent application. The **USPTO Patent Application** guidelines are important to consider when applying for a patent. See Table 11.1 for steps for obtaining a patent. The following sections outline the main sections of the application and are meant to convey information and influence the patent committee: (1) Title (2) Abstract (3) Drawings (4) Background of the Invention (5) Summary of the Disclosure (6) Brief Description of the Drawings (7) Detailed Description of the Invention (8) Claims.

Table 11.1 USPTO patent application process.

Determine whether a patent has already been granted for a substantially similar product, design, or plant. Care must be taken not to infringe upon another patent. A review of all patents also requires attention to priority—which product was first?

Asses the invention: is it eligible, novel, or nonobvious? will the applicant be the owner?

Assess monetization. Mere licensing may be a means, or sales or another method are possibilities.

Make a prototype or model. At this stage, a business plan should be available. Investor or Patent Resource Centers may be helpful.

A "how to" approach is acceptable. Include the scope of the intellectual aspects of the project and explain usage. Consulting the information about eligibility on the USPTO site is helpful. The same section furnishes a discussion of the Broadest reasonable interpretation of the claim as a whole.

Complete an initial patent application; electronically is best. The Application Assistance Unit or the Patent-process-overview section of the USTPO may be of interest.[4] A marketing plan needs to be developed.

An attorney or agent is not required; the applicant may file him/herself. ("pro se"). If any portion of the above is complex, professional help is advised.

Examination of the application will follow. An estimate of the time this will take is given on the USTPO site, Chapter 700.

The examiner may contact the applicant to supply or modify information to complete the application. Most examiners will schedule an interview. If rejected, reasons are given, enabling a revised response. A second rejection may be appealed. If accepted, fees, including maintenance fees, are due; the patent will then be published. The patent will expire in 20 years.

A project summary must be available before the patent application and at each step mentioned above. A complete description, function, features, and all other aspects should be within the contents. A larger scope of the description with a clear definition of terms and details favors approval. In contrast, a brief, narrow scope with unclear content is limiting, favors rejection, and cannot be defended or upheld in court.

Strategies for commercializing novel ideas without a patent

Patents are intended to protect novel ideas, but most granted patents never make it to the marketplace. Due to the high cost of producing new products and starting new businesses, many patents are simply framed and mounted on walls. While waiting for patent approval, licensing an invention to a reputable company in exchange for royalties is a viable option. Big businesses always look for new ways of thinking and doing things to stay competitive. Large corporations would not wait around for patent applications to be processed or for intellectual property to be protected before releasing a new product. There is a short shelf life for products. Whatever was trending today may fall short of expectations tomorrow. As product lifespans get shorter and intellectual property theft becomes more common, businesses are becoming more receptive to licensing innovations from outside sources and paying royalties to creators. Prominent investors who are willing to invest in novel ideas are (1) *The Shark Tank*, the most well-known television show about business innovation, entrepreneurship, and partnership negotiation. The program facilitates communication between start-ups and investors, (2) *Quirky*, an organization that oversees a million-strong online community of people who submit new ideas, tweet their opinions (product names, labels, and pricing), and ultimately take part in deciding which ideas make it to market. (3) *Crowdfunding* refers to the practice of obtaining financial backing for a venture by soliciting a large number of individuals for monetary contributions, typically through the use of the Internet. Sites like Kickstarter, Indiegogo, and Fundable are a few examples. (4) Some other great platforms for open innovation are as follows: United Genomes Project, AstraZeneca Open Innovation, Open Health Innovations, Merck, HeroX, OpenIDEO, Wazoku, Brightidea, HYPE Innovation, HYVE, Planview Spigit, Nosco, Chaordix, Qmarkets, Lilly, NTTData, Air99, Health++, Opioid360, and many other.

Patenting pitfalls to avoid

There are a few things to keep in mind for inventions that could potentially earn a patent: (1) Delaying applying for a patent is a major drawback in the patenting process. Since the AIA's "first to file" law, it is no longer relevant who invented something first. The first-come, first-served policy will be enforced. If an application is delayed, it gives competitors time to submit proposals with similar features. (2) When deciding whether or not to grant a patent, the inventor's logbook takes precedence over all other sources of information. Delays in the patenting process can be avoided if inventors keep thorough notes, drafts, codes, and correspondence. (3) Validating Uniqueness: Many applicants overlook the initial patent search. To search for a patent, look through all published patents. Legally, it is required when applying. Before taking on an enormous project, it is prudent to check the patent records to see if the invention is truly original. (4) Examiners and courts often reject a patent application for lack of written description or enablement. To be "enabled," an invention must be usable without much practice or experimentation. (5) Patents are rejected for incomplete disclosure. All claims must be addressed in the specification section. (6) When submitted, the ideas and applications of the invention should be concrete, physical, and tangible. If they are solely abstract, the chances of acceptance will decrease significantly. A software program or mobile telephone "app" is less likely to succeed. (7) Finally, stay alert for any possible scams. Individuals, businesses, and other organizations may provide inaccurate, misleading, or unnecessary patent services. Accepting these items can cause application delays, additional fees, or even rejection.

Examples of patents in anesthesiology

In 1982, British anesthesiologist Archie Brain of London's Royal Hospital developed the first laryngeal mask airway. Dr. Brain's desire to overcome ventilation limitations drove him to conduct extensive research into the anatomy and physiology of the upper airway. Without his discovery, doctors had only two options for managing the airway of an unconscious patient: inserting a nasal tube or oral tracheal tube or using a facemask with an oral or nasopharyngeal airway. At home, Dr. Brain experimented with the Goldman Dental Mask to see if he could successfully connect an artificial tube to his trachea. After some initial gagging and hacking, he was able to resume normal breathing, and the experiment was deemed a success.

John Allen Pacey, a general and vascular surgeon from Canada, created the Glidescope in 2001, which was the first video laryngoscope to be sold commercially. Lower rates of esophageal intubation, decreased laryngeal/airway trauma, and increased success in establishing a secure airway with fewer attempts were all made possible by the high-resolution digital camera.

Inventor Shawn Schumacher of Maho Med Tech, LLC was granted a United States Utility "Pending Patent" (US20220265983A1)

US 20220265983A1

(19) **United States**

(12) **Patent Application Publication** (10) **Pub. No.: US 2022/0265983 A1**

Schumacher (43) **Pub. Date: Aug. 25, 2022**

(54) **CATHETER SYSTEM WITH SUBCUTANEOUS, IMPLANTABLE PORT AND ULTRASOUND-GUIDED PLACEMENT METHOD**

(71) Applicant: **Maho Med Tech, LLC**, (US)

(72) Inventor: **Shawn Schumacher**, Corvallis, OR (US)

(21) Appl. No.: **17/518,815**

(22) Filed: **Nov. 4, 2021**

Related U.S. Application Data

(60) Provisional application No. 63/200,204, filed on Feb. 21, 2021.

Publication Classification

(51) **Int. Cl.**

A61M 39/02	(2006.01)
A61M 5/158	(2006.01)
A61M 25/01	(2006.01)
A61M 19/00	(2006.01)

(52) **U.S. Cl.**

CPC ***A61M 39/0208*** (2013.01); ***A61M 5/158*** (2013.01); ***A61M 25/0108*** (2013.01); ***A61M 19/00*** (2013.01); *A61M 2025/0166* (2013.01); *A61M 2005/1587* (2013.01)

(57) **ABSTRACT**

A catheter system includes a medication dispenser located external to a patient. A subcutaneous port displaced internal to the patient and receives a quantity of medication. The port can be filled from the medication dispenser using a Huber needle, or some other suitable needle, fluidically connected to the medication dispenser by tubing. The catheter can be placed in proximity to a patient's nervous system using ultrasound imaging guidance. A method of administering a nerve block to a patient includes the steps of placing a subcutaneous port using ultrasound imaging for guidance and administering medication via a catheter connected to the port.

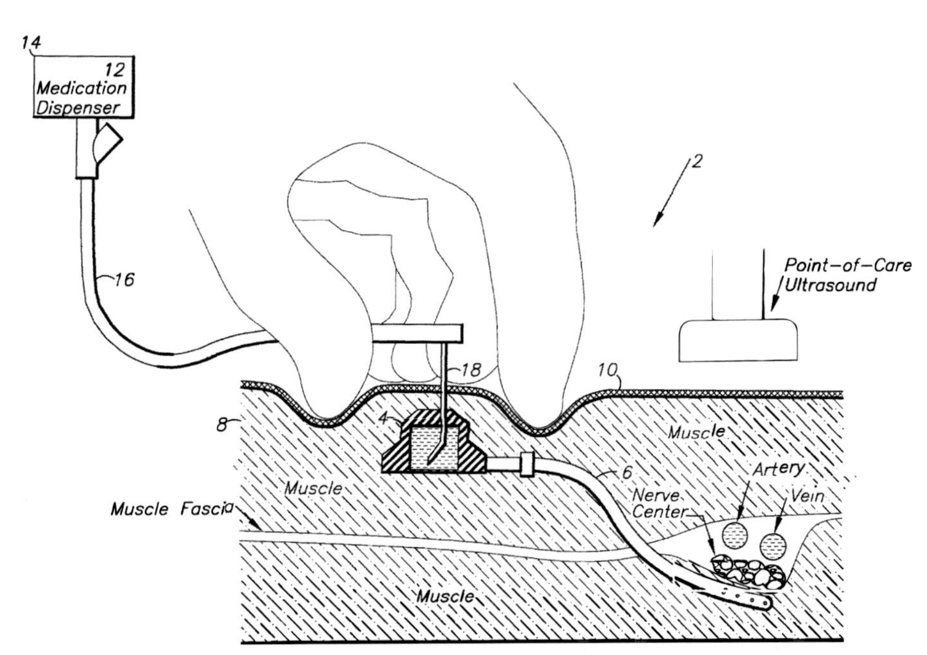

Figure 11.2 Catheter system with subcutaneous, implantable port and ultrasound-guided placement method. The Patent can be accessed at https://patents.google.com/patent/US20220265983A1/en?oq = US20220265983A1.

US010124021B2

(12) **United States Patent**
Gostine

(10) **Patent No.: US 10,124,021 B2**
(45) **Date of Patent: Nov. 13, 2018**

(54) **INTRAVENOUS FLUID**

(71) Applicant: **Andrew L. Gostine**, Chicago, IL (US)

(72) Inventor: **Andrew L. Gostine**, Chicago, IL (US)

(*) Notice: Subject to any disclaimer, the term of this patent is extended or adjusted under 35 U.S.C. 154(b) by 0 days.

(21) Appl. No.: **15/443,054**

(22) Filed: **Feb. 27, 2017**

(65) **Prior Publication Data**

US 2018/0177821 A1 Jun. 28, 2018

Related U.S. Application Data

(60) Provisional application No. 62/438,491, filed on Dec. 23, 2016.

(51) **Int. Cl.**

A61K 33/00	(2006.01)
A61K 9/00	(2006.01)
A61K 9/08	(2006.01)
A61K 33/20	(2006.01)
A61K 31/19	(2006.01)
A61K 31/191	(2006.01)
A61K 33/06	(2006.01)
A61K 47/02	(2006.01)
A61K 47/12	(2006.01)

(52) **U.S. Cl.**

CPC ***A61K 33/00*** (2013.01); ***A61K 9/0019*** (2013.01); ***A61K 9/08*** (2013.01); ***A61K 31/19*** (2013.01); ***A61K 31/191*** (2013.01); ***A61K 33/06*** (2013.01); ***A61K 33/20*** (2013.01); ***A61K 47/02*** (2013.01); ***A61K 47/12*** (2013.01)

(58) **Field of Classification Search**

CPC A61K 33/00; A61K 9/0019; A61K 9/08; A61K 31/19; A61K 31/191; A61K 33/06; A61K 33/20

USPC 424/663

See application file for complete search history.

(56) **References Cited**

U.S. PATENT DOCUMENTS

3,993,750	A	11/1976	Fox, Jr.	
5,443,848	A	8/1995	Kramer et al.	
2007/0135343	A1 *	6/2007	Webb	A61K 9/0019 424/680
2008/0125488	A1 *	5/2008	Leverve	A61K 31/19 514/557
2011/0189091	A1 *	8/2011	Bachwich	A61K 9/08 424/9.1
2011/0318431	A1 *	12/2011	Gulati	A61K 9/0019 424/681
2013/0274340	A1 *	10/2013	Jeffs	A61K 9/0019 514/569

OTHER PUBLICATIONS

Liujiazi et al., Multiple Branch and Block Prediction. Hypertonic Saline for Brain Relaxation and Intracranial Pressure in Patients Undergoing Neurosurgical Procedures: A Meta-Analysis of Randomized [online],Jan. 30, 2015 [retrieved on Aug. 18, 2017]. Retrieved from the Internet:< URL:http://journals.plos.org/plosone/article?id=10.1371/journal.pone>.*

Strandvik, Anaesthesia, 2009,64, pp. 990-1003.*

Infusion Nurse Blog, Is there a difference? Osmolarity vs. Osmolality . . . [online],May 2010 [retrieved on Mar. 16, 2018]. Retrieved from the Internet:< https://infusionnurse.org/2010/05/14/osmolarity-vs-osmolality/>.*

* cited by examiner

Primary Examiner — Johann R Richter
Assistant Examiner — Courtney A Brown
(74) *Attorney, Agent, or Firm* — Butzel Long; Gunther J. Evanina

(57) **ABSTRACT**

Disclosed are intravenous hypertonic electrolyte solutions for treating intracranial hypertension while reducing the risk of inducing hyperchloremic metabolic acidosis. The solutions are characterized by a ratio of sodium-to-chloride (Na:Cl) ions of 1.2-1.6 and a total osmolarity of 310-400 mEq/L for a maintenance solution, and a total osmolarity greater than 1000 mEq/L for an initiation solution.

8 Claims, No Drawings

Figure 11.3 Intravenous hypertonic electrolyte solutions. The Patent can be accessed at https://patents.google.com/patent/US10124021B2/en?oq = US10124021B2.

on August 25, 2022, for his work on a catheter system for nerve block pain management that features an implantable subcutaneous port and a method of ultrasound-guided placement (Fig. 11.2). To perform a nerve block, a subcutaneous medication port with a catheter is inserted in proximity to the nerve of interest under ultrasound guidance. A controlled medication dispenser dispenses the prescribed medication to the patient at predetermined intervals and in the prescribed quantity. The subcutaneous port is typically filled with a Huber needle or a similar needle that is attached to a medication dispenser via tubing.

On November 13, 2018, Andrew L. Gostine was awarded a United States Utility Nonprovisional or Active Patent (US10124021B2) for developing two novel intravenous fluid compositions to treat intracranial hypertension (Fig. 11.3). The inventor disclosed hypertonic electrolyte solutions for intravenous administration with a lower risk of inducing hyperchloremic metabolic acidosis. An initiation solution has an osmolarity greater than 1000 mEq/L, while a maintenance solution has an osmolarity in the range of 310–400 mEq/L and a Na:Cl ion ratio of 1.2–1.6. Traditionally, a 3%, 7.5%, 10%, 23.4% saline, or 20% mannitol is the standard of care.

Further reading

The Federal Register. n.d. US PO, Washington, D. C. Retrieved from https://www.uspto.gov/patents/basics.

USPTO Patent Application Initiatives Timeline. n.d. USTPO, Washington, D Retrieved from https://www.uspto.gov/patents/initiatives/uspto-patent-application-initiatives-timeline/.

USPO 2106 Patent subject matter eligibility [R-10.2019], 2019. Retrieved from https://www.uspto.gov/web/offices/pac/mpep/s2106.html.

Patent Process Overview. n.d. Retrieved from https://www.uspto.gov/patents/basics/patent-process-overview#step1.

Ji CD, Pan X, Xiong YC, et al. An analysis of patents for anesthetic laryngoscopes. J Zhejiang Univ Sci B 2017;18(9):825–32. Available from: https://doi.org/10.1631/jzus.B1600259.

Protect Yourself from Invention Promotion Scams. n.d., USPTO. Retrieved from https://www.uspto.gov/sites/default/files/documents/ScamPrevent.pdf.

The life cycle of a patent. Reproduced from the U S Patent Office, 2022. Fig. 1 is in the public domain. USPTO. Retrieved from https://www.uspto.gov/patents/basics.

Seger C, Cannesson M. Recent advances in the technology of anesthesia. F1000Res 2020;18:9. Available from: https://doi.org/10.12688/f1000research.24059.1 F1000 Faculty Rev-375.

Key Stephen, Janice Kimball Key. Sell your ideas with or without a patent. QuarkXPress; 2015. Kindle Edition.

IdeaConnection: Share your open innovation success stories with inventors and innovators. 2022, https://www.ideaconnection.com/open-innovation-success/. Accessed 13 Dec. 2022.

12

Concept testing and selection—where and why will it fail

P.K. Benson
Department of Anesthesiology, HRH, Secaucus, NJ, United States

Chapter outline

Abstract

Tesla had a unique strategy when it launched its Model 3 in 2017. It presented the concept and various features and allowed consumers to give a deposit. It was a huge success and raised over $400 million. In one stroke, Tesla attained invaluable feedback and secured the financial backing necessary to ensure a successful launch. Fast forward to 2022, and the NY Times headlines read: "Tesla's online success is forcing auto industry to rethink how it sells cars."

Concept testing gauges customers' needs, acceptance, and willingness to buy a new product. There are various methods to gain in-depth insights into the customers' opinions about the different aspects of the product. By determining their level of interest regarding specific features, how the product looks/feels, and the price this process strives to optimize the product before launch. It is an opportunity to iron out flaws and add new features. This enhancement in product development will in turn fine-tune marketing strategy, which could shorten the time to market and could confer a first mover's competitive advantage.

Keywords: Concept test; concept selection; prototyping; screener questions; sympathy bias

Innovation in Anesthesiology. DOI: https://doi.org/10.1016/B978-0-12-818381-6.00013-9

Glossary terms and definitions (a list of 5–10 terms with their definitions—this list may be identical to your keywords but needs to include the definitions)

Concept test The response of the potential customers in the target market to a conceptual description of a product.

Concept selection Usually the step that precedes concept testing, a judgment call by the developers that narrows down a set of concepts to one or two that can be tested.

Prototyping Physical representation of a product.

Screener questions The first few questions of a survey that ensure that the respondents represent the target market.

Sympathy bias When respondents fake liking the product to appease an anxious product developer.

Chapter outline

The following are the seven generally recommended steps of concept testing according to Ulrich and Eppinger:

1. Explicitly articulate the questions that the team wants answered.
2. Select the survey population that closely represents the target market
3. Select the appropriate survey format
4. Promote the product/benefits in a manner that matches the survey format
5. Accurately gauge customer response
6. Analyze their response and glean factors that impact their intent to buy
7. Use their feedback to synthesize qualitative insights about the product [1]

Key points

- To estimate the sale potential of a product, customers from the target market are solicited to choose between concepts that describe the closest approximation to an ideal product.
- Their input plays a critical role in designing and or improving a potential product.
- A prototype, a physical representation of the conceptually designed product, is utilized to assess potential demand, improvements, and the customer's intent to purchase.

Insight must precede application.

—Max Plank.

One of the most complicated yet exciting areas in the specialty of Anesthesiology has been the management of pain. The challenges involved have evolved to such a level that there are now national and international guidelines, and they keep getting updated. About 15.3 million patients who undergo surgery every year in the U.S. experience moderate to severe postoperative pain, which needs to be better controlled [2,3]. Less than satisfactory pain control in the 21st century has been associated with significant postoperative complications, longer hospital length of stay, and the development of chronic pain conditions [3–7]. The historic gold standard modality of pain control with opioids has created an epidemic of devastation, addiction, and death which costs $78.5 billion annually [8,9]. This is in addition to the commonly observed untoward effects such as ileus, somnolence, pruritis, postoperative nausea and vomiting, and urinary retention. The most vulnerable are middle-aged women who undergo colectomy or knee replacement. From this cohort, 13% become newly persistent opioid users and have the highest opioid-related death rates [10,11]. Consequently, there has been a shift to sharply reduce or altogether eliminate opioid use in the perioperative setting and replace it with a multimodal approach that results in pain control that is acceptable to the patient [12,13]. Many protocols essentially involve preoperative use of oral acetaminophen, gabapentoids, and Non Steroidal Anti-Inflammatory Drugs preoperatively as well as regional blocks with local anesthetics when applicable [14,15]. Such infiltration remained a valuable yet underutilized option due to the short duration of action of the local anesthetics. This evanescent duration of short-acting local anesthetics led to the search for long and ultra-long-acting anesthetics. Liposomes are lipid vesicles encasing aqueous compartments into which a local anesthetic can be loaded, which are ideally suited as a carrier molecule. Formation of complex utilizing liposomes as a carrier for bupivacaine led to prolongation of its release at the injection site. Grant et al. provided the proof of concept in a seminal article in *Anesthesiology Journal* in 2004 [16–20]. Once a favorable release profile was demonstrated, Pacira Pharmaceuticals, Inc. further developed, marketed, and greatly profited from liposomal bupivacaine (Exparel).

The Exparel saga further defined important lessons to ensure that the product launch is one of the 5% that is successful. A full 95% fail in the highly regulated healthcare ecosystem. Ineffective market segregation is an important cause of failure identified by Dr. Clayton Christensen, who teaches Marketing Strategy at HBS. As consumers do not execute their purchasing decisions by dividing into target demographics, this means that product developers must have a significant and current understanding of the competitive landscape. In Exparel's case, the team focused their efforts on not only the anesthesiologists, who could use it for regional blocks but also other proceduralists including podiatrists who could use it for wound infiltration. Marketing the product according to a product's target use, Christensen notes, has the added advantage that it is difficult to copy [21]. This may be more pertinent to innovations that are not protected by patents. The caveat is that product planning usually involves analyzing existing data, which is organized by consumer demographics or product categories. This is where creativity and including an expanded target customer increases the probability of a successful and sustained successful launch (Fig. 12.1).

As concept testing is in essence an experiment, it is vital to ask the pertinent questions that define the purpose of the product. It should discuss alternative concepts and to what extent they need to be explored. Additionally, potential improvements that enhance how the product could address the unfulfilled customer needs are addressed. A projection of the number of units likely to be sold is made, as well as the decision as to whether there is a need for enhanced product development.

The population surveyed during concept testing should accurately reflect the target market, otherwise, the inferences that will be made will be biased. Thus, the survey population should closely resemble the target market. The screener questions are usually posed to ensure that the respondent fits the

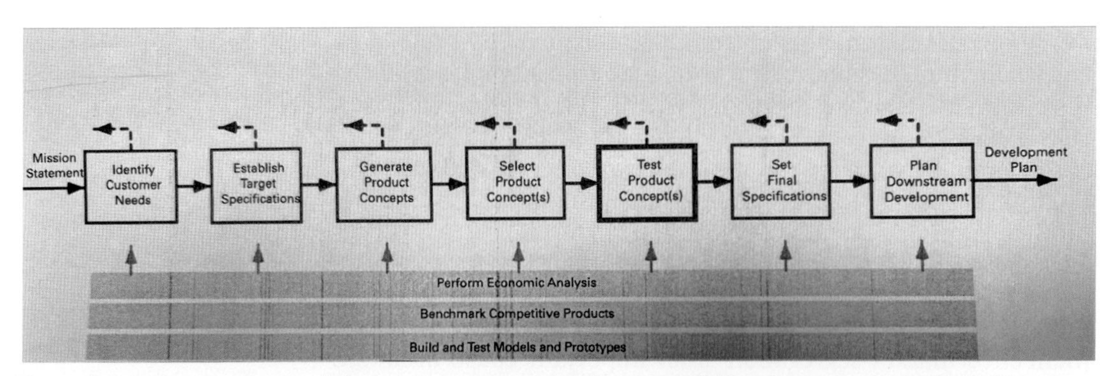

Figure 12.1 Concept testing in relation to other concept development activities. Concept testing by Ulrich KT, Eppinger SD. Product design and development. Seventh Edition McGraw-Hill Education; 2019.

attributes of the target market. To accurately guide decision-making, a sufficient and representative sample size is essential to ensure that the development team has confidence in the obtained results. Although there are not any scientifically proven formulas to accurately determine the adequacy of a sample size, some of the pertinent factors are included in Fig. 12.2. It is recommended that various objectives be explored with a multitude of surveys. Each survey can seek to answer questions about a different aspect of the product, such as attractiveness or purchase intent. As a demand forecast will be created based on purchase intent, the higher cost of a larger sample survey is often justified to ensure that future financing decisions will be made most accurately.

Most surveys incorporate several modalities of open-ended interactive format. A face-to-face format benefits product developers by enabling direct observation of the consumer's reaction to the product in rich detail. It is often used when exploring alternative concepts or suggestions for product improvement. A virtual concept testing site is often developed where specific individuals are invited to a phone interview while on the virtual website. Various concepts, product details, and attributes are presented, and responses are recorded and later analyzed. With the advent of Zoom meetings due to the COVID-19 pandemic, this idea has progressed to having virtual meetings where customer experience could be augmented with tools created in Virtual Reality. The consumer can have direct input to a multitude of options and provide real-time feedback that can be recorded for future review and decision-making. Such realistic representation has been more

Factors Favoring a Smaller Sample Size

- Test occurs early in concept development process.
- Test is primarily intended to gather qualitative data.
- Surveying potential customers is relatively costly in time or money.
- Required investment to develop and launch the product is relatively small.
- A relatively large fraction of the target market is expected to value the product (i.e., many positively inclined respondents can be found without a large sample).

Factors Favoring a Larger Sample Size

- Test occurs later in concept development process.
- Test is primarily intended to assess demand quantitatively.
- Surveying customers is relatively fast and inexpensive.
- Required investment to develop and launch the product is relatively high.
- A relatively small fraction of the target market is expected to value the product (i.e., many people have to be sampled to reliably estimate the fraction that values the product).

Figure 12.2 Factors leading to relatively smaller or larger survey sample sizes. Concept testing by Ulrich KT, Eppinger SD. Product design and development. Seventh Edition McGraw-Hill Education; 2019.

favorably received [22]. To ensure a fair comparison, alternative concepts need to be presented in the same media and at the same level of detail. It is critically important to understand what information could encourage the consumer to make a purchase decision and to relay the concept description in a manner that checks most if not all the boxes (Fig. 12.3).

Upon effective presentation of the product as a concept, customer response is usually measured by asking them to choose from two or more alternatives. It is paramount to analyze the consumer's reaction to the product and then focus on potential suggestions for improvement. Their intention to purchase the product could be assessed using a verbal scale (from definitely would buy to definitely would not buy) or a numerical probability. Upon a clear understanding of the differences between the various concepts, the responders' preferred concept should become evident. If the results are inconsistent, the team may choose it based on cost or may decide to offer different tiers or versions of the product. Forecasting models for a new long-acting local anesthetic, for instance, should consider rates of trial and likelihood of future repeat purchases when estimating sales potential to estimate demand following launch.

While forecasts often correlate with actual future sales, several considerations account for the difference observed in the "real world." Concept testing does not capture the importance of word-of-mouth. It could significantly affect the demand for the product and account for the difference. In addition, to the degree that the concept description is reliably replicated in the final product, there

	Telephone	Electronic Mail	Postal Mail	Internet	Face-to-Face
Verbal description	•	•	•	•	•
Sketch		•	•	•	•
Photo or rendering		•	•	•	•
Storyboard		•	•	•	•
Video				•	•
Simulation				•	•
Interactive multimedia				•	•
Physical appearance model					•
Working prototype					•

Figure 12.3 Appropriateness of different survey formats for different ways of communicating the product concept. Concept testing by Ulrich KT, Eppinger SD. Product design and development. Seventh Edition McGraw-Hill Education; 2019.

will be a corresponding positive concordance between the forecast and actual sales. Disclosure of the price, and by extension, cost breakdowns as a form of "intimate disclosure" remains controversial. Research has demonstrated that voluntary cost transparency enhances customer attraction and increases product sales. However, most authorities do not recommend including it. Instead, they suggest gauging customer's expectations and explicitly inquiring about how much the consumer is willing to pay. Since most insurers and third-party payers base their reimbursement on Medicare rates, it would be prudent to assess how much CMS would pay for a new medication, for instance. As experience with Aduhelm (aducanumab) has demonstrated, if the price is perceived to be too high, even if the drug receives FDA approval, it will not generate the expected ROI. This would also result in a discrepancy from the forecast. For most new products, advertising and other promotional activities can positively influence future demand. This influence is weakly correlated in the forecasting model by minimal awareness of the consumer's potential future desire for the product and the material used to make the prototype, which could account for the difference as well.

Concept testing, especially when performed early and in an iterative fashion in the development process, will yield valuable qualitative insights from open-ended discussions. This feedback from potential future customers provides the most important evidence for a blueprint of future product development. The team should consider alternative markets to ensure the inclusion of all future customers. Advertisements and promotions could increase customer awareness of a new product and subsequent sales. Modification of product design could enhance the attractiveness of the product and thereby increase the customers' intent to purchase. The degree to which the communication about the product would elicit the true intent of the customer to purchase the product is very important. For example, if the primary benefit of an iPhone was in its esthetic appeal, the team must present it as such clearly to the respondents. Furthermore, the resulting forecast should be congruent with observed sales rates of similar products as a benchmark.

[Potential Pitfalls]

(3–5 pitfalls to avoid).
Sample bias.
Sympathy bias.
Choosing the inappropriate survey format.

References

[1] Ulrich KT, Eppinger SD. Product design and development. Seventh Edition McGraw-Hill Education; 2019.

[2] Apfelbaum JL, Chen C, Mehta SS, et al. Postoperative pain experience: results from a national survey suggest postoperative pain continues to be undermanaged. Anesth Analg 2003;77:201.

[3] Svensson I, Sjostrom B, Haljamae H. Assessment of pain experiences after elective surgery. J Pain Symptom Manage 2000;20(3):193–201.

[4] Shea RA, Brooks JA, Dayhoff NE, et al. Pain intensity and postoperative pulmonary complications among the elderly after abdominal surgery. Heart Lung 2002;31(6):440–9.

[5] Tsui SL, Law S, Fok M, et al. Postoperative analgesia reduces mortality and morbidity after esophagectomy. Am J Surg 1997;173(6):472–8.

[6] Pavlin DJ, Chen C, Penaloza DA, et al. Pain as a factor complicating recovery and discharge after ambulatory surgery. Anesth Analg 2002;95(3):627–34.

[7] Perkins FM, Kehlet H. Chronic pain as an outcome of surgery: a review of predictive factors. Anesthesiology 2000;93(4):1123–33.

[8] Portenoy RK, Foley KM. Chronic use of opioid analgesics in non-malignant pain: report of 38 cases. Pain 1986;25(2):171–86.

[9] Florence CS, Zhou C, Luo F, Xu L. The economic burden of prescription opioid overdose, abuse, and dependence in the United States, 2013. Med Care 2016;54(10):901–6.

[10] Centers for Disease Control and Prevention. Opioid Overdose: Data Overview," 16 December 2016. [Online]. Available: <https://www.cdc.gov/drugoverdose/data/index.html> [accessed 22.05.17].

[11] Rudd RA, Seth P, David MS, Scholl L. Increases in drug and opioid-involved overdose deaths—United States, 2010–2015. CDC: Morbidity Mortal Wkly Rep 2016;65(50–51):1445–52.

[12] Prabhaker A, Cefalu JN, Rowe JS, Kaye AD, Urman RD. Techniques to optimize multimodal analgesia in ambulatory surgery. Curr Pain Headache Rep 2017;21(5).

[13] Carr DB, Goudas LC. Acute pain. Lancet 1999;353:2051–8.

[14] Helander EM, Menard BL, Harmon CM, Homra BK, Allain AV, Bordelon GL, et al. Multimodal analgesia, current concepts, and acute pain considerations. Curr Pain Headache Rep 2017;21(1).

[15] Buvanendran A, Kroin JS. Multimodal analgesia for controlling acute postoperative pain. Curr Opin Anaesthesiol 2009;22(5):588–93.

[16] Grant GJ, Bansinath M. Liposomal delivery system for local anesthetics. Reg Anesth Pain Med 2001;26:61–3.

[17] Planas ME, Gonzalez P, et al. Noninvasive percutaneous induction of topical analgesia by a new type of drug career, and prolongation of local pain insensitivity by anesthetic liposomes. Anesth analg 1992;75:615–21.

[18] Lafont ND, Legros FJ, Boogarets JG. Use of liposome associated bupivacaine in a cancer pain syndrome. Anaesthesia 1996;51:578–9.

[19] Boogarets JG, Lafont ND, et al. Epidural administration of lisosome-associated bupivacaine for the management of post surgical pain: a first study. J Clin Anesth 1994;6:315–20.

[20] Grant GJ, Barenholtz Y, et al. A novel liposomal bupivacaine formulation to produce ultalong-acting analgesia. Anesthesiology 2004;101:133–7.

[21] Christensen CM, Hall T, et al. Competing against luck: the story of innovation and customer choice. Illustrated edition. Harper Business; 2016.

[22] Macomber B, Yang M. The role of sketch finish and style in user responses to early design concepts. ASME Int Des Eng Technical Conf 2011;.

13

What to prototype?

Vittorio Mottini[1], Bryan Smith[1], Lucy Xu[1] and Joshua Younger[2]

[1]Department of Biomedical Engineering, Michigan State University, East Lansing, MI, United States [2]Henry Ford Hospital, Henry Ford Innovations, Detroit, MI, United States

Chapter outline

Abstract

This chapter highlights the importance and breadth of prototyping in the Biodesign paradigm within the context of anesthesiology. The processes before the prototyping phase are briefly described to summarize the previous steps. Instead of providing detailed instructions on how to construct different types of prototypes, we outline a general approach for innovators to maximize the effectiveness of concept exploration and testing to refine and enhance what to prototype. Readers will become familiar with translating concepts into functional blocks and generating technical specifications related to product feasibility. We emphasize derisking to transform a vision into an increasingly complex prototype as the Biodesign process evolves through various rounds of iterative testing.

Keywords: Biodesign; innovation; prototyping; derisking; testing; validation; need identification; need screening

Innovation in Anesthesiology. DOI: https://doi.org/10.1016/B978-0-12-818381-6.00017-6

Key points

1. Prototyping is guided by questioning and addressing possible pitfalls
2. Functional block selection is critical to the progress of the prototyping phase
3. Testing begins from the most critical components of the concept to reduce the risk of failure

Glossary

Iterations repetitive review and testing of design to improve each time
Ideation concept of creating new concepts or ideas
Derisking identifying and addressing possible risks that can be encountered while developing the prototype
Functional blocks one aspect of a concept that can be prototyped and tied to a unique engineering discipline based on its characteristics
Brainstorming generating ideas individually or as a team to solve clearly defined problems
Continuous opportunity discovery an ongoing commitment to seek out problems and challenges around oneself before identifying solutions
Testing points of evaluation and feedback as innovators move from prototype to product; specifications undertaken should be based on characteristics of the product
Validation measures taken at each design step to check quality, performance, and reliability before continuing to the next phase
Concept map a conceptual diagram that illustrates relationships between different ideas
Prototyping the process of creating early experimental versions of the product

Introduction/why it matters

The Biodesign process, as described in [1] and summarized in Fig. 13.1, aids in the challenge of bridging engineering's approach and innovation to the field of medicine and healthcare. The prototyping phase is an essential step for all clinical innovators, with the potential to reveal unforeseen challenges, receive feedback from end users, and identify improvements with iterative testing. Through first-hand observation of clinical problems at Henry Ford Health System's intensive care unit, Michigan State University's Biomedical Engineering graduate students had the opportunity to integrate Biodesign skill sets in identifying and then providing answers to pressing clinical

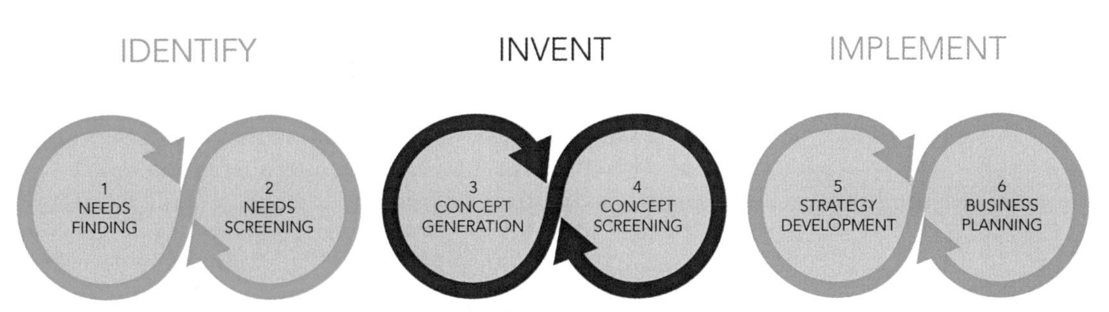

Figure 13.1 Summary of the three-step iterative Biodesign process. Adapted from [1].

issues. This chapter provides an overview of the Biodesign and prototyping guidelines to help innovators decide what to prototype, with examples in the context of anesthesiology.

Processes before prototyping

Before the prototyping phase, several necessary steps must occur to narrow down to a highly impactful clinical need and generate an appropriate solution concept to address the issues.

As discussed in previous chapters, the Biodesign innovation process begins with the identification of a problem or opportunity that needs to be addressed. For this reason and sustainability, innovators must actively seek need areas in which solving a problem can potentially generate improved economic outcomes. Continuous opportunity discovery through ethnography/observations and taking note of structured conversation can be used to identify the most pressing hypothesis and ask what underlies the problem before seeking solutions.

After identifying problems and needs, Biodesign innovators focus on fewer unmet clinical needs through a process of needs finding and screening. Following that, primary research can be performed by reading widely to uncover essential information and understand why something is occurring. Several needs can be found for a single problem; while all clinical needs are to be considered, several aspects must be taken into consideration to screen the identified needs and facilitate the best solution.

Needs are identified through direct observation of actual clinical situations, and, to be considered a real problem deserving further investigation, the innovator has to identify: (1) a problem, (2) the population affected by it, and (3) the intended outcome of a possible solution to the identified problem.

Identifying these three factors allows for formulating a Need Statement. This phrase clearly states the innovator's goal in the form of "A way to address (problem) in (population) that (outcome)." Well-formulated need statements must not be too general or specific and to not include a solution within the statement. Including or directing towards a specific solution would cause the innovators to have too many ideas to explore or not enough and introduce biases. This ultimately determines which needs are in the greatest alignment with the interests and priorities of the inventors. In-depth research into understanding the disease state fundamentals, existing solutions, stakeholder analysis, and market analysis can help drive innovators toward an option where significant risks can be mitigated.

Once needs have been screened, and a need statement has been formulated, the second step in the Biodesign process focuses on selecting possible solutions worth pursuing. Innovators teams will start by brainstorming possible ways to address the need. During this process, careful attention should be placed on:

- judgment deferral: criticism of a proposed solution to be made later in the process
- encouragement of wild ideas
- building on ideas from others
- aiming for quantity over quality
- having one conversation at a time
- being visual
- staying focused on the topic being discussed

At the end of brainstorming, it is essential to capture the results and organize them by clustering the ideas into groups of similar concepts through a mind map. Organizing concepts allows innovators to identify gaps, biases, and synergies among the ideas. Identifying gaps among groups of solutions is necessary so that additional ideation can be performed to address any opportunities that may have been missed. Innovators can also identify commonalities and complementarities between concepts so they can be merged to create synergistic, combined concepts.

In determining which activities to undertake during concept exploration and which questions to answer through prototyping, innovators should strategically prioritize which elements of a concept present the most significant risk to the technical viability of an idea and thus need to be addressed early on. Concepts must be screened after thorough research of existing solutions and patents, the regulatory pathways for the tentative solution, its reimbursement basics, and business models.

Prototyping

After identifying the most promising solutions, the prototyping phase is the next step in the Biodesign process. The goal is to translate a promising concept from an idea into a preliminary design and then ultimately into a working form. Early-stage prototyping must focus on simple, inexpensive, and quick processes to allow for fast iterations and weaving out possible bottlenecks in the product development process. The process of prototyping is a highly iterative endeavor, where each successive prototype assembled should be used to answer critical questions about the device and its functions. This allows for the improvement of the design while addressing deficiencies in the product. As the project progresses, the technical requirements, designs, and models call for more robust prototypes. Refinement of functionality and features requires a combination of user, bench, simulated use, and tissue testing to meet essential design requirements more comprehensively.

When developing a prototype, one must first define specific questions and issues to address throughout the process. This helps to identify what critical components are required at each stage and eliminates extraneous features that could detract from testing or unnecessarily increase the time/cost of testing. Technical feasibility is an important starting point, followed by user feasibility, novelty and nonobviousness for patentability, and value proposition. The questions will evolve as the prototypes progress based on the lessons learned from the previous iterations. Several combinations of models could be used to help innovators address issues that may arise:

- Works-like model: demonstrate the technical feasibility and gather user feedback
- Feels-like model: uses surrogate material to demonstrate ergonomics, weight, size, and tactical features
- Is-like model: performs desired functions as intended and gathers user experience
- Looks-like model: looks like the final product in terms of shape, size, color, and packaging
- looks-like/is-like model: functions and looks-like final device

It is advisable to start with the simplest model that has a narrow focus to answer specific questions. More complicated models should be saved for the later stages to prove the total concept.

Functional blocks

The ability to break down the selected concept into smaller functional blocks is a critical import to the prototype in a focused and iterative manner. A functional block can be defined as a subset of principles, skills, technologies, and expertise needed for the solution to perform a characteristic function. Many functional blocks can be identified for systems and equipment in anesthesiology. These may be divided based on physicochemical principles at work in the specific situation or skills required to assemble the prototype (mechanical engineering, biomaterials science, electrical engineering, and computer science/software engineering).

A good functional block selection allows the innovator to prototype and test the feasibility of each of them, independently of the others. Selecting functional blocks appropriately allows for evaluating different processes and elements of the envisioned solution without needing a working prototype beforehand. Moreover, addressing functional blocks individually optimizes the time, resources, and skills required. An example of a concept breakdown into functional blocks is shown in Fig. 13.2.

Derisking

Derisking refers to the active process of identifying and addressing possible risks that can be encountered while developing the prototype. Innovators should focus on prototyping a specific functional block with straightforward questions in mind. Derisking and functional block are two intrinsically related concepts: functional block identification should help answer specific questions to remove risks of failure, and derisking helps better identify functional blocks.

In deciding which activities to undertake during functional block prototyping, innovators should prioritize the elements of a concept that present the most significant risk to the technical viability of an idea and thus need to be addressed early on. Through this iterative process, innovators can quickly identify which functions of a concept can or cannot be realized, considerably speeding up the product development stage by moving forward with promising solutions and discarding unfeasible ones. Several types of tests can be done to derisk a candidate solution. These are, in order of complexity, regulations involved, and quality of information that can be obtained: user testing, bench testing, simulated use testing, acute animal testing, human cadaver testing, chronic animal testing, in-human testing, and clinical trials.

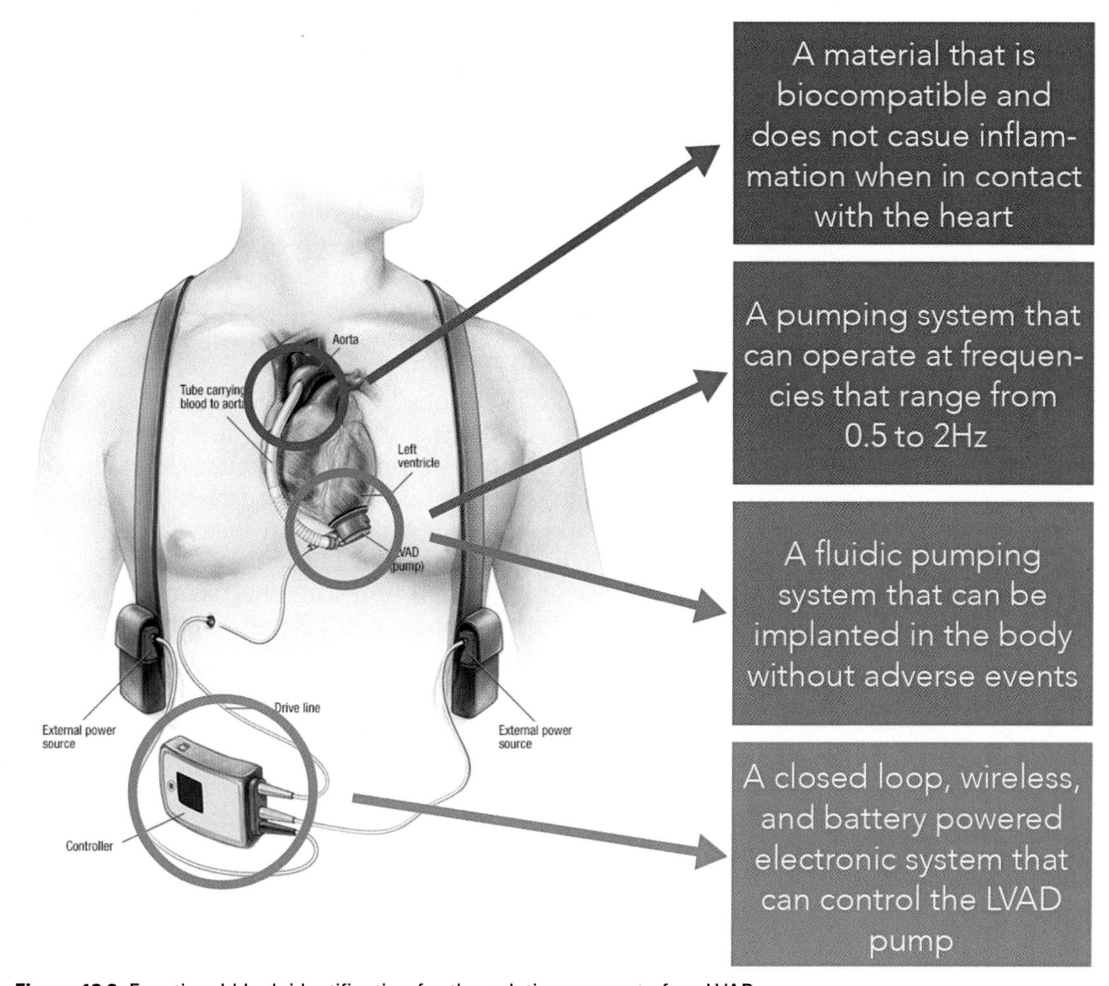

Figure 13.2 Functional block identification for the solution concept of an LVAD.

Examples

Biodesign group example #1

A Biodesign team from MSU Bioengineering rotated through the Henry Ford Health ICU and made 100 initial observations. Upon further reflection, a top need was identified. The need statement in this scenario was: "reduce blood clots entering the body for ICU ECMO patients to decrease clotting complications following device insertion." Following need identification, the top solution identified was to build integrated filters at the end of the tubing before blood re-enters the body to remove any clots formed inside the ECMO machine and prevent complications. This could also provide a solution for scenarios where anticoagulants are contraindicated. Altering the tubing or pump system might decrease the shear force that leads to clotting. However, it will not be a significant change nor eliminate the problem as the interaction between blood platelets and the ECMO surface will always be present. The structure of the filtering device will resemble an existing IVC filter. However, an opening at the tip connected with tubing to an external reservoir will be added for blood clots to be collected. When no blood clot is detected, the opening at the tip will be closed. A duplex ultrasound imaging system will be aimed at the tip of the filter to detect any clots that have been captured before changes in pressure or flow can occur. Once detected, the bypass tubing will be switched to the other channel to prevent the stopping of blood flow. Then, a syringe can be attached to the outer tube connected to the filter, and the clot can be removed by suction when the opening is unblocked. From there, the team broke down the design into three functional blocks:

1. A design with big enough pores for the blood to pass with minimal interference of flow but small enough to capture any clots formed in the stream

   Design and results: The first test involved the rigid paperclip design of the filter. It was tested by placing simulated blood clots into the ½" tubing and pumping them through to see if they could be captured. When the clots were pumped through the tubing, they traveled freely through them until they reached the filter, which caught them and housed them towards the middle of the filter, which was considered a positive result.

2. A way to detect when the filter captures blood clots.

   Design and results: Using a laser cutting design, a microfluidic device was created to mimic the lung-human alveolus. The device successfully accommodated rates matching blood flow up to 26 mL/min. Flow rates under conditions with similar viscosities as human blood will be tested.

3. A way to remove captured blood clots without stopping blood flow.

Design and results: The model was printed using elastic and hard-clear plastic materials. The elastic one was used at the top of the bypass to easily pinch one side of the model to act like a switch. The hard plastic one was used at the bottom end of the bypass. The glucose water was introduced into the tube. Pinching the elastic model successfully redirected the flow from one stream to the other. This indicated the feasibility of the idea of having a bypass with a switch.

Biodesign group example #2

A second team of graduate students identified that catheter-associated urinary tract infections (CAUTI) are a common, costly, and morbid complication for ICU patients. The need statement characterizing this problem is "A way to detect urinary tract infections in ICU patients requiring the use of an indwelling catheter." Following the Biodesign process, the team identified as a candidate solution the "Smart-catheter": a foley catheter carrying a biosensing module able to detect the development of free-flowing infection-inducing bacteria in urine while monitoring several other environmental factors such as temperature, pH, and salinity to prevent infections from developing.

The team divided the concept into four different functional blocks, each characterizing different functions of the thought design:

1. A device having a form factor that allows its placement on already available catheters, that does not add weight to the catheter, can be easily handled and placed in ICU rooms;

   Design and results: for the innovation to be implementable in current medical practice, it had to be readily manufacturable, used, transported, and attached to commercially available catheter tubing. The team focused on designing a catheter attachment that could easily be hung from the bedside. Through several iterative designs and 3D printing, the innovators' team produced a catheter attachment that can be placed along any catheter tubing through standard Luer fittings.
2. A Sample collection method for accurate and reproducible measurements;

   Design and results: To conduct reliable, accurate, and continuous bio- and environmental sensing in flowing urine, the flow rate has to be kept constant—the team identified as a possible risk the inability to produce consistent and reliable sensing data. Prototyping centered on developing disposable flow-controlled cartridges that could store a set volume of urine and discharge it upon reaching a fixed volume.
3. Sensing technologies that can measure significant biomarkers and environmental factors in flowing urine;

   Design and results: to remove the risk of not being able to detect the infection onset, the team researched possible infection biomarkers and sensing techniques for their identification. After extensive research and bench- and invitro testing, they developed a system that could detect physiologically significant nitrite concentrations through cyclic voltammetry.
4. An electronic system that collects, processes, stores, and sends data to an external monitor and displays relevant information.

Design and results: to monitor infections and report the data collected, the developed device had to send the measurements reliably and continuously to an external display that could easily access and monitor healthcare providers in the ICU. The team focused on assembling an electronic system based on open-source microcontrollers that could sense, store, and process data and send the information through Bluetooth communication protocols. The Smart-catheter prototype successfully underwent a series of tests using synthetic urine to validate its functionalities. Not only was it leak-proof, but it also demonstrated the capability to detect known concentrations of analytes within physiologically relevant ranges wirelessly. Although promising, it is important to note that the device is still in its prototyping phase. The next steps in its development include animal testing and real urine sample analysis, with data confirmation to be done through mass spectrometry.

Real-world example

To better clarify what to prototype for a candidate solution, we report here an example of innovators developing a solution to the need for "a way to prevent strokes in patients with atrial fibrillation caused by left atrial appendage (LAA) thrombus."

A team of Biodesign innovators is focused on preventing instances of LAA thrombus by filling the LAA with an engineered material that can solidify after being injected in liquid form to eliminate the space where thrombi can form. The starting concept can be divided according to its specific functions or fields involved. Although possible solutions can involve a single engineering field, they will likely involve multiple functions that can be separated and prototyped individually. Another possible approach in this scenario could involve classifying all the blocks related to materials into a single block with multiple properties and prototyping phases. In this case, there are mechanical engineering and biomaterials science components, as shown in Fig. 13.3.

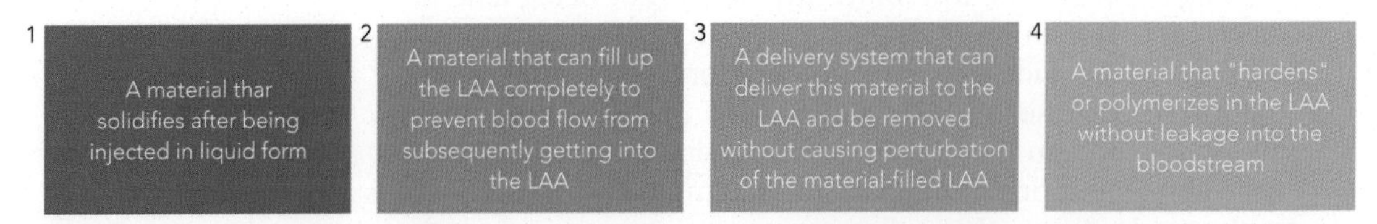

Figure 13.3 Functional block subdivision of a concept solution for the need statement "a way to prevent strokes in patients with atrial fibrillation caused by left atrial appendage (LAA) thrombus." Adapted from [1].

However, dividing the concept into precise and distinct functional blocks, each trying to answer critical questions in the prototyping phase, can significantly help innovators in derisking the solution while avoiding delays or bottlenecks in the testing process and helps in identifying the most critical and challenging components to be developed

Get started

1. Previous steps:
   a. Problem identification: through research and first-hand experience identify a need that can be addressed.
   b. Need statement: formulate a sentence clearly stating the problem the innovator team is trying to solve, in what population, and with which foreseeable outcome.
   c. Concept generation/screening: generate multiple solution concepts and screen for the most promising few.
2. Functional blocks selection: divide the solution concept into multiple units characterized by different critical questions to be answered with their prototyping
3. Derisking question formulation: formulate a question, directed by the functional block generation, that if answered could reduce the prototype risk of failing
4. Prototyping and testing: assemble the prototype and use one of the testing methods to answer the critical questions
5. Has the question been answered?
   a. Yes: move forward
   b. No: adapt the design and iterate

Potential pitfalls

- Trying to answer too many questions with initial prototyping models
- Not dividing prototypes into manageable functional blocks
- Questions and issues are too broad and not specific enough
- Not identifying and derisking the most pressing issues initially

Reference

[1] Paul, S.A. Zenios, J. Makower, G. Yock. Biodesign: the process of innovating medical technologies, Cambridge University Press, 2009.

14

Medical device development

Ruth Segall[1], Harry Burke[2], Zach Frabitore[2] and Trent Emerick[2]

[1]University of Pittsburgh, Pittsburgh, PA, United States [2]Department of Anesthesiology and Perioperative Medicine, University of Pittsburgh Medical Center, Pittsburgh, PA, United States

Chapter outline

Abstract

Medical device development is an iterative process during which problems and new information constantly arise, forcing innovators to make improvements and validate those changes before a prototype is ready for a pilot study and the next steps of commercialization. From beginning to end of the medical device development process, innovators will find their clinical need, concept, and final solution in an iterative loop based on feedback from internal and external stakeholders as well as the clinical and competitive landscape. The flow of changes to the design of the device will result in refinement to ensure the product addresses the problem of a specific population with a targeted change in the outcome. The final version of the product will be validated with the approval of internal and external stakeholders, most importantly the FDA in the US and EU-notified bodies.

Keywords: Iterative prototyping; iterative product development; design controls; design verification and validation; regulatory approval

- What is the iteration process?
- How are improvements defined, refined, and validated?
- How do you know a version is good enough?

Innovation in Anesthesiology. DOI: https://doi.org/10.1016/B978-0-12-818381-6.00007-3

Glossary terms and definitions

Design controls
Design and development planning
Design input
Design output
Design review
Design verification
Design validation
Design transfer
Design changes
Design history file

Chapter outline

- Iteration occurs through every step of medical device development, as new information constantly arises, creating a constant loop until the product is successfully launched.
- Improvements must be made to the device based on the definition of problems, refinement of needs to design inputs and validation of design outputs.
- The medical device is deemed "good enough" when it passes design validation and design verification protocols and can be translated into product specifications.

Medical device development is an iterative process where problems and new information constantly arise, forcing the innovators to make improvements and validate changes before the prototype is ready for a pilot study and the next steps of commercialization. From beginning to end of the medical device development process, innovators will find their clinical need, concept, and final solution in an iterative loop based on feedback from internal and external stakeholders as well as the surrounding medical landscape. The flow of changes to the design of the device will result in refinement to ensure the product addresses the problem of a specific population with a targeted change in the outcome. The final version of the product will be validated with the approval of internal and external stakeholders, most importantly the FDA in the US and EU-notified bodies. An understanding of the design controls for Medical Device Manufacturers is crucial for successful medical device development to ensure innovators achieve product commercialization [1].

Medical device development is an iterative process shaped by a multitude of stakeholders, depending upon the medical landscape of which the innovators and product are a part. Most innovators consider taking an idea from concept to minimum viable product (MVP) as the biggest hurdle in device development. With proper resources and education, physicians can ensure their product is prepared for the next steps of a pilot study and commercialization.

The iteration process begins well before a concept turns into a product, from the development of the clinical need statement to initial commercialization through the end of a product's life cycle. After developers identify the unmet clinical need their device will address, they create a prototype version of the needs statement that will become more descriptive and refined through needs scoping and validation. The team must conduct needs validation early on in the development process by assessing existing and potential competitor solutions to the clinical problem, exploring each with in-depth research. Validation of the clinical need should be conducted with a variety of practicing physicians or end-users to ensure all assumptions are tested before further iterations. Additionally, through market analysis and identification of the target market where the innovators can develop the product to meet the clinical need, the process of selecting needs is repeated until there is a set of 1–10 priority needs. The clinical need statement that the innovators present to the regulatory bodies might be further iterated when new information is obtained from the market and medical environment of use [2].

The exploration of the concept and subsequent prototype testing is another constant iterative process that innovators must visit as they progress along the path to commercialization. Every failure that innovators face in medical device development should trigger a brainstorming session, in which digressions can be encouraged to pivot the product to an enhanced concept that will address market needs. A good version of the MVP triggers new ideas to be explored and tested in the prototyping phase of product development. The crucial aspects of product commercialization lie in regulatory, intellectual property, and reimbursement, so innovators must constantly iterate on those respective strategies and allow those to enhance the MVP. Scoping the patent landscape is an iterative process, requiring innovators to make progressively deeper dives into a patent similar to their product to ensure the design specifications of their product do not infringe on claims in other patents. The same iterative process applies to regulatory and reimbursement about consistent research into the FDA database for similar products and premarket clearance pathways along with reimbursement pathways. Innovators will absorb feedback and information from developing these strategies that they can implement into the next iteration of their prototype [2].

In successful iterations of prototyping, the product can be divided into smaller components that correspond to certain device functions. It is helpful if the innovators focus on making essential components of the product operable and ensure the technical feasibility of those parts before testing them in the whole system. With the method of component iterations, certain aspects of the design will be refined individually, so fewer problems arise when components of the device are brought together as a complete product. There must be multiple rounds of user, bench, simulated use, and tissue tests during the iterative process for the complete product to be ready for animal, cadaver, and human testing later in the development timeline. Therefore, conducting iterative prototype testing of specific aspects of the product design will help optimize the device as a whole to ensure the concept is the effective solution to the desired clinical need [2].

Medical device development is not a linear process through the iterative development that has been previously discussed, but the FDA does set out a linear model for device manufacturers through the FDA Quality System Regulation (QSR) that is contained in the Code of Federal Regulations, Title 21 Part 820 [3]. The FDA aims to help innovators with a framework to develop and implement design controls, but these vary in practice as device manufacturers grapple with the iterative process of development. Shluzas et al. studied in-depth the development process of six different devices, whose developers ranged from start-ups to large companies, and various regulatory pathways. The linear device development model identifies five phases of product development defined by four different gates [2]. These phases are defined by the FDA's QSR: design input, design output, development, testing, risk analysis, and process qualification for manufacturers to develop and implement design controls [3]. However, iterative medical device development is not as smooth as the path outlined by the FDA, and loops often occur across various development phases (see Fig. 14.1). The variations that disrupt the linear model of development tend to stem from a user feedback loop that is repeated multiple times until the device can fulfill the user's needs and functionality requirements. When the innovators receive user feedback, typically physicians and other medical personnel, they must implement improvements and then re-test the device with users and other key stakeholders (i.e., industry thought leaders) until the device inputs meet the outputs.

Medical Device Development
ITERATIVE PROCESS

Clinical need identified
Isolate the problem and how it affects patient care.

Concept design
The design takes into account the surrounding medical landscape, including current patented prototypes.

Targeted change in outcome
The design is molded to maximize the effect on a specific population.

Final design solution
Varying internal and external stakeholders (i.e., FDA Design Control) further contour the design to its final form.

Figure 14.1 Linear medical device development.

Innovators will continue to define, refine, and validate improvements to their product design as they move further along the phases of testing. Benchtop and animal testing represent the first step of preclinical testing to assess for unexpected outcomes that will require further redesign. The prevalidation (performance) testing might need to be repeated several times if the device fails to move to the next phase of verification and validation (V&V) testing. To enter formal V&V testing, the product must reach a design freeze, so if the device fails testing at this phase, then the engineers can identify the root causes of the failure and implement changes. However, after these improvements are made, the device could end up returning to this testing further down the path of product commercialization. In parallel to V&V testing, innovators should be utilizing their respective regulatory agency presubmission programs to get feedback on the safety and effectiveness of the device before a formal submission. If clinical trials are not needed, a formal submission can be made to the FDA, but the FDA will most likely respond to the manufacturers with deficiency requests surrounding product testing and results. This will trigger additional testing by the innovators, requiring an iterative process of product refinement and improvement until the product is granted formal regulatory approval [4].

To ensure the product gains regulatory approval, innovators should closely follow the FDA's Design Control Guidance for Medical Device Manufacturers. The design controls process begins when the innovators define the design inputs but does not end with the transfer of the design to manufacturing, as the postmarket launch period could result in changes made to the product design. The design control process can be viewed as a waterfall loop (see Fig. 14.1 from FDA Guidance), in which the user needs to consistently redefine the design inputs and resulting outputs for product verification and validation. The validation of the design encompasses verification that the designed product satisfies the user's needs and intended use. Risk management should also be implemented early in the design process to ensure necessary changes are made before the cost and impact are too severe. Innovators should read through every section (A–J) of the design controls guidance and consistently refer to the document throughout the medical device development process [1]. When design controls are properly defined and executed, the medical device should not face any major issues with its regulatory clearance to market and can be deemed a safe and effective product [1].

Overall, innovators will find themselves in an iterative loop of medical device development full of product refinement and validation based on design controls to meet the clinical need they intend their product to solve. To ensure the medical device is good enough to move on to the next phase of commercialization (pilot study), identification and reduction of risk should occur. Marešová et al. [5] conducted in-depth research of the medical device development process worldwide and concluded that the most important factor is stability. Innovators must practice good manufacturing protocols for every version of their product to mitigate failure risk and present consistent evidence to regulatory bodies worldwide over the entire product life cycle, making certain that every version of the device is safe and effective until the product is retired from the market.

[Get Started]

1. Read FDA Design Control Guidance for Medical Device Manufacturers [3] (Table 14.1)

Table 14.1 Glossary terms and definitions, as defined by the Food and Drug Administration (Title 21, Vol 8., Sec 820.3).

Terms	Definitions
Design Controls	Creating and sustaining tasks to gaurantee design goals are appropriate.
Design and Development Planning	Creating and sustaining a system that details the design and development tasks.
Design Input	The physicial goals of the device that serve as the construct for the device design.
Design Output	The consequence of a design task at each design stage.
Design Review	A detailed summative evaluation of a design to determine the validity of the design tasks.
Design Verification	Verification by inspection of objective data that particular tasks have been completed.
Design Validation	Creating by objective data that device criteria comply with user specifications.
Design Transfer	Creating and sustaining methods to gaurantee that the device design is appropriately conferred into manufacturing descriptions.
Design Changes	Creating and sustaining tasks for the recognition, authentication, analysis, and confirmation of design changes prior to their application.
Design History File (DHF)	A document that holds or cites the evidence required to show that the design was created analogously with the confirmed design plan.

Source: *Food and Drug Administration. Sec. 820.3, Definitions. Subchapter H: Medical devices. In: Code of Federal Regulations, Title 21, vol. 8. < https://www.accessdata.fda.gov/scripts/cdrh/cfdocs/cfcfr/CFRSearch.cfm?CFRPart = 820&showFR = 1> [accessed 15.08.22].*

[Examples of...]

A. St. Francis Medical Technologies; acquired by Kyphon for X-STOP product [2]

  1. Began with two orthopedic spine surgeons, Jim Zucherman and Ken Hsu, who faced problems with the standard laminectomy procedure for lumbar spinal stenosis.

    a. The CEO of St. Francis, Kevin Sidow, found the surgeons in need of a less invasive way to treat spinal stenosis

    b. Surgeons noticed some patients with cognitive changes following anesthesia for the procedure, so they sought to only use local anesthesia

    c. Surgeons developed a minimally invasive implant to solve the problem

  2. Iterations of business model and regulatory development by the team led to medical device development

    d. Going down the regulatory path of a brand-new spinal implant was difficult. The FDA rejected the PMA application for the X-STOP

    e. Innovators were forced to regroup their clinical testing strategy of the product to present more straightforward data to the FDA

      – Reimbursement pathway was iterated as a new implantable spinal implant would not be adopted

      – Large companies did not want this implant following their reimbursement code

  3. Reimbursement strategy had to be iterated multiple times and therefore, the earlier in the development process this is addressed, the more likely the product makes it to market

    f. The iterations of these strategies and pathways resulted in FDA approval for the X-STOP as the first surgically implanted minimally invasive procedure requiring under an hour

    g. First-year post-FDA approval led to $58 M in worldwide sales and Kyphon bought St. Francis for $725 M

  4. Overall, the medical device development for the X-STOP was driven by failures that resulted in new iterations of regulatory and reimbursement strategies to ensure the product was good enough to get to market

B. Gradian Health Systems [2]

  5. Dr. Paul Fenton, a British anesthesiologist, was frustrated with unnecessary surgical death caused by unreliable anesthesia equipment, so he designed the Universal Anesthesia Machine (UAM)

- **h.** Can deliver safe, reliable anesthesia even during a power outage by generating its oxygen from an integrated oxygen concentrator and using room air with integrated oxygen monitoring
- **i.** The device worked, but insufficient interest from investors or buyers existed to bring the product to market
- **j.** Dr. Fenton persisted, and eventually, a private philanthropy fund helped to find a manufacturer to develop the product
- **k.** The foundation revamped itself to become Gradian Health Systems, LLC. to bring the product to market
  - Identified a niche market in low-resourced hospitals and used the foundation to support operations
  - Capital anesthesia equipment usually has a slow sales cycle, so the innovators needed to build a network to support the product
- **l.** Gradian developed a novel, low-resource market for this capital equipment that did not exist prior
  - Peer-reviewed publications, key opinion leaders, similar users, large-scale companies, and research donor organizations were utilized
  - Forced to untangle deep problems in the anesthesiology capital equipment space for low-resourced hospitals, adding another challenge to development

C. Emphasys Medical [2]

- **6.** Transferred between branches of the FDA to gain regulatory clearance, had to reiterate clinical strategy and conduct a pivotal trial to gain clearance
- **7.** Because a similar product had an FDA recall, the government and scientists were hesitant, resulting in a House Committee Investigation.

References

[1] FDA. Design control guidance. <https://www.fda.gov/media/116573/download?attachment>.

[2] Zenios S, Makower J, Yock P, Brinton T, Kumar U, Denend L, et al. Biodesign: the process of innovating medical technologies. Cambridge: Cambridge University Press; 2009. Available from: http://doi.org/10.1017/CBO9780511757853.

[3] Food and Drug Administration. Sec. 820.3, Definitions. Subchapter H: Medical devices. In: Code of Federal Regulations, Title 21, vol. 8. <https://www.accessdata.fda.gov/scripts/cdrh/cfdocs/cfcfr/CFRSearch.cfm?CFRPart = 820&showFR = 1> [accessed 15.08.22].

[4] Shluzas L.A., Pietzsch J.B., Paté-Cornell M.E., Yock P.G., Linehan J.H. (2009). The iterative nature of medical device design. In: DS 58-1: proceedings of ICED 09, the 17th international conference on engineering design, vol. 1, design processes, Palo Alto, CA, USA, September 24–27, 2009. p. 85–96.
[5] Marešová P, Klímová B, Honegr J, Kuča K, Ibrahim WNH, Selamat A. Medical device development process, and associated risks and legislative aspects—systematic review. Front Public Health 2020;8:308. Available from: https://doi.org/10.3389/fpubh.2020.00308 PMID: 32903646; PMCID: PMC7438805.

15

Pilot study

Odmara L. Barreto Chang
Department of Anesthesia and Perioperative Care, University of California San Francisco, San Francisco, CA, United States

Chapter outline

Abstract

This chapter explains the components of a pilot study and the value of medical device development. It explains why a pilot study is crucial before preparing for larger trials and why it is one of the first steps used in developing medical devices. The chapter provides the methodology to follow and reviews the CONSORT 2010 guidelines for conducting pilot studies, including a detailed table of the information that should be reported for pilot studies. Finally, we discuss practical considerations and include examples of pilot studies highlighting the importance and role of pilot studies in preparing for more extensive clinical trials.

Keywords: Pilot study; feasibility; randomized clinical trials (RCT); Consolidated Standards of Reporting Trials (CONSORT); Institutional Review Board (IRB)

What is a pilot study?

A pilot study is a small study or prestudy that can help assess the feasibility of the research, allowing the investigators to refine, improve, and modify the methods and anticipated

Innovation in Anesthesiology. DOI: https://doi.org/10.1016/B978-0-12-818381-6.00004-8

analyses intended for a full-scale study [1,2]. There are a variety of pilot study designs that can differ in size and methodology (randomized vs nonrandomized).

Why pilot studies are performed? Considerations and advantages

Pilot studies are performed to consider potential pitfalls in product design and assess the proposed project's practicality. One of the main goals of the pilot study is to determine the feasibility of the proposed project to outline the clinical and product use strengths, weaknesses, and potential areas of improvement. Further, define objectives, the research question, how to recruit subjects, how to fund the study, and obtain research personnel.

This also allows testing the hypothesis in a small number of patients to evaluate the logistics of conducting the study. It provides the opportunity for the investigator to refine the research hypothesis while identifying potential barriers and assessing if the methods used for the proposed experiments are feasible. Pilot studies can also provide data for estimates of parameters needed to calculate the sample size and estimates of missing data and dropout. When no preliminary data is available about the variables and outcomes of interest, the data from the pilot study can provide initial measurements of the effect size and statistical variability [1]. On some occasions, pilot studies have enabled the investigators to narrow or broaden the focus of the research based on data gathered from the analyses [2]. The reasons for conducting pilot studies have been classified into four categories: process, resources, management, and scientific [3]. The process is referred to as the study feasibility. Resources consider the logistics of performing the research; this includes time, materials, personnel, and participants. Management involves assessing data acquisition, quality, and sources. Scientific reasons for conducting a pilot study include evaluating treatment dose, time, safety, treatment response, and variability of effect [3].

How to conduct a pilot study?

Given the growing number of pilot studies, the Consolidated Standards of Reporting Trials (CONSORT) statement assembled a 26-item checklist to improve the transparency and quality of pilot study reporting. Specifically, they focused on randomized pilot studies performed before randomized clinical trials (RCT). The 26-item checklist was developed by experts, including trialists, methodologists, statisticians, funders, journal editors, and the CONSORT executive committee [4]. When reporting a pilot study, the following components should be included: title and abstract, introduction, methods, results, and discussion. It is recommended that the word pilot or feasibility is included in the title of any manuscript reporting the data of the study. There should be an explanation for the study's rationale in the introduction, including the objectives or hypothesis. The CONSORT has developed extensive guidelines for the methods, including trial design, participants, interventions, outcomes, sample size, randomization, sequence generation, allocation, implementation, blinding, and analytical techniques [4]. When reporting the results, it is recommended to follow a flow diagram that includes the participants, recruitment, follow-up, and the number of participants included in the analysis (Fig. 15.1). The statistical methods used and the outcomes of the study should be reported. The discussion should contain an interpretation of the results, applicability, and limitations of the study. Finally, the study should have approval by a group that has been formally designated to review and monitor biomedical research involving human subjects, such as an institutional review board (IRB) or ethical board approval. The IRB will review the research and protocols to ensure the protection of the rights and welfare of human subjects. For clinical trials, they should be in the study registry such as clinicaltrials.gov.

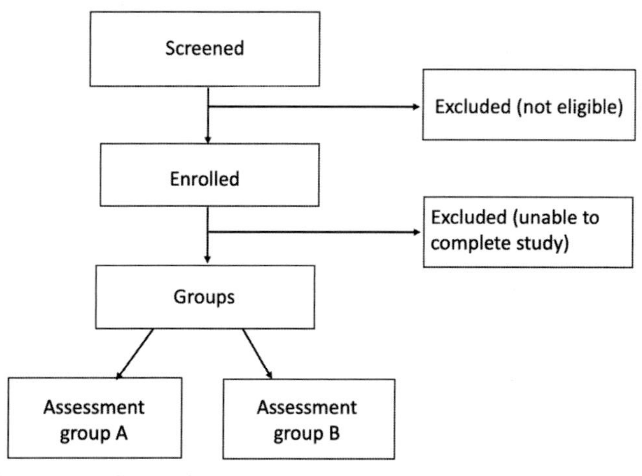

Figure 15.1 Flowchart of a sample pilot study.

Given the growing number of pilot studies, the Consolidated Standards of Reporting Trials have put together a checklist that can serve as a guideline when planning pilot trials (Tables 15.1 and 15.2).

Table 15.1 CONSORT checklist of information to include when reporting a pilot trial.

Section/topic and item No	Standard checklist item	Extension for pilot trials	Page No where item is reported
Title and abstract			
1a	Identification as a randomised trial in the title	Identification as a pilot or feasibility randomised trial in the title	
1b	Structured summary of trial design, methods, results, and conclusions (for specific guidance see CONSORT for abstracts)	Structured summary of pilot trial design, methods, results, and conclusions (for specific guidance see CONSORT abstract extension for pilot trials)	
Introduction			
Background and objectives:			
2a	Scientific background and explanation of rationale	Scientific background and explanation of rationale for future definitive trial, and reasons for randomised pilot trial	
2b	Specific objectives or hypotheses	Specific objectives or research questions for pilot trial	
Methods			
Trial design:			
3a	Description of trial design (such as parallel, factorial) including allocation ratio	Description of pilot trial design (such as parallel, factorial) including allocation ratio	
3b	Important changes to methods after trial commencement (such as eligibility criteria), with reasons	Important changes to methods after pilot trial commencement (such as eligibility criteria), with reasons	
Participants:			
4a	Eligibility criteria for participants		
4b	Settings and locations where the data were collected		
4c		How participants were identified and consented	

(*Continued*)

Table 15.1 (Continued)

Section/topic and item No	Standard checklist item	Extension for pilot trials	Page No where item is reported
Interventions:			
5	The interventions for each group with sufficient details to allow replication, including how and when they were actually administered		
Outcomes:			
6a	Completely defined prespecified primary and secondary outcome measures, including how and when they were assessed	Completely defined prespecified assessments or measurements to address each pilot trial objective specified in 2b, including how and when they were assessed	
6b	Any changes to trial outcomes after the trial commenced, with reasons	Any changes to pilot trial assessments or measurements after the pilot trial commenced, with reasons	
6c		If applicable, prespecified criteria used to judge whether, or how, to proceed with future definitive trial	
Sample size:			
7a	How sample size was determined	Rationale for numbers in the pilot trial	
7b	When applicable, explanation of any interim analyses and stopping guidelines		
Randomisation:			
Sequence generation:			
8a	Method used to generate the random allocation sequence		
8b	Type of randomisation; details of any restriction (such as blocking and block size)	Type of randomisation(s); details of any restriction (such as blocking and block size)	
Allocation concealment mechanism:			
9	Mechanism used to implement the random allocation sequence (such as sequentially numbered containers), describing any steps taken to conceal the sequence until interventions were assigned		

(*Continued*)

Table 15.1 (Continued)

Section/topic and item No	Standard checklist item	Extension for pilot trials	Page No where item is reported
Implementation:			
10	Who generated the random allocation sequence, enrolled participants, and assigned participants to interventions		
Blinding:			
11a	If done, who was blinded after assignment to interventions (eg, participants, care providers, those assessing outcomes) and how		
11b	If relevant, description of the similarity of interventions		
Analytical methods:			
12a	Statistical methods used to compare groups for primary and secondary outcomes	Methods used to address each pilot trial objective whether qualitative or quantitative	
12b	Methods for additional analyses, such as subgroup analyses and adjusted analyses	Not applicable	

Table 15.2 CONSORT checklist of information to include when reporting a pilot trial.

Section/topic and item No	Standard checklist item	Extension for pilot trials	Page no. where item is reported
Results			
Participant flow (a diagram is strongly recommended):			
13a	For each group, the numbers of participants who were randomly assigned, received intended treatment, and were analysed for the primary outcome	For each group, the numbers of participants who were approached and/or assessed for eligibility, randomly assigned, received intended treatment, and were assessed for each objective	
13b	For each group, losses and exclusions after randomisation, together with reasons		

(*Continued*)

Table 15.2 (Continued)

Section/topic and item No	Standard checklist item	Extension for pilot trials	Page no. where item is reported
Recruitment:			
14a	Dates defining the periods of recruitment and follow-up		
14b	Why the trial ended or was stopped	Why the pilot trial ended or was stopped	
Baseline data:			
15	A table showing baseline demographic and clinical characteristics for each group		
Numbers analysed:			
16	For each group, number of participants (denominator) included in each analysis and whether the analysis was by original assigned groups	For each objective, number of participants (denominator) included in each analysis. If relevant, these numbers should be by randomised group	
Outcomes and estimation:			
17a	For each primary and secondary outcome, results for each group, and the estimated effect size and its precision (such as 95% confidence interval)	For each objective, results including expressions of uncertainty (such as 95% confidence interval) for any estimates. If relevant, these results should be by randomised group	
17b	For binary outcomes, presentation of both absolute and relative effect sizes is recommended	Not applicable	
Ancillary analyses:			
18	Results of any other analyses performed, including subgroup analyses and adjusted analyses, distinguishing prespecified from exploratory	Results of any other analyses performed that could be used to inform the future definitive trial	
Harms:			
19	All important harms or unintended effects in each group (for specific guidance see CONSORT for harms)		
19a		If relevant, other important unintended consequences	

(*Continued*)

Table 15.2 (Continued)

Section/topic and item No	Standard checklist item	Extension for pilot trials	Page no. where item is reported
Discussion			
Limitations:			
20	Trial limitations, addressing sources of potential bias, imprecision, and, if relevant, multiplicity of analyses	Pilot trial limitations, addressing sources of potential bias and remaining uncertainty about feasibility	
Generalisability:			
21	Generalisability (external validity, applicability) of the trial findings	Generalisability (applicability) of pilot trial methods and findings to future definitive trial and other studies	
Interpretation:			
22	Interpretation consistent with results, balancing benefits and harms, and considering other relevant evidence	Interpretation consistent with pilot trial objectives and findings, balancing potential benefits and harms, and considering other relevant evidence	
22a		Implications for progression from pilot to future definitive trial, including any proposed amendments	
Other information			
Registration:			
23	Registration number and name of trial registry	Registration number for pilot trial and name of trial registry	
Protocol:			
24	Where the full trial protocol can be accessed, if available	Where the pilot trial protocol can be accessed, if available	
Funding:			
25	Sources of funding and other support (such as supply of drugs), role of funders		
26		Ethical approval or approval by research review committee, confirmed with reference number	

*Reprinted from Eldridge SM et al. CONSORT 2010 statement: extension to randomised pilot and feasibility trials with permission from BMJ Publishing Group (PMID: 27777223).

Practical considerations

When conducting a pilot study, one of the first steps is to define what the study aims to answer. If the pilot study is designed to evaluate a medical device, it is important to consider any previous studies done with the device or similar devices and conduct an in-depth review of the literature. Consulting experts in that area can provide input in creating the research question and the study. While performing the pilot study, it is important to remember that the research question must be feasible, interesting, novel, and relevant. A framework that can be used when formulating research questions for clinical studies is the PICOT (Patient, Intervention, Comparison, Outcome, and Time) format [5]. This can be used as a guide to develop the research question and address an area of interest or topic in which there are gaps in the current knowledge; thus, the study may contribute to expanding that area.

Another critical consideration for pilot studies is the recruitment and retention of subjects. The elements involved in recruiting subjects include engaging the participants by explaining the study, prescreening participants, discussing informed consent, and providing incentives when appropriate [6]. Studies can often be posted online (academic centers' websites, laboratory webpages, companies' webpages), in newspapers, newsletters, or by companies specializing in recruitment for clinical trials.

Furthermore, financial considerations need to be considered when doing a clinical study, including the study's cost and the required research personnel. In general, pilot studies tend to be smaller and serve as the first step in preparation for a larger study. However, the cost can be staggering, depending on the cost of developing/testing the new medical devices. The funding for the study can come from different sources, including institutions, private industry, voluntary groups, private and nonprofit foundations, and federal agencies, such as the National Institutes of Health, the US Department of Defense, and the US Department of Veterans Affairs among others.

Examples of pilot studies

Pilot studies can be divided into nonrandomized and randomized study processes. In nonrandomized studies, the participants are not randomized. In contrast, randomized studies are those in which the participants are randomly assigned to an experimental group or a control group. The advantage of conducting a randomized pilot is that it decreases bias and confounding that is observed when the subjects are assigned to specific groups in a nonrandom manner. For example, Garcia et al. developed a pilot study before implementing a larger randomized trial to investigate the feasibility, safety, and initial efficacy of acupuncture for uncontrolled pain among cancer patients [7]. With the pilot data, they were able to demonstrate that acupuncture was feasible, safe, and a helpful treatment adjunct for cancer patients experiencing uncontrolled pain in this study. This pilot data serves as the initiation point for the future development of a randomized placebo-controlled trial [7].

Pilot studies play an important role in medical device development. The effect of medical devices is often tested by conducting pilot studies. For example, the use of percutaneous peripheral nerve stimulation (neuromodulation) was studied using a randomized sham-controlled pilot study [8]. The investigators conducted a pilot study to determine (1) the feasibility and optimize the protocol for a subsequent clinical trial and (2) estimate the treatment effect of percutaneous peripheral nerve stimulation on postoperative pain and opioid consumption. By conducting this pilot study, the investigators determined that the study was feasible, and the findings served as power estimation for subsequent studies. In this case, their treatment effect was much greater than what they expected, as opioid consumption was reduced by 80% and pain scores by more than 50% in the group that received neuromodulation. The pilot's data also allows them to adjust the number for a subsequent trial; in this case, they could reduce the initially planned number of 528 to 250 based on effect size. Other input collected from the study included removing centers that lack enrollment, calling participants before the surgery to review the protocol and answer questions, and adding additional long-term time points for evaluation of long-term benefits and adverse events [8].

Medical devices in anesthesia play an important role in diagnosis and guiding treatment decisions. Device reliability and use case feasibility in the perioperative setting are often tested by doing a pilot study. For example, a pilot study was designed to evaluate the reliability of a wireless electrocardiogram monitor in the operating room [9]. Wireless monitoring has many advantages, such as low cost and reducing the number of cables attached to patients, thus improving anesthesia ergonomics and facilitating patient management. In this study, the authors tested if wireless electrocardiograms using Bluetooth can reliably be used in the operating room. They showed that the Bluetooth electrocardiogram prototype could provide reliable electrocardiogram monitoring in the operating room. The results from this pilot study provided initial data about the reliability of intraoperative wireless electrocardiograms and encouraged manufacturers to develop new devices for wireless monitoring based on their results [9].

Common medical devices that have been used in clinical applications can also be often explored as diagnostic tools in other clinical settings. Pilot studies can be done to test the modalities of these diagnostic tools and the feasibility of these devices in different environments. For example, a study by Monastesse and colleagues aimed to investigate the role of lung ultrasonography in the assessment of perioperative atelectasis [10]. They conducted a prospective observational pilot study that included 30 patients undergoing laparoscopic surgeries. The study involved doing a systematic ultrasound evaluation of the patient's thorax. The investigators identified 2 capnothoraces, 1 endobronchial intubation, and 1 episode of subclinical pulmonary edema [10]. They showed that lung ultrasonography in the perioperative period is feasible and allows tracking of perioperative atelectasis. Given the limited diagnostic tools available for intraoperative evaluation of lung atelectasis, the authors suggested that this pilot study should encourage larger studies to further research the role of lung ultrasonography as a modality to monitor perioperative atelectasis.

Lastly, there are occasions when the pilot study allows the investigators to make protocol modifications that can strengthen the larger study. For example, Buse et al. did a pilot trial to determine the feasibility of a trial comparing accelerated care versus standard care among patients with hip fractures. They found the study was feasible, and identified design issues they were able to overcome through modifications in the trial protocol, thus supporting a larger definitive trial [11]. These examples illustrate how the results from pilot trials can be used to decide if a larger study should proceed or if any modifications should be made to the protocol before embarking on a larger study.

Summary

Pilot studies are considered initial studies to determine if the study design, measures, procedures, recruitment, and resources under consideration will be feasible and provide an adequate sample size for a more extensive study. They allow the investigator to test novel approaches and at the same time, gather preliminary data that can serve to improve or modify methods of a larger trial. The ability to start with a pilot study allows the investigator to invest fewer resources initially and, at the same time, consider the possibility of modifying the proposed protocols to facilitate the success of a larger study. The CONSORT 2010 statement provides appropriate guidelines, including a checklist that should be used when preparing and designing a pilot study. The CONSORT guidelines allow the investigator to have a solid framework and facilitate the organization of the data, resulting in improvements in the reporting of pilot trials. A successful pilot study requires a solid framework that includes study design, analysis plan, and considers sample size. Furthermore, when designing the pilot study, the investigator should carefully consider the overarching research question addressed with the larger study.

References

[1] Hulley SB, Cummings SR, Browner WS, et al. Designing Clinical Research. 4th edition Lippincott Williams & Wilkins (LWW); 2013. ISBN:978-1-60–831804-9.

[2] Eldridge SM, Lancaster GA, Campbell MJ, Thabane L, Hopewell S, Coleman CL, et al. Defining feasibility and pilot studies in preparation for randomised controlled trials: Development of a conceptual framework. PLoS One 2016;11(3):e0150205. Available from: https://doi.org/10.1371/journal.pone.0150205. PMID: 26978655; PMCID: PMC4792418.

[3] Thabane L, Ma J, Chu R, Cheng J, Ismaila A, Rios LP, et al. A tutorial on pilot studies: the what, why and how. BMC Med Res Methodol 2010;10:1. Available from: https://doi.org/10.1186/1471-2288-10-1. PMID: 20053272; PMCID: PMC2824145.

[4] Eldridge SM, Chan CL, Campbell MJ, Bond CM, Hopewell S, Thabane L, et al. PAFS consensus group. CONSORT 2010 statement: extension to randomised pilot and feasibility trials. BMJ 2016;355:i5239. Available from: https://doi.org/10.1136/bmj.i5239. PMID: 27777223; PMCID: PMC5076380.

[5] Riva JJ, Malik KM, Burnie SJ, Endicott AR, Busse JW. What is your research question? An introduction to the PICOT format for clinicians. J Can Chiropr Assoc 2012;56(3):167–71. PMID: 22997465; PMCID: PMC3430448.

[6] Thoma A, Farrokhyar F, McKnight L, Bhandari M. Practical tips for surgical research: how to optimize patient recruitment. Can J Surg 2010;53 (3):205–10. PMID: 20507795; PMCID: PMC2878987.

[7] Garcia MK, Driver L, Haddad R, Lee R, Palmer JL, Wei Q, et al. Acupuncture for treatment of uncontrolled pain in cancer patients: a pragmatic pilot study. Integr Cancer Ther 2014;13(2):133–40. Available from: https://doi.org/10.1177/1534735413510558. Epub 2013 Nov 25. PMID: 24282103.

[8] Ilfeld BM, Plunkett A, Vijjeswarapu AM, Hackworth R, Dhanjal S, Turan A, et al. PAINfRE Investigators Percutaneous peripheral nerve stimulation (neuromodulation) for postoperative pain: a randomized, sham-controlled pilot study. Anesthesiology 2021;135(1):95–110. Available from: https://doi.org/10.1097/ALN.0000000000003776. PMID: 33856424; PMCID: PMC8249357.

[9] Ariès P, Bensafia K, Mansour A, Clément B, Vincent JL, Nguyen BV. Design and evaluation of a wireless electrocardiogram monitor in an operating room: a pilot study. Anesth Analg 2019;129(4):991–6. Available from: https://doi.org/10.1213/ANE.0000000000003972. PMID: 30540614.
[10] Monastesse A, Girard F, Massicotte N, Chartrand-Lefebvre C, Girard M. Lung ultrasonography for the assessment of perioperative atelectasis: a pilot feasibility study. Anesth Analg 2017;124(2):494–504. Available from: https://doi.org/10.1213/ANE.0000000000001603. PMID: 27669555.
[11] Hip Fracture Accelerated Surgical Treatment and Care Track (HIP ATTACK) Investigators. Accelerated care versus standard care among patients with hip fracture: the HIP ATTACK pilot trial. CMAJ 2014;186(1):E52–60. Available from: https://doi.org/10.1503/cmaj.130901. Epub 2013 Nov 18. PMID: 24246589; PMCID: PMC3883849.

16

Clinical research

Rachel Wang Hoffman[1,2]
[1]*Clinical & Regulatory Affairs, Medcura, Inc, Riverdale, MD, United States*
[2]*Syneos Health, Morrisville, NC, United States*

Chapter outline

Abstract

This chapter introduces medical device development and the role of clinical research in medical innovation. Advancing new and existing approaches to patient treatment not only includes ongoing research to increase understanding of disease states but also developing new techniques based on this increased understanding. Medical devices are a vital part of patient care, ranging from imaging to equipment such as anesthetic vaporizers, to products permanently implanted into patients. They can be designed to prevent, diagnose, or treat diseases and disorders. Continued development and innovation of devices are critical to improving patient outcomes and involve healthcare practitioners, regulators, manufacturers, Clinical Research Organizations, and more. This chapter defines clinical studies and describes their purpose and basic elements of design and conduct.

Innovation in Anesthesiology. DOI: https://doi.org/10.1016/B978-0-12-818381-6.00002-4

Keywords: Clinical research; randomized controlled trial; clinical investigation; protocol; study design; medical device development; medical technology

What is clinical research?

Clinical research is a systematic investigation undertaken in one or more live subjects to assess the clinical performance, effectiveness, or safety of a treatment. The Belmont Report summarizes the basic ethical principles that should underlie the conduct of biomedical and behavioral research involving human subjects and defines clinical research as "an activity designed to test a hypothesis, permit conclusions to be drawn, and thereby to develop or contribute to generalizable knowledge…" A narrative review based on a course of clinical trials reported by Umscheid CA, et al. notes that clinical research is "designed to observe outcomes of human subjects under "experimental" conditions controlled by the scientist" [1]. The ethical and quality conduct of well-designed clinical studies is a rigorous approach to demonstrating causal associations in the practice of evidence-based medicine and can contribute significantly to improving the effectiveness, safety, and efficiency of healthcare practices [1].

Purpose of conducting clinical research

Clinical research can be performed to improve knowledge about a disease state, develop diagnostic methods, or assess the safety and performance of a specific treatment or technology, which includes medications and medical devices. This chapter focuses on medical device clinical research but also include drug research.

Different types of medical device studies are shown in Table 16.1.

Registrational studies are performed to provide evidence for Regulatory Authorities, such as the United States Food and Drug Administration (FDA), to grant clearance or approval to market the device. Registrational studies can be first in human, feasibility, pilot, pivotal, bridging, or label expansion. Postmarketing studies are performed after Regulatory Authorities have granted clearance or approval and generally provide data on the continued performance and safety of devices, sometimes in a real-world setting.

Table 16.1 Examples of medical device study types.

Medical device registrational studies	Postmarketing and research studies
First in human	Nonsubmission research
Feasibility or pilot	Scale development and validation
Pivotal	Human factors
Bridging (supplemental study)	Postmarket patient registry
Label expansion (expanded indication)	Prospective postmarketing
	Retrospective postmarketing
	Patient preference
	Surveys

Research studies, which can be performed prior to or after marketing authorization, are intended to support the development and validation of scales or algorithms or to evaluate human factors or device usability.

Though the specific purposes can vary from study to study and be tied to a commercial goal, clinical research is generally performed to gather data to advance healthcare and improve patient outcomes.

Clinical study design

This section will focus on the design of a pivotal randomized clinical trial, considered the most rigorous type of medical device study. Study design is critical in:

- the protection of the safety, rights, and welfare of human subjects [2]
- conducting scientifically sound research that soundly tests a hypothesis, including identification and mitigation of effects of confounding variables on data analysis
- ensuring that collected data is representative of and applicable to, the targeted patient population and disease state
- preserving operational feasibility

In a pivotal study, the main goal is to demonstrate that a new product has better or noninferior efficacy, safety, and/or usability than another product or standard of care procedure. These studies test the impact of an intervention or technology on pre-specified outcome(s). The variable—use of the device—is presented, and the effects of that variable on outcomes are assessed.

Key factors to consider in study design are the endpoints, patient population, reproducibility of results, and operational efficiency. The study should not only be designed to prove that the product is safe and efficacious, but should also consider the patient population, risk/benefit profile, and the logistics of executing the study in compliance with applicable regulations and guidance.

Study objectives and endpoints

Clinical studies should have well-defined objectives and endpoints. Objectives of the study describe its purpose and identify what questions need to be answered or what needs to be proven. Endpoints define the success of the study and determine the assessments that need to be performed. A well-designed study has clearly defined endpoints that are measurable—a measure of how a patient feels, functions, or survives. Measurable parameters can be quantitative, semiquantitative, or qualitative, but should always be reportable and clinically meaningful. In other words, an endpoint is an event or outcome that can be measured, ideally in an objective manner, to determine whether the product being studied has a certain effect.

There can be multiple endpoints for a single study. The primary endpoint drives the statistical methods and sample size and addresses the main objective of the study. Secondary endpoints further support the objectives of the study and provide information on the product's performance or safety. Examples of endpoints are listed below in Table 16.2.

Endpoints should tie directly to the objectives of the study; both should be driven by the indication for use, target patient population, regulatory requirements, and commercial considerations.

Study population and eligibility criteria

Participants in a clinical study should be a subset of patients from the population that adequately represents the population from which it is drawn, such that true inferences about the patient population can be made from the study results. The patient population is determined by the intended use and device labeling. The patient population for the study may be further restricted to balance the risk/benefit ratio of the study (safety concerns), limit confounding variables, take into account input from regulatory authorities, and protect vulnerable populations (e.g., pediatric, cognitively impaired, or incarcerated). Participants in a clinical study are selected using pre-defined

Table 16.2 Different types of endpoints.

Objective	• Measurable disease state or measurable variable • Examples include the occurrence of a clinical event (myocardial infarction, stroke, or death), time to occurrence of a clinical event, or another measurable variable such as percent stenosis (stent) or wound surface area (wound dressing)
Subjective	• Questionnaires or patient-reported outcomes • Validated instruments or assessment tools should be utilized whenever possible to increase the reliability, repeatability, and/or validity of the outcome measure
Surrogate	• A surrogate endpoint could be a laboratory measure or physical sign that is intended to be used as a substitute or indication of a clinically meaningful endpoint or outcome • Examples include urine leakage being a physical sign of incontinence, or elevated cholesterol to be an indicator of heart disease
Composite	• A composite endpoint is a single measure of effect but is based on a combination of individual endpoints or variables • Example: thromboembolic event + stenosis + repeat revascularization → stent failure

eligibility criteria. In order to be considered eligible for study participation, a patient must meet all inclusion criteria and none of the exclusion criteria.

Control group and randomization

A control group is a comparison group used to determine if a new product has better performance and/or safety. Control groups are utilized in controlled studies, and can be nonrandomized or randomized. Random allocation of patients into test or control groups helps reduce selection bias resulting frosm an imbalance of confounding variables; in other words, it helps ensure comparability of patient populations between groups (Umscheid, 2011). This means factors that could potentially affect the study outcomes or assessments are similarly and evenly distributed between groups [3]. Similar characteristics of the patient population enrolled into each group enhance the validity of different treatment effects seen between groups—often tied to study endpoints. These factors could include, but are not limited to, severity of disease state, age, sex, and comorbidities.

Control groups could be historical, placebo/sham, active treatment, or dose-comparator. A historical control is external to the clinical study, where patients in the control group were treated in an earlier study or different setting, which is an important distinction from studies involving contemporaneous patient enrollment groups. A placebo or sham is a form of medical therapy or intervention designed to simulate medical therapy, without specificity or activity for the condition being treated and assessed. Placebos are usually used to show a difference in efficacy and equivalence in safety. An active treatment comparator provides medical therapy for the condition being treated; in these cases, the study can be designed to show equivalence, noninferiority, or superiority in both efficacy and safety. A dose comparator uses different doses or regimens of the same treatment; the intent is to establish a relationship between dose and efficacy or safety.

The selection of the control used in a clinical study is critical. The selected control should:

- have the same, or very similar, intended use or targeted indication for use of the product being studied
- have a similar mechanism of action as the product being studied and/or

be the gold standard, or prevalently used method, for the treatment of the condition.

Justification of the control group should be provided within the protocol and included in submissions to regulatory authorities.

Factors to consider in establishing the control group include, but are not limited to:

- current standard of care practices;
- minimizing bias and confounding variables;
- enrollment;
- acceptability and applicability of results;
- safety profile;
- trial execution and logistics; and
- ethics (e.g., serious illnesses).

Statistical methods

Randomized controlled trials are designed to test a hypothesis. Statistical methodology includes information about the hypothesis(es) and how they will be tested.

Sample size calculation is an integral piece of statistical methodology. The number of patients should have sufficient power to detect the effect of scientific importance [4]. Factors to consider in determining sample size include, but are not limited to:

- baseline incidence—if an outcome occurs infrequently, more patients are needed to detect a difference
- treatment effect size—if the difference is small, more subjects are needed to detect a difference

- population variance—the higher the standard deviation, the more subjects needed to demonstrate a difference
- attrition rate—patients may drop out of the study due to withdrawal, death, loss of follow-up, etc.
- allocation ratio—patients may be randomized at different ratios to test or control

The type of test impacts sample size calculation. A test for superiority has the primary objective of showing that the response to the test treatment is superior to the control. A test for noninferiority has the primary objective of showing that the response to the test treatment is not clinically inferior to the control. A baseline equivalence test is aimed toward showing that the response to the test treatment is neither superior nor inferior to the control. These different tests may apply to efficacy, safety, usability, and other variables indicative of the effectiveness of treatment.

Randomized controlled pivotal trials for medical devices usually involve several hundred patients, but this number can vary based on the factors above, with some large studies involving more than one thousand.

Key parameters of statistical methodology are described in Table 16.3.

Clinical study flow

Fig. 16.1 depicts the general flow for a randomized controlled trial. This flow may differ depending on the number of study groups, cross-over study design, and study-specific procedures. The flow of a study and the patient journey are described in the trial's clinical investigational plan (16.3.6).

Clinical investigational plan

The Clinical Investigational Plan (CIP) is a document that states the rationale, objectives, design, pre-specified analyses, methodology, organization, monitoring, conduct, and record-keeping of the study. The CIP is often referred to as the protocol, though the precise definition of this term may differ from country to country. The International Organization for Standardization (ISO) has published a standard with which randomized controlled trials should comply: ISO14155:2020—Clinical Investigation of Medical Devices for Human Subjects—Good Clinical Practice (GCP).

Components of the CIP are detailed in Annex A of the standard. Note that not all components are applicable to every

Table 16.3 Statistical methodology considerations.

Design and analytical methods	In addition to the relative performance claims described above, the specific type(s) of test (s) should be described. Examples include *t*-tests, analysis of variance (ANOVA), and chi-square tests. Different analytical methods exist for different endpoints.
Analysis considerations	Variable types, such as continuous or categorical, and their associated calculations, should be detailed. Software used for analyses should also be considered. Analysis populations, subgroup analyses, data transformations, and handling of missing data should be described.
Success	The conditions of study success and individual patient success should be defined.
Sample size justification [5]	Parameters and methods for sample size calculation should be specified, including, but not limited to: • assumptions about test and control performance, usually related to the primary endpoint and null and alternative hypotheses • statistical power (β)—the probability that a hypothesis test correctly infers that a sample effect exists in the population; this is usually set at 0.80 or 0.90 for randomized controlled trials • significance (α)—the probability of rejecting the null hypothesis when it is true, usually set at 0.05 for randomized controlled trials • margin for noninferiority tests; this is usually set at 0.10 or 0.15 and should be justified based on risk and clinical significance [5]

study; additional content may also be warranted based on country-specific regulations and guidance. The CIP should provide relevant background information and justification of the study design and risk-benefit ratio. This document governs the conduct of the study, so procedures should be described in enough detail to ensure compliance and consistency in study conduct across patients and investigational sites.

Ethics in clinical research

Ethics are a key piece of clinical research. The protection of human subjects has a long and complex history. The Belmont Report, Nuremberg Code, and Declaration of Helsinki are the foundations for ethical principles in the protection of research subjects.

The Nuremberg Code is the direct result of the Nuremberg trials, which were held at the end of World War II for Nazi war crimes, including those for German physicians who carried out

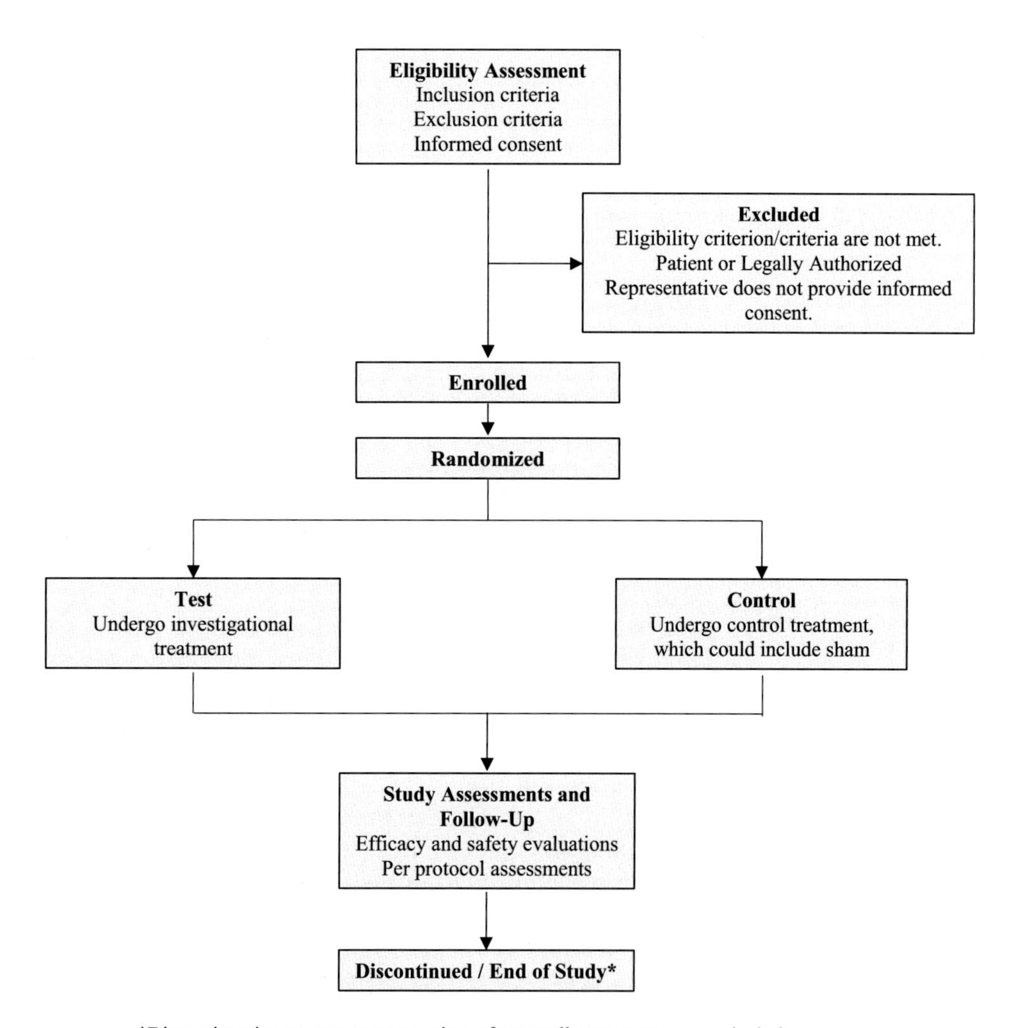

Figure 16.1 Randomized controlled trial flow.

unethical human experimentation during the war. The verdict included "Permissible Medical Experiments" describing the ten points of the Nuremberg Code:

1. Voluntary consent of the human subject
2. Experimentation should be necessary and justified
3. Design of the experiment such that anticipated results justify the conduct
4. Experiment should be conducted in a way to avoid unnecessary physical and mental harm

5. No experiment should be conducted where there is an a priori reason to believe that death or disabling injury will occur
6. Risk and benefit justification
7. Protection of the subject against the possibility of injury, disability, or death
8. Experiments should be conducted by scientifically qualified persons
9. Subject right to withdraw
10. Ability to terminate the experiment upon the belief that continuation would likely result in injury, disability, or death

The ten principles in the Nuremberg Code are further elaborated in The Declaration of Helsinki, considered the cornerstone of ethical human research, which was originally adopted in 1964 and has undergone seven revisions, the most recent in 2013. The principles presented in The Declaration center around human research subject respect, right to self-determination, right to make informed decisions regarding participation in research, and subject welfare taking precedence over interests of science and society. Further, ethical considerations also take precedence over laws and regulations. Vigilance around the participation of vulnerable populations is also addressed.

The Declaration describes the responsibilities of the physician in terms of ethics, with the key point that the physician will act in patients' best interest and safeguard their health, well-being, and rights. Physicians who are involved in medical research have a duty to protect the life, health, dignity, integrity, right to self-determination, privacy, and confidentiality of personal information of research subjects. The Declaration also presents operational principles for conducting clinical research, with a focus on risk/benefit analysis, independent ethical review, and justification of study design.

The Belmont Report (1978) was written by the National Commission for the Protection of Human Subjects of Biomedical and Behavioral Research, created as a result of the National Research Act of 1974. The Commission was responsible for the identification of the ethical principles for conducting research on human subjects. The Report presents three basic principles:

1. respect for persons, requiring acknowledgment of autonomy and requiring protection of persons with diminished autonomy
2. beneficence, not only in the ethical treatment of human subjects but also in protecting them from harm through efforts to maximize benefits and minimize potential harms
3. justice, or equal treatment of equals.

Standards and regulations also describe parameters of human subject protection and required elements of informed consent, which stem from the principles in these three foundational documents. Included in this list are ISO14155 and United States Codes of Federal Regulations (CFR), specifically 45 CFR Part 46 and 21 CFR Part 50. Compliance with ethical principles, standards, and regulations is critical for approval to conduct research, marketing clearance, and acceptability of data for publication.

Conclusion

Clinical research is critical to the advancement of healthcare. As new technologies are developed, meaningful and quality data are required to demonstrate the effectiveness, performance, and safety of new treatments. The basic concepts of clinical studies and their design are described in this chapter, along with considerations for establishing a scientifically sound approach to research. Although scientific rigor is critically important, conducting clinical research must put the safety and welfare of human patients first.

References

[1] Umscheid CA, Margolis DJ, Grossman CE. Key concepts of clinical trials: a narrative review. Postgrad Med 2011;123(5):194–204.

[2] World Medical Association. Declaration of Helsinki. 1964, Retrieved from: https://www.who.int/bulletin/archives/79(4)373.pdf.

[3] Lim CY, In J. Randomization in clinical studies. Korean J Anesthesiol 2019;72 (3):221.

[4] Barkan H. Statistics in clinical research: important considerations. Ann Card Anaesth 2015;18(1):74.

[5] Shan G, Banks S, Miller JB, Ritter A, Bernick C, Lombardo J, et al. Statistical advances in clinical trials and clinical research. Alzheimer's Dementia Transl Res Clin Intervent 2018;4:366–71.

17

Licensing—how to, terms, expectations, examples for university technology transfer

Steven Kubisen[1] **and Brian Coblitz**[2]

[1]InnoComm, LLC, Annapolis, MD, United States [2]Technology Commercialization Office, The George Washington University, Washington, DC, United States

Chapter outline

Abstract

The process for developing and licensing a life science invention out of a university is unique as compared to development of an invention in a private company. This chapter covers the unique culture in a university, which can lead to very innovative inventions and the particular process that is followed by leading research universities to commercialize their inventions. Areas that are discussed include key steps in the development, creating a win-win deal, when to license or option technology, key terms in the license and how the best technology commercialization offices are structured to increase the value of the invention and the probability of successful licenses.

Keywords: License; Licensee; Option; negotiation; reimbursement; regulatory; partners; technology; commercialization; diligence; milestone; intellectual property; royalty; equity; exclusive; patent; university; development; startup

Innovation in Anesthesiology. DOI: https://doi.org/10.1016/B978-0-12-818381-6.00032-2

How to maximize the value of licensing technologies

Licensing a university technology involves developing a licensing agreement (a "License"), but more importantly requires developing relationships with the business entrepreneur and connecting to commercialization funding sources. The legal document of a License together with the process of connecting to external and internal resources is how the technology is transferred and commercialized. We will cover the key elements of the actual licensing agreement in later sections, but here we will discuss the important external relationships that need to be developed for an effective License and effective commercialization process.

Universities conduct basic research and some early-stage development work in some cases. They are not set up to actually commercialize a product. Thus the technology commercialization offices (TCO) strive to find a commercialization partner; either an existing company or a startup venture (a "startup"). The key role of the TCO is to connect technology, people and money to move the innovation forward through the commercialization process. Fig. 17.1 depicts the process and various

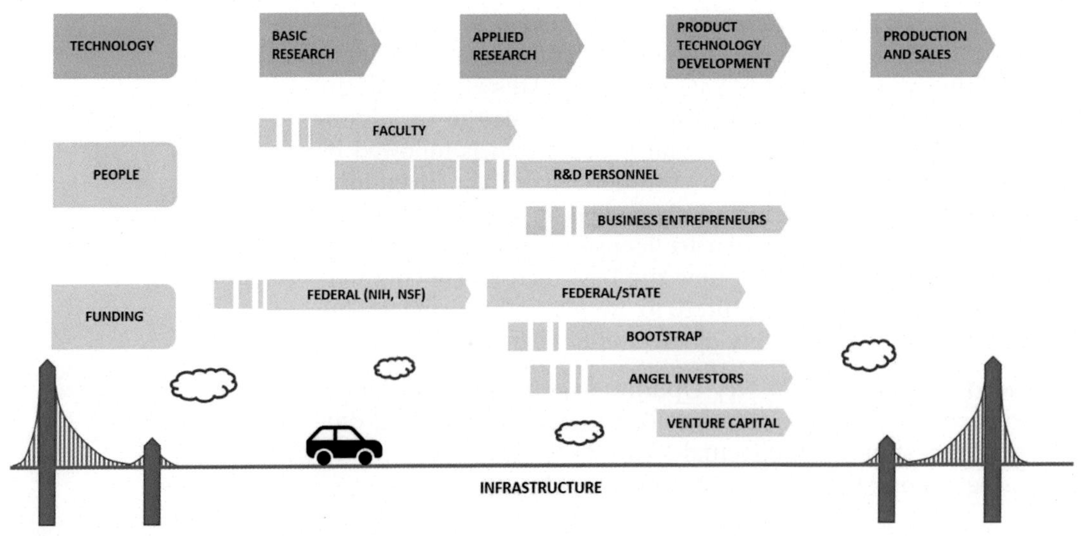

Figure 17.1 Technology commercialization process.

steps in moving early stage technology forward to final commercialization. Universities have technology, but it is early stage. The technology needs to be productized, manufactured and sold. The two elements outside the university, which must be brought in are the people and money. The key people resource is the business entrepreneur who will form the company, build out the team and raise the money to conduct the commercialization process. For life science product commercialization, the business entrepreneur is usually an experienced entrepreneur. An experienced entrepreneur is generally needed in life science projects, since the funding requirements are high and time to commercialization is longer. Investors will want an experienced leader to minimize the risk of their large investment in a complicated process that may take 5–15 years before a return is realized with a commercial product.

The early money that is needed for commercializing a life science product can come from some early-stage federal funding such as Small Business Innovation Research (SBIR) and Small Business Technology Transfer (STTR) programs, but it is usually from early-stage angel investors or professional venture capital firms. Some early-stage state or regional funding sources are also possible in the early phases of development.

Value of technology commercialization to key stakeholders

The License is between a university that has the rights to the innovation, and a business entrepreneur or company that has the commercialization expertise (the "Licensee"). This partnership is key to moving the process forward. There is also an important internal relationship with the internal university stakeholders (inventors and administrators) who need to see the value of commercializing the technology and entering into a License. The university administration needs to understand the value of broadening the organization's capability to do licensing and partner to commercialize their innovation.

Besides providing innovation into the market and also new products to save lives and improve healthcare, university technology commercialization has other benefits to the university stakeholders as well as the Licensees.

The main benefit of technology commercialization to university inventors is that they are able to see their invention have a positive impact on society by solving problems. More and more university researchers want to see their work have an impact on society, going beyond publishing in well-respected journals to share their new knowledge. These new products may save patient lives or improve quality of life of patients. There can also be financial rewards to the inventors, since universities always share the compensation that is derived from licensing an invention. So, inventors can "hit the trifecta" by doing great science, impacting society and also making some money.

For the university the benefit of technology commercialization is often seen as helping fulfill one of their main missions of societal impact or patient care. Most universities have three main missions: education, research, and societal impact. One can see where technology commercialization actually impacts all three prime missions. In terms of education, new knowledge is developed when new innovations occur and new products are commercialized. Also, students are sometimes offered positions with the company that is commercializing the new innovation, since they have firsthand knowledge of the new innovation. Postgraduate employment is a great benefit for the students and if it is with a startup, it provides entrepreneurial experience in addition to product development experience.

Technology commercialization also benefits the research mission. Not only when the initial research is conducted that results in the innovation, but also when sponsored research funding may come back to the university from the Licensee to further develop the technology.

Many of the new breakthrough technologies in the world have seen their origin in university research. Communities around major research universities like Stanford with Silicon Valley and MIT/Harvard with Cambridge are great examples of how university technology commercialization has had impact on society and also the regional, if not the global economy. Of course, these innovation zones are not just due to the technology from the universities. Graduating students provide an educated, entrepreneurial workforce to help these regions grow.

The Licensees also benefit from university technology commercialization. The unique culture of the university with deep expertise, open communications, longer time horizons, and broader vision often lead to these breakthrough innovations that are so valuable to Licensees. It is very difficult to replicate this culture in a company, where time frames are shorter and researchers need to have a broader view of the total commercialization process. Bridging these two disparate cultures is an interesting challenge and is one of the key responsibilities of university technology commercialization offices.

Examples of best practices in innovation marketing

There are a number of best practices in organizing and operating a university technology commercialization office. The most important element is the staff. TCO staff who are technically trained can help understand the inventors' technologies and also assist with the patent claims drafting. TCO staff who are business savvy can run smoother License negotiations with businesses to the advantage of all parties. When many university technology offices were first formed in the 1980s, the licensing staff typically had Ph.D.'s and little if any company experience. This makes it difficult for them to bridge the cultural gap between university researchers and administration and corporate and startup executives. So, more and more universities have hired more licensing executives into the technology commercialization offices, who have business experience. Not just research experience in a business, but also sales and marketing and general management experience. The staff must also establish norms to respond at the pace of business to their business partners who are their customers and partners in the commercialization process. The differences in culture between universities and businesses, differences in values and difference in the time constant (time at which results are expected to be achieved). So, the technology commercialization staff needs to bridge this cultural difference.

The staff also needs to be science savvy so they can gain the respect of the researchers and also be able to translate the deep science into a description of the technology that shows its potential benefit to the business. These science savvy business entrepreneurs are a rare breed for a university. Progressive universities understand that increasingly the public and the federal funding agencies are expecting impact on society from their universities. The progressive universities thus understand the importance of the technology commercialization mission. Progressive universities also understand the value of the science savvy business entrepreneurs in helping them achieve that mission in collaboration with researchers and outside business entities.

Identifying potential commercial partners to take the university innovation to the market is a complex process. Corporate partners are usually visible through various business listings. Finding the right person in the company can be difficult. A number of corporations have technology scouts, who look to find new innovations in universities that may answer some of their corporate technology needs. For cases where a startup is the original Licensee, the startup entrepreneur may be more difficult to find. Once a startup entrepreneur has had a successful sale or exit of their business, they will often look to invest in other startup opportunities. So, they often join angel investment groups. Having connections with angel investment groups can be an excellent source of startup entrepreneurs. Also, various organizations will hold venture competitions, where universities may connect to startup entrepreneurs. Another location the startup entrepreneurs frequent is venture capital firms, so this is another group to target. Once universities have a few successfully commercialized innovations that are known, startup entrepreneurs and funders (venture capital firms and angel investor groups) will visit the campus. So, it is important to publicize your successes to attract business entrepreneurs and funders to your campus.

A key element of connecting to outside business entrepreneurs is to network. Join organizations where funders and business entrepreneurs join. They also are looking for technologies to commercialize, so they are always willing to network with science savvy business entrepreneurs at a good research university. Networking is a skill and starts with building relationships versus just collecting business cards. One should try to develop relationships and help the other person, so they may want to help in return. These symbiotic relationships will be lasting. Also securing connections and marketing innovations is difficult and time consuming. Once a university has executed a licensing deal with an organization or an entrepreneur, they will look back to them when another opportunity comes out of the university with the purpose of either the prior Licensee to get involved in the commercialization or to secure a referral.

A key step in commercializing university innovations is the negotiation of the License terms agreement with the Licensee. This step occurs after identification of a potential Licensee. There are three key requirements in a License negotiation; (1) never do a deal with someone you do not trust, (2) do your homework and (3) have options. When entering a License where a valuable asset (intellectual property) will be entrusted to an outside party, it is important to develop a trusting relationship. The agreement will last up to many years and will undoubtedly have highs and lows. If during the negotiation process, the potential licensor and Licensee end up mistrusting their counterparty, they should not do the deal. It is not possible to write into the agreement all the contingencies for averting situations where the parties may act in poor faith and not be a trusted partner. So, if trust is lacking, it is best not to do the deal. Next each party needs to do its homework and understand the strategic or operational value the technology will offer to the Licensee and this should be translated into economic value to the licensor. To know the value to the Licensee, the TCO needs to understand the commercial value of the innovation and the value proposition it can have to this particular Licensee. It is often wise to secure external valuation of the innovation, from an entity that performs such analysis, and is not connected to either the entrepreneur or the investor. Such efforts can move the negotiation process forward expeditiously and fairly. This is the basis of defining a "win-win" License for both parties. In a successful negotiation both parties win something and both parties probably will have to give up something. But the end result is a synergistic creation of more value to both parties than exists without the deal. The final rule is to have options. Since university innovations often don't have more than one bidder, how can one have options? The option that is always available is "to walk." Each party must have the mindset that they can walk away from an agreement, even when they do not have another bidder with whom to negotiate. If someone has the mindset that they must do a deal with the current potential business partner, they will tend to make suboptimal decisions and have to live with a troublesome License and partner for years.

So, the three rules are:

1. Never do a deal with someone you don't trust
2. Do your homework
3. Have options

This approach will lead to win-win agreements for both parties and also lead to business partners who may develop other innovations together in the future.

University technology commercialization offices typically have a number of metrics they collect to determine if they are being successful in commercializing the university IP. The typical metrics are:

1. Inventions disclosures
2. Patents filed
3. Patents issued
4. Deals done
5. Licensing income

While these are useful metrics, they tend to be trailing indicators of success. In order to determine if the office is operating to build a successful program, I would suggest two other metrics be added:

1. Speed of execution
2. Breadth of network

You might consider to rate the university commercialization with data that is available from a data base such as AUTM's annual survey: You might want to consider calculating an ROI of the university, following businesses practices. The amount of:

1. Approximate income generated per license/option
2. Licenses/options generating >$1M revenue
3. Disclosures/$10M research expenditures
4. Issued patents/invention disclosures

One of the key functions of the technology commercialization office is connecting to the outside business world that will be their partners in commercializing the innovations. The outside business world operates at a much faster pace than universities. In order to have effective business relationships, the office must operate at the speed of business. Therefore the office must have the authority to negotiate the Licenses and not have to send it through multiple university administration levels that tend to slow the speed of execution. Speed and "time constant" match are key to being able to execute a smooth transition of licensing the innovation to the business.

The other key forward looking metric is how large the business network is. The outside business network requires the technology commercialization staff to have good relationships with businesses, entrepreneurs and funders; those who will refer the office staff to other people in the business world and thus continue to grow the office's network. Since it can be a challenge to identify a particular business or entrepreneur that may be interested in partnering to commercialize the technology, the larger the pool of candidate Licensees, the better.

So, if an office operates at the speed of business and effectively networks into the business world, they are likely to produce great results in the technology commercialization process metrics. Thus these are valuable forward-looking metrics, especially when a university is building their offices to achieve the next level of performance.

To option or to license?

When a startup company and university initiate discussions on licensing a technology, they often start with the question of which deal better meets each parties' needs, a License or an option agreement (an "Option"). The License is like a marriage, where the parties say, "til death do us part." In this case the death is the death of the licensed patent portfolio, or the startup company, whichever comes first (usually not the death of the university). In many cases the License relationship lasts for 10–20 years. Startup Licenses are also like a marriage, because they are generally exclusive Licenses, whereby the university agrees not to License to third parties. To secure funding for product development, startups require the potential high profitability that comes from a period of market exclusivity.

In contrast, a an Option is like having a boyfriend or girlfriend. It is a relationship with commitment, but also a trial period that may not last for a long time. Startup Options are for evaluative purposes, not for making sales. Options allow for the startup to validate the technology, regulatory path, reimbursement path, customer needs, changing market trends, competitive landscape, value proposition, and interest of development funding sources. Options also allow for the university to assess the ability of the startup team to meet development goals and financial obligations. If at the end of a set period of months or years, both parties are satisfied with each other and the potential for success, then the startup will ask to execute the Option to license the technology (or request the university's hand in marriage). Given the comparison with marriage one might think that Options are always put in place before a License, but that is not the case. Requirements of venture capitalists, high levels competition in the field, or short development timelines may lead to rapid licensing. Many venture capital firms do not feel comfortable investing their money without a long-term commitment on the exclusivity of the intellectual property. Their money would be wasted if a startup took an Option, evaluated the technology and market to find a path to success and for one reason or another (like corporate interest, different ideas of fair market value, or legal terms) the university and startup could not agree on a full License when the time came. There is also the fact that as the startup establishes the path to market and identifies potential customers, the value of the intellectual property increases, so it may be more expensive to take a License later than sooner. One solution is to skip the Option stage and take a License right away. As illustrated in Fig. 17.2, the decision on when to Option or License may be an iterative process that continually utilizes new data on the technology, market, and finances.

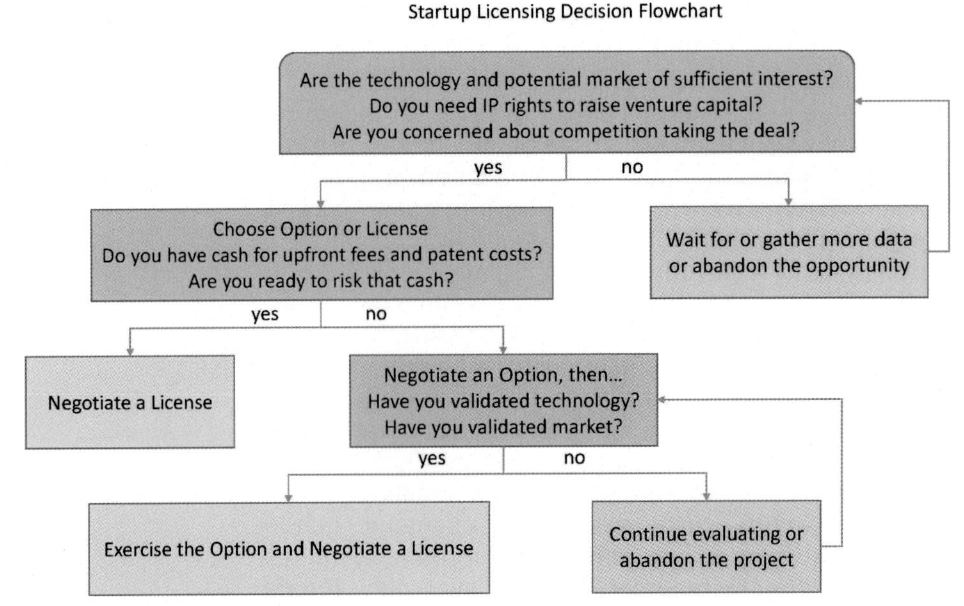

Figure 17.2 Licensing decision flowchart.

Taking a License does come with greater up-front costs than an Option. Usually, universities require past patent costs to be fully reimbursed at the time a License is signed, whereas Options only require payment of ongoing patent costs, if at all. Both Licenses and Options also have up-front fees that startups pay and those fees are usually significantly higher for Licenses (often five or six figures for Licenses versus four figures or low five figures for Options). Another early piece of License consideration startups provide in a License is equity, which is not the case for Options. Because startups are cash-poor they cannot pay License initiation fees in the hundreds of thousands of dollars like corporates, so they offer equity instead. But just like cash fees, the startups won't get back the equity they give away if the development of the technology fails.

A frequent solution that balances the risk and commitments on both sides is to prenegotiate some or all of the License terms when doing an Option. The amount of negotiation that a startup wants to do ahead of time depends on the amount of time and money it can spend on negotiating complex legal documents prior to evaluating the technology. On the low end of complexity, parties may just want to agree to a range for some key business terms, like the royalty rate, within which they will negotiate when it is time to execute the Option and negotiate the License. In the middle ground, parties may agree ahead of time on specific key terms, like royalty rate, fees, development resources, and development milestones. On the high end, Licenses can be fully negotiated ahead of time and appended to an option.

Key terms in a license

Here is a brief overview of key terms in a startup technology License and how they factor in short- and long-term risk for the startup.

Intellectual property, exclusivity, and sublicensing. The most fundamental terms in a License define the intellectual property rights to be licensed, what is allowed to be done, made, or sold under those rights, and whether the rights will be exclusive to the startup or nonexclusive. Generally due to the extremely high costs and amount of time involved in getting an early-stage life science technology to market, startups will require exclusive Licenses. The intellectual property is most often in the form of patent rights, but may be patents and copyrights in the case of software or medical devices. The intellectual property may also include know-how and physical materials. It is critical to identify all necessary pieces of intellectual property to commercialize the technology. In addition to defining how the startup is allowed to use the intellectual property rights, the License also defines whether, how, and to whom the startup is allowed to establish sublicense agreements. Sublicense agreements from startups to corporate pharmaceuticals are sometimes made in life science as an exit strategy instead of selling the entire startup. Sublicenses can also be an important way to have a business partner execute manufacturing, distribution and/or sales in countries where the startup has no presence. More and more startups are virtual and don't do any manufacturing themselves, so sublicensing is critical.

Term and termination. The amount of time that the License will remain in effect is the "Term" and when licensing patents it the standard License is until the patents expire. The Term for know-how and copyrights can be much longer. One may set a time as a number of years after the patents expire, or any other time during which the intellectual property rights remain in effect. Know-how is protected by trade secret laws on a state-by-state basis and remains proprietary as long as the know-how stays secret. Aside from the natural death when the intellectual property expires, Licenses include terms for early termination.

Some examples of reasons for early termination include: when the company no longer wants to make the product, when the company fails its obligations under the License, and when the company is going bankrupt.

Diligence requirements. The role of the startup in the License relationship is to develop and commercialize the university's technology. Diligence Requirements ensure that the startup performs the activities necessary and makes the necessary investments to successfully develop the early-stage technology into a market-ready product. For life science technologies this often means getting the technology through preclinical development and clinical trials. Diligence Fees specify the amount of money that the startup must invest in development each year until the first sale of a product based on the licensed technology. If the company fails to invest the full amount, then they must pay the difference to the university. Rarely do universities collect Diligence Fees, rather they provide a means to terminate Licenses that are not performing. Diligence Fees should be set at a relatively low bar compared to the requirements of a business plan, to account for challenges and delays that startups may meet while working in good faith to develop technologies. Diligence Milestones are development goals that must be met by specific dates, for instance: building a first prototype, filing an application for an Investigational New Drug (IND), first treatment of a human, initiation of a Phase II clinical trial, initiation of a Phase III clinical trial, acceptance of a New Drug Application, application for an IDE (Investigational Device Exemption for a device to collect safety and effectiveness data) or first Sale. Again, these dates should follow the startup's business plan, but allow plenty of wiggle room, since plans change and the parties do not want to have to amend the License often and startups do not want to take too much risk of having a License terminated for failure to meet Diligence Milestones. Diligence Requirements should motivate startups, and allow for termination of failed startup Licenses, but should be readily achievable in the case of ongoing development efforts.

Intellectual property management. Startups are responsible for the costs of obtaining intellectual property protection for technology licensed from a university. Startups reimburse the university for expenses universities incur prior to the License and pay ongoing costs, such as for patenting. Given that the startups are paying the bills, universities are happy to take significant direction from startups regarding decisions around patenting, such as when to file in what jurisdictions and what rights to maintain. For the majority of the relationship, the parties' goals are aligned. Both the startup and university want to establish strong intellectual property rights to create a barrier to entry for competition and support utilization of the technology customers of the startup. However, sometimes startup and university goals may diverge, such as when a company starts to develop a competing technology. Very often, as the patent owner, the university maintains control of the patent prosecution and may take advice from the startup's attorneys. This maintains the ability of the university to act as steward of the innovative ideas. However, in some cases, universities may be willing to allow startups to control patent prosecution under a joint representation agreement whereby one law firm will represent both the startup and the university. A university may see the advantage of the law firm billing the startup directly for services. A startup may see the advantage of cost savings from utilizing one law firm as opposed to one for each party. These advantages may balance some of the risk from the potential for misaligned goals. Startups should be prepared to accept that some universities are not flexible on intellectual property management. In addition to establishing and maintaining intellectual property rights, the License will also set ground rules for enforcement of intellectual property rights against infringing parties and how to share any related costs or revenues.

Fees. Usually, life sciences Licenses to startups involve an assortment of fees that range from short term to long term and from low value to high value. Usually, the higher fees come further in the future as the value of the technology increases due to becoming closer to generating sales revenues. The License Initiation Fee is due at the time of executing the License. Even though startups are usually cash-poor, universities see the License Initiation Fee as essential to ensure that the entrepreneur has "skin in the game." Someone that has invested their own money in a new venture is much more motivated and committed than someone who has received a License for free. As the company meets development milestones, the value of the startup company increases, the amount of money that the startup can raise increases, and accordingly the size of the fees increases. Milestone Fees are fees the startup pays to the university when a Development Milestone is achieved. They are a way for the university to share in the increased value of the company/technology. This can be similar to bonuses that the entrepreneurs may receive upon meeting milestones or raising money due to meeting milestones regardless of whether the company or technology ever ends up successfully reaching customers. In addition to development milestones, startups may receive cash payments when they establish sublicensing agreements and universities share in that income through Sublicensing Fees. Often License Maintenance Fees are charged on an annual basis to ensure continued interest on the part of the startup in developing the technology and to compensate the university for keeping the intellectual property off the market. License Maintenance Fees generally end when the first commercial sale is made and their function is then taken over by Minimum Royalties. Minimum Royalties are a set amount of royalties to be paid regardless of whether the startup makes the sales that would generate those royalties. They are not intended to be paid, but rather ensure that the startup does continue to maintain interest in selling the products made with the licensed technology. If the startup is unable or unwilling to effectively tap the potential market then these fees provide impetus for return of the technology and licensing by the university to a new company.

Royalties. Most commonly royalties are defined as a percentage of Net Sales. This is distinct from Net Profits which are too easily manipulated by accounting practices. The royalty ensures sharing in the success of the venture by the university and represents a real win-win assuming it is set at a level which leaves room for profits for the startup. This is why royalties for pharmaceuticals or medical devices are generally in the low single digits for university technologies. Royalties for technologies with lower development costs and timelines can be much higher, even as much as 50%–80% for some software or copyrighted works. The more work needed before getting a pharmaceutical to the clinic, the lower the royalty rate will be. If a university has already tested a drug in humans, it can obtain a royalty in the mid single digits to low double digits. Another factor is the strength of the intellectual property. In other words, composition of matter patents and broad patents with great ability to block competition will have higher value that gets reflected in a higher royalty rate (and higher fees).

Equity. Sharing of equity in the startup aligns equity owners' goals. One of the first things that the License does is enable the startup to raise venture capital. As the provider of one of the key ingredients to the technology startup, the university should rightly share in some of the value created by raising the venture capital even if the startup achieves success through products that do not use the licensed technology. The value in the equity given to the university won't be realized until the startup reaches a liquidity event that provides opportunity for the university to sell its equity. Liquidity events could include sale of the company, a merger, or the startup becoming publicly traded through an initial public offering (IPO). Until such liquidity event, the university and company will maintain greater goal alignment as the university shares in increased company value (on paper) when development milestones are reached. Furthermore, provision of equity is a way that cash-poor startups can provide fair compensation at the time of a License deal despite having no way to pay six or seven figure up front License Initiation Fees. The amount of equity provided to the university depends on the stage of development of the technology, the anticipated development costs and time, the size of the potential market, the strength of the intellectual property, and anticipated future needs of the company to sell further equity or to use it as compensation for employees. The equity compensation to the university should be balanced with the fees and royalties, so that the university receives some value early and during development and a majority of value once the technology is generating sales revenues. In other words, the risks and rewards are distributed over the life of the License. However, different universities have different risk tolerances and will seek different weights of fees, royalties, and equity.

18

Funding approaches—why, from where, how much, terms

Junichi Naganuma
Department of Anesthesiology, University of California, San Diego, CA, United States

Chapter outline

Innovation in Anesthesiology. DOI: https://doi.org/10.1016/B978-0-12-818381-6.00001-2

Abstract

You have a brilliant idea that might change the world, but you do not have enough money to bring your idea to the market. Without external financing, your company will likely not succeed. Unless you are an experienced entrepreneur, you may not know where or how to start. This chapter aims to provide an overview of different funding approaches—why, from where, how much, and terms, and explain the steps necessary in raising money to make your startup a success. Different sources of funding available and the progressive stages of funding rounds will first be discussed in this chapter. Essentials of a business pitch with a brief discussion of financial modeling will then follow. The chapter will conclude with a primer on important terms and clauses commonly included in a term sheet.

Keywords: Bootstrapping; angel investor; venture capital; seed capital; series A, B, C funding; initial public offering (IPO); business idea pitch; financial model; term sheet

Key points

- Having a great idea alone is usually not enough for a startup to succeed; external financing is often necessary to grow and scale your company.
- Successful fundraising requires time, effort, and careful planning.
- An overview of different funding approaches is presented in this chapter.

Introduction/why it matters?

You have a brilliant idea that might change the world, but you do not have enough money to bring your idea to the market. Without external financing, your company will likely not succeed. Unless you are an experienced entrepreneur, you may not know where or how to start. This chapter aims to provide an overview of different funding approaches—why, from where, how much, and terms, and explain the steps necessary in raising money to make your startup a success. Different sources of funding available and the progressive stages of funding rounds will first be discussed in this chapter. Essentials of a business pitch with a brief discussion of financial modeling will then follow. The chapter will conclude with a primer on important terms and clauses commonly included in a term sheet.

Why do you need funding?

Starting and maintaining a startup company costs money, and self-funding is usually not enough. Without external financing, most startups will not succeed.

"From where?" Sources of funding

Founders/self-funding/bootstrapping

To "bootstrap" means to launch your startup without seeking help from any outside investment [1]. By bootstrapping, you are putting your own money into the startup, that is, "putting your skin in the game." In addition to retaining full ownership and decision-making authority in your company, bootstrapping sends a positive signal to future investors during subsequent fundraising rounds [2]. On the other hand, by bootstrapping and growing your startup slowly, you risk depleting your cash reserve more quickly and potentially having competitors emerge. With these two downsides combined, you may run out of time before successfully launching your product or service [1].

Family and friends

This is a quick way of getting cash, often without interest, though it is often just a loan and not a gift.

Crowdfunding

Crowdfunding raises funds from a group of people on the Internet and allows you to keep full ownership of your startup. In exchange for their money, crowdfunders may just expect perks such as discount codes. Crowdfunding has become a popular funding source for early-stage startups because even if your business fails, you usually do not have to pay them back [3]. Indiegogo, Kickstarter, and GoFundMe are examples of crowdfunding platforms.

Accelerators and incubators

While accelerators and incubators both offer networking opportunities and mentorship from seasoned entrepreneurs and founders, there are important differences to consider when choosing between the two.

Accelerators are competitive, fixed, short-term programs for startups that are ready to scale but in need of cash. In exchange for enrolling startups in their program and providing them with capital typically ranging from $20,000 to $80,000, most accelerators require equity from the startups [4].

Incubators, on the other hand, are more suitable for early-stage startups not yet ready to grow quickly. Aside from providing office space for startups, incubators offer guidance as needed over a longer time period compared to accelerators. Incubators often do not require equity in exchange for their support, but they also usually do not offer capital [4,5].

Small business loan

Traditional debt financing, for example, loan, is for those who want to keep complete ownership of the startup but do not have enough cash. Banks, however, usually do not approve loans for startups that are considered too risky [3].

Small business administration-backed loan

If a traditional business loan is declined by the bank, consider SBA-backed loan via Lender Match, as your startup may have a better chance of getting the business loan approved through this type of loan [3].

SBA's Small Business Innovation Research (SBIR) and Small Business Technology Transfer (STTR) Programs

Also known as America's Seed Fund, Small Business Innovation Research (SBIR) and Small Business Technology Transfer (STTR) programs are U.S. federal programs that encourage domestic small businesses to engage in Research/Research and Development (R/R&D) with the potential for commercialization [6].

There are three phases of SBIR/STTR programs: Phase I (concept development) awards are generally $50,000 to $250,000 for 6 months (SBIR) or 1 year (STTR). Phase II (prototype development) awards are generally $750,000 for 2 years. Phase III (commercialization) is not funded by either SBIR or STTR programs. STTR program requires the small business to collaborate with a nonprofit research institution in Phase I and Phase II, but STIR program does not [6].

Healthcare and biomedical professionals should know that National Institutes of Health (NIH), a part of the Department of

Health and Human Services, is one of the federal agencies that participates in these SBIR/STTR programs. According to NIH's website, NIH's SBIR/STTR programs invest over $1 billion in promising small businesses in the biomedical innovation space [7].

Angel investors

Angel investors are wealthy individuals who invest their own money in startups in exchange for equity. Angels typically invest from $500,000 to $1 million. In addition to offering money, angels often provide networking opportunities and sector-specific knowledge [2].

Venture capitalists

Venture capitalists are professional investors who invest in early-stage startups with other people's money. In exchange for capital, they expect equity and decision-making authority in the company. Venture capitalists typically focus on high-growth startups, take higher risks hoping for much higher returns, and have a longer investment horizon than traditional financing [3].

"How much?" Stages of funding rounds

You should aim to raise as much money as you need to either reach profitability or get to the next "fundable" milestone [8]. Progressive stages of funding rounds will be discussed below (see Fig. 18.1).

Preseed funding (bootstrapping)

Founders develop prototypes of product or service hoping to build a business case by using their own money, money from family and friends, and/or money from crowdfunding.

Seed funding

Seed funding is the initial external capital raised. In addition to raising capital, one of the purposes of this funding round is to make the startup look good to investors in future rounds. How much to raise will depend on the startup's financial situation and its growth trajectory, but it is important to keep in mind that the more you raise, the more you dilute your ownership and give away your decision-making authority in your startup.

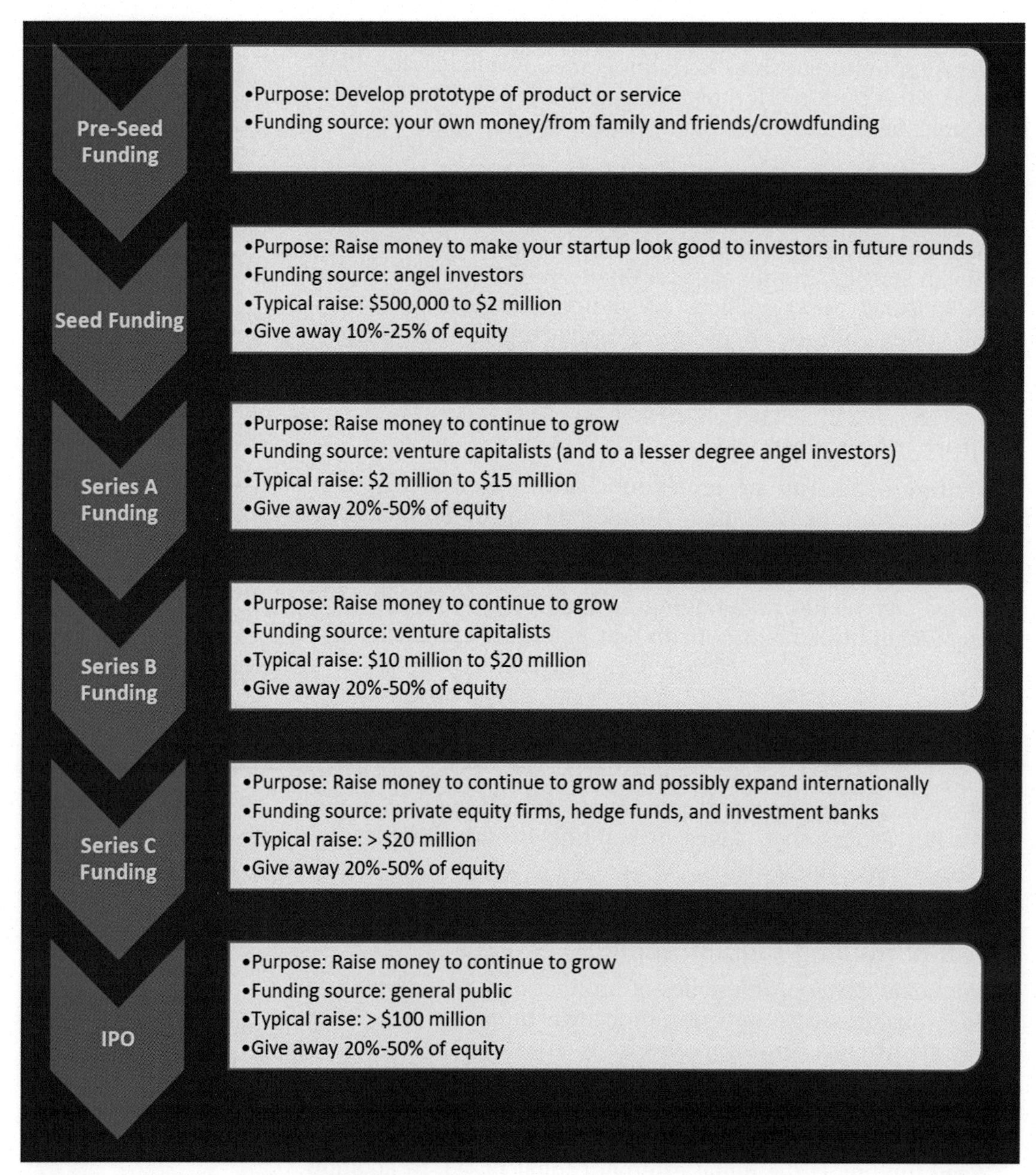

Figure 18.1 Stages of funding rounds.

Investors for this stage are typically angel investors, and a typical seed funding can raise between $500,000 and $2 million [5]. A startup raising seed capital should usually expect to give away between 10%–25% of equity [5].

When asking yourself when to raise money with this round, you should raise before you get down to 6 months of cash left or start thinking about raising when you have 12 months of cash left, because fundraising takes time and effort [9].

Funding in this round is commonly structured as convertible debts or convertible equity.

1. Convertible debt

   Convertible debt is a type of convertible security that is like a loan with a principal amount, an interest rate, and a maturity date, but it "converts" to equity when the company does equity financing. It will usually have a "Cap" or "Target Valuation" and/or a discount. A cap is the maximum effective valuation that the owner of the note will pay. A discount defines a lower effective valuation via a percentage off the round valuation [8,10].

2. Convertible equity

   Convertible equity is a type of convertible security that allows investors to make an investment before a company establishes valuation, and later converts to equity based on predefined parameters. Simple Agreements for Future Equity (SAFE) and Keep It Simple Security (KISS) are some examples of this type of security commonly used [11].

Series A funding

Your startup will have reached appropriate milestones and should have a large customer or user base by this round, but you need more money to scale further [5].

Investors for this stage are primarily venture capitalists, and to a lesser extent, angel investors. Series A Funding can raise between $2 million and $15 million [5], and most companies should expect to give away 20%–50% of equity after this funding round [10]. Data show that only up to 30% of founders who raise seed funding will successfully raise Series A funding [9].

Series B funding

Venture capitalists remain as the primary investors for this next funding round, and capital raised in most Series B funding is between $10 million and $20 million [10].

Series C funding

Private equity firms, hedge funds, and investment banks are the primary investors for this funding round. Capital raised in this round is often used to expand the company internationally. Valuation of companies that make it to this round is typically around $100 million and typical capital raised is more than $20 million [10].

Initial public offering

Initial Public Offering (IPO) is the first time a company offers its shares of capital stock to the general public. IPO is one of the exit strategies of a startup along with "merger" and "acquisition" [12]. Typical capital raised in a successful IPO is over $100 million but companies should expect to give away 20%–50% of equity.

Pitching your business idea

Every time you try to convince investors to give you money, you will need to "pitch" them your business idea. Simply having a great idea will not suffice. Present a clear and succinct pitch that tells investors why your startup will be a big success. Prepare a tight slide deck following the "10–20–30 Rule": 10 slides, no more than 20 minutes, and no font smaller than 30 points [13]. Successful pitches tell a compelling story with passion and make investors want to join you.

Elements of a successful pitch—what should a pitch include?

A pitch should include the following information about the product, the market, and the team [14–16]:

1. Problem: What is the problem you are solving?
2. Product: What is different about your product or service?
3. Customer: Who are your customers and how big is the market?
4. Competitor: Who is your competitor and can you create barriers to competition?
5. People: Who are the founders?
6. Plan: How will you reach those customers?
7. Profit: What is your revenue model?
8. Plan: What is your exit strategy?
9. Ask: How much money are you asking for?

Financial model basics

As part of your startup's financial projections, by building a financial model using revenue and cost calculations based on actual data, you are better able to forecast profits. When pitching your business idea to potential investors, it is obviously more convincing to have real numbers and not simply estimates to support your business case and your ask.

There are two forecasting methods used in building a financial model [17]:

1. Top down forecasting method (macro to micro view):

With the top down method, estimate the entire global market first and work your way down to your potential market share.

The "TAM SAM SOM" model is a useful framework to explain the three levels of market size (see Fig. 18.2):

1. TAM (total available market): total worldwide market for your product or service
2. SAM (serviceable available market): the part of TAM you address with your product or service
3. SOM (serviceable obtainable market): the part of SAM you can realistically capture; this is the market share you aim to capture

The pitfall of this approach is that the forecast may be too optimistic, since you are taking a somewhat arbitrary percentage of the market in your forecast.

2. Bottom up forecasting method (micro to macro view):

With the bottom up method, estimate revenues, costs, expenses, and investments based on actual company data, and work your way up.

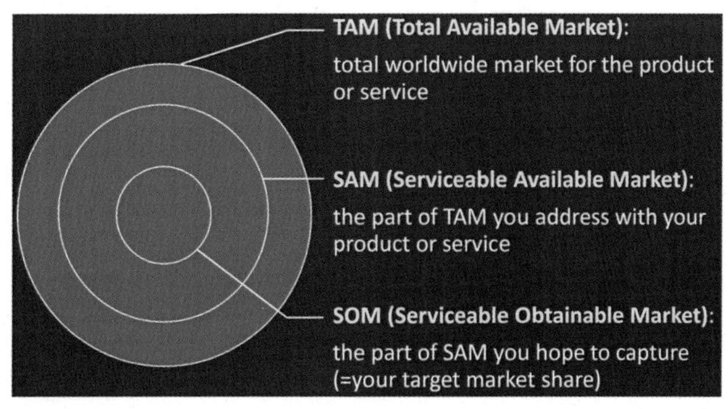

Figure 18.2 "TAM SAM SOM" model.

While this approach provides a more realistic number, the pitfall is that it may fail to reflect the fast growth trajectory of your startup.

For better forecasting, some recommend using the bottom up approach for the short term forecast (1–2 years ahead) and the top down approach for the longer term forecast (3–5 years ahead) [17].

"Terms?" The term sheet

Term sheet basics

A term sheet is "a nonbinding agreement that shows the basic terms and conditions of an investment" [18]. It is basically an agreement between the founder and the investor on how to split the future upsides and downsides of the startup under consideration.

In any emerging industry or business sector such as that in the startup space, standards do not yet exist. As such, a wide spectrum of what a "typical" term sheet might look like can be found on the internet. For illustration purposes only, readers can refer to the Y Combinator website (https://www.ycombinator.com/library/4P-a-standard-and-clean-series-a-term-sheet) for an example of a "concise" term sheet. This should serve as a primer on important basic terms and clauses in a term sheet that founders must understand. For a more "comprehensive" example, readers should refer to the sample term sheet on the National Venture Capital Association (NVCA) website (https://nvca.org/).

Different types of securities

Before discussing individual terms and clauses in a term sheet, an understanding of the different types of securities is a prerequisite.

Securities are "fungible and tradable financial instrument used to raise capital in public and private markets" [19].

There are basically two types of securities: equity (ownership rights to shareholders by means of stock) and debt (loans).

Equity can be broken down into the following three categories:

1. common stock: shares issued to the founders, employees, and future employees of the company.

2. preferred stock: shares often issued to investors in Series A and subsequent funding rounds; preferred shares will have more claim to the company's assets than common shareholders.
3. stock option: gives an investor the right to buy or sell a stock at an agreed upon price and date [20].

Capitalization or "Cap" table

A capitalization table is a "table showing the equity ownership capitalization for a company." It should include all of the company's equity ownership capital, such as common stock, preferred stock, stock options, and convertible equity [21].

Four categories of terms and clauses in a term sheet

Terms and clauses included in a term sheet can be grouped into the following four categories [22]. The glossary section at the end of the chapter will provide a description of each term and clause. The investopedia website (https://www.investopedia.com/) is a useful resource to be used as reference.

1. Economics of the deal
   a. Investment amount
   b. Valuation
   c. Liquidation preference
   d. Dividends
   e. Conversion rights
2. Investor rights
   a. Antidilution rights
   b. Right of first refusal (ROFR) and Co-sale rights
   c. Prorata rights
3. Governance and control
   a. Voting rights
   b. Board rights
   c. Information rights
   d. Founder (and employee) vesting
4. Exits
   a. Drag-along

Getting started

Equipped with the preliminary knowledge of different funding approaches, readers (future entrepreneurs and founders) should now be more comfortable when considering external funding beyond bootstrapping. No matter which path you decide to pursue, always keep in mind that the more you raise the more you dilute your ownership and give away control of your company. Finally, the funding process does not end when you sign the term sheet. There is still a long list of legal documents which needs to be signed before you can start receiving money from your investor.

Glossary of terms and clauses in a term sheet

1. Investment amount

   Amount is specified per investor (lead, nonlead).
2. Valuation
   a. How much the company is thought to be worth by the founders and investors.
   b. Premoney valuation: value of the company excluding the funding you are raising.
   c. Postmoney valuation: value of the company including the investment you are receiving.
3. Liquidation preference

   Determines who gets paid first in the event of a liquidation, bankruptcy, or sale [23].
4. Dividends

   Distribution of the company's earnings to a class of its shareholders [24]. This is included to guarantee a certain return to investors [22].
5. Conversion rights

   Gives investors right to convert preferred stock into common stock, where preferred stock is more valuable [22]. It also usually includes the conversion ratio, which is the number of common shares into which the preferred shares can be converted [25].
6. Antidilution rights

   Protects investors from their equity ownership becoming diluted or less valuable [26]. Dilution occurs when a company issues more shares and the existing shareholders' ownership percentage therefore decreases [27]. There are two types of anti-dilution: weighted average anti-dilution (broad-based and narrow-based) and ratchet based anti-dilution.
7. Right of first refusal (ROFR) and co-sale rights

   ROFR protects investors in the case of a secondary offering where existing shares are sold. ROFR offers the investor the right to buy the stock before it can be sold to a third party [22].

   Co-sale rights protect a minority shareholder. If a majority shareholder has negotiated a sale of their shares at a certain price, holders of the rights can join the sale and sell their stock at the same deal terms [22,28].

8. Pro-rata rights
   Give investors the right to maintain their level of ownership throughout subsequent financing rounds by allowing holders of these rights to participate proportionally (pro-rata) in any future issues of common stock prior to nonholders [22].
9. Voting rights
   Give a shareholder the right to vote on matters of corporate policy. "Approval of a preferred majority is required" means that your preferred shareholders have a veto on the items that follow [22].
10. Board rights
   This is an important clause since the composition of the board can potentially result in the founders' loss of company control. A board of directors is a group of individuals chosen to represent the interests of the shareholders in the company and to establish policies for corporate management and make corporate decisions [22].
11. Information rights
   The company is obligated to share the company's financial and business condition with its investors on a regular basis [22].
12. Founder (and employee) vesting
   Vesting makes it difficult for a founder to leave the company early by putting their shares at risk [22].
13. Drag-along rights
   Enables a majority shareholder to force a minority shareholder to join in the sale of a company. The majority shareholder doing the dragging must give the minority shareholder the same price, terms, and conditions as any other seller [29].

References

[1] Ghosh, S., Westner, M.M. Should you bootstrap? https://www.hbsaccelerate.org/fundraising/fundraising-pitching/should-you-bootstrap/, 2021 (accessed 08.04.21).
[2] Bobbink, W. Thirteen sources of finance for entrepreneurs: make sure you pick the right one! https://www.ey.com/en_nl/finance-navigator/12-sources-of-finance-for-entrepreneurs-make-sure-you-pick-the-right-one, 2020 (accessed 08.04.21).
[3] U.S. Small Business Administration. Fund your business. https://www.sba.gov/business-guide/plan-your-business/fund-your-business, 2021 (accessed 08.04.21).
[4] Richards, R. Accelerators vs incubators: how to choose the right one. https://masschallenge.org/article/accelerators-vs-incubators, 2020 (accessed 08.04.21).
[5] Hanlon, W. How to get seed funding and how it is different from series A. https://masschallenge.org/article/How-Get-Seed-Funding-Different-Series-A, 2019 (accessed 08.04.21).
[6] SBIR and STTR programs. https://www.sbir.gov/, 2021 (accessed 08.04.21).
[7] NIH SBIR and STTR programs. https://sbir.nih.gov/, 2020 (accessed 08.04.21).

[8] Ralston, G. A guide to seed fundraising. https://www.ycombinator.com/library/4A-a-guide-to-seed-fundraising, 2021 (accessed 08.04.21).
[9] Harris, A. and Tam, J. Series A guide. https://www.ycombinator.com/library/14-series-a-guide, 2021 (accessed 08.04.21).
[10] Hanlon W. Understanding the differences of series A, series B, and series C funding. https://masschallenge.org/article/understanding-difference-series-a-series-b-series-c-funding-examples, 2018 (accessed 08.04.21).
[11] FundersClub. Convertible Securities. https://fundersclub.com/learn/guides/understanding-startup-investments/convertible-securities/, 2021 (accessed 08.04.21).
[12] U.S. Securities and Exchange Commission. Investor bulletin: investing in an IPO. https://www.investor.gov/introduction-investing/general-resources/news-alerts/alerts-bulletins/investor-bulletins-17, 2013 (accessed 08.04.21).
[13] Kawasaki, G. The 10/20/30 rule of powerpoint. https://guykawasaki.com/the_102030_rule/, 2005 (accessed 08.04.21).
[14] Dance, A. Develop the perfect pitch to launch a start-up. https://www.nature.com/articles/d41586-019-02252-w, 2019 (accessed 08.04.21).
[15] Landry, L. How to effectively pitch a business idea. https://online.hbs.edu/blog/post/how-to-pitch-a-business-idea, 2020 (accessed 08.04.21).
[16] Ghosh, S. and Westner, M.M. How to build a great pitch that hooks investors. https://www.hbsaccelerate.org/fundraising/fundraising-pitching/how-to-build-a-great-pitch-that-hooks-investors/, 2021 (accessed 08.04.21).
[17] Bobbink, W. The ultimate guide to financial modeling for startups. https://www.ey.com/en_nl/finance-navigator/the-ultimate-guide-to-financial-modeling-for-startups, 2019 (accessed 08.04.21).
[18] Ganti, A. Term sheet. https://www.investopedia.com/terms/t/termsheet.asp, 2020 (accessed 08.04.21).
[19] Kenton, W. Security. https://www.investopedia.com/terms/s/security.asp, 2021 (accessed 08.04.21).
[20] Chen, J. Stock option definition. https://www.investopedia.com/terms/s/stockoption.asp, 2021 (accessed 08.04.21).
[21] Young, J. Capitalization table. https://www.investopedia.com/terms/c/capitalization-table.asp, 2021 (accessed 08.04.21).
[22] Hermans, F. The ultimate term sheet guide – all terms and clauses explained. https://blog.salesflare.com/term-sheet-guide, 2021 (accessed 08.04.21).
[23] Twin, A. Liquidation preference. https://www.investopedia.com/terms/l/liquidation-preference.asp, 2020 (accessed 08.04.21).
[24] Hayes, A. What is a dividend? https://www.investopedia.com/terms/d/dividend.asp#what-is-a-dividend, 2021 (accessed 08.04.21).
[25] Chen, J. Conversion in finance. https://www.investopedia.com/terms/c/conversion.asp, 2020 (accessed 08.04.21).
[26] Kenton, W. Anti-dilution provision. https://www.investopedia.com/terms/a/anti-dilutionprovision.asp, 2020 (accessed 08.04.21).
[27] Ganti, A. Dilution. https://www.investopedia.com/terms/d/dilution.asp, 2021 (accessed 08.04.21).
[28] Chen, J. Tag-along rights. https://www.investopedia.com/terms/t/tagalongrights.asp, 2020 (accessed 08.04.21).
[29] Tarver, E. Drag-along rights. https://www.investopedia.com/terms/d/dragalongrights.asp, 2020 (accessed 08.04.21).

19

Personnel—teams needed, roles

Junichi Naganuma
Department of Anesthesiology, University of California, San Diego, CA, United States

Chapter outline

Abstract

Having an idea that addresses an unmet market need that is timely, cost-effective, and can be rapidly developed is only the beginning of your startup journey. Before you consider building a team, however, you might ask yourself, “Do I really need a team? Why not attempt this development on my own?” In short, having a high-functioning team is a prerequisite for any successful startup. Before stakeholders confirm their interest in investing, they review many aspects of a startup including product, market, and people. It is therefore crucial that you build a team with the requisite personnel with complementary expertise. Team composition may undergo fine-tuning at different stages of the startup

Innovation in Anesthesiology. DOI: https://doi.org/10.1016/B978-0-12-818381-6.00023-1

as it scales, but successful startups all share similarities in the essential roles and responsibilities required.

Keywords: Solopreneur; founder; co-founder; founding team; Chief Executive Officer (CEO); Chief Technology Officer (CTO)

Key points

- For any startup to succeed, a product must address a clinical or market need, a market of sufficient size with a confirmed interest, and a team to support the development from several vantage points.
- Having a co-founder is better than being a solopreneur.
- During the early stages of a startup, C-level executive titles are not as important as the roles and responsibilities of those critical personnel in your development team.

Introduction/why it matters?

Having an idea that addresses an unmet market need that is timely, cost-effective, and can be rapidly developed is only the beginning of your startup journey. Before you consider building a team, however, you might ask yourself, "What is the value of having a team in this development effort?" Why not attempt this development on my own? In short, having a high-functioning team is a prerequisite for any successful startup. Unless you are a seasoned solopreneur with a proven track record of success, studies argue against founders going solo and recommend that they identify at least another co-founder who can complement their skill set. Reviewing several medical start-ups such as EPIC of electronic health records, Intuitive Surgical of Da Vinci Surgical System, and Doximity of a digital platform for physicians confirms all had co-founders. Founding, maintaining, and scaling a startup successfully requires a diverse set of skills that can more efficiently be provided by a team.

The reality of startups is that up to 75% of them fail, and not having an aligned team is one of the top reasons identified [1]. Before investors agree to fund development, they investigate three primary aspects of a startup: product, market, and people. It is therefore crucial that you build a team with complementary expertise. Team composition may undergo fine-tuning at different stages of the startup as it scales, and the addition of essential personnel (full-time vs. part-time) should be considered on a case-by-case basis. However, as described in this chapter, successful startups all share similarities in the essential roles and responsibilities required.

Stay a solopreneur or look for a co-founder?

A founder eager to launch the startup may be tempted to remain a solopreneur, but studies support the anecdotal evidence that having a co-founder is better than being a solopreneur for the following reasons [2]:

1. Increased productivity and efficiency

   Having a co-founder allows the development team to divide critical tasks based on established and proven skills.
2. Moral support

   A co-founder may also provide moral support during developmental tasks.
3. Track record of success

   Identifying team members with the proven requisite skills can ensure that development tasks will progress forward more rapidly and integrate with the overall objectives.

According to the Startup Genome Report [3] which analyzed over 650 internet startups, solo founders took 3.6 times longer to scale when compared to startup teams of two or more. The Report [3] also found that teams were more likely to attract investors and experience success in comparison to solo founders. The optimal number of co-founders appears to be two [4]. When seeking a co-founder, select individuals with proven experience applicable to the development and able to work under pressure.

Essential roles in a founding team

You have identified a co-founder who has skills that complement yours. But what does a successful founding team look like? Who else besides a co-founder do you need in your startup team? Research shows that at a minimum, a startup's founding team should have personnel with the following three roles [1,4,5]:

1. The Leader

   The Leader is the central figure who others follow and is usually the founder and/or the Chief Executive Officer (CEO). This role should be held by a passionate individual able to communicate the vision and capabilities required for the development efforts of the startup's product.
2. The Innovator

   The Innovator ensures that the execution goes forward and has the technical skills to execute. The Innovator is often the Chief Technology Officer (CTO).

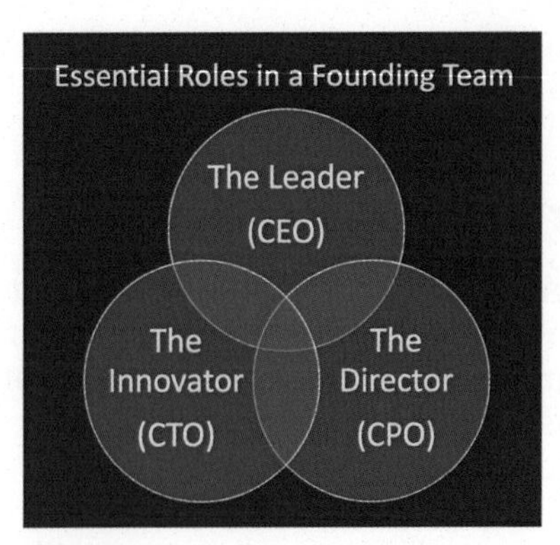

Figure 19.1 Essential roles in a founding team.

3. The Director
The Director should have a clear vision for the future of the startup's product and is usually the Chief Product Officer (CPO), This product person has operational skills and is responsible for the product strategy and development (Fig. 19.1).

Essential startup company personnel

Chief Executive Officer

In the early stages of a startup, the Founder, or one of the co-founders, is usually the CEO. The CEO is responsible for controlling the direction, vision, and culture of the company [6]. The CEO should be the leader of the group who others follow [5,7].

Leaders create trust in their team, leading, motivating, attracting, and retaining a talented team [8]. Furthermore, proven leaders generally have the following three characteristics [9]:

1. Clarity of Thought and Communication
Leaders can clearly and succinctly communicate a compelling vision of the future of their product or service.
2. Judgment about People
Leaders have proven experience at staffing selections to drive the execution of the company vision.
3. Personal Integrity and Commitment
Leaders are passionate about their product and their company and are not only interested in their personal gains. Leaders whom others look up to are willing to go above and beyond what is expected, oftentimes sacrificing their time and energy.

The CEO's main responsibilities include the following [10]: (1) Manage revenue, expenses, and external financing; (2) Refine the vision of the product; (3) Build a successful team with appropriate and capable personnel; (4) Keep investors informed regularly; (5) Maintain a strong brand image; (6) Corporate development.

The CEO's responsibilities evolve as the startup progresses and the CEO must be flexible and ready to adapt. In the early stages of the startup, the CEO is responsible for building an attractive product for its users and investors. A founder having the dual title and responsibilities of a CEO may suffice at this stage of the startup, but as the demand for the product increases and the team grows to about 20–25 employees, the job of the CEO transitions from building a product to building a company, requiring an additional set of skills [11]. According to Wasserman [12], when startups were three years old, 50% of founders were no longer the CEO; in year four, only 40% were still the CEO; and less than 25% were the CEO at their companies' initial public offerings (IPO) (Fig. 19.2).

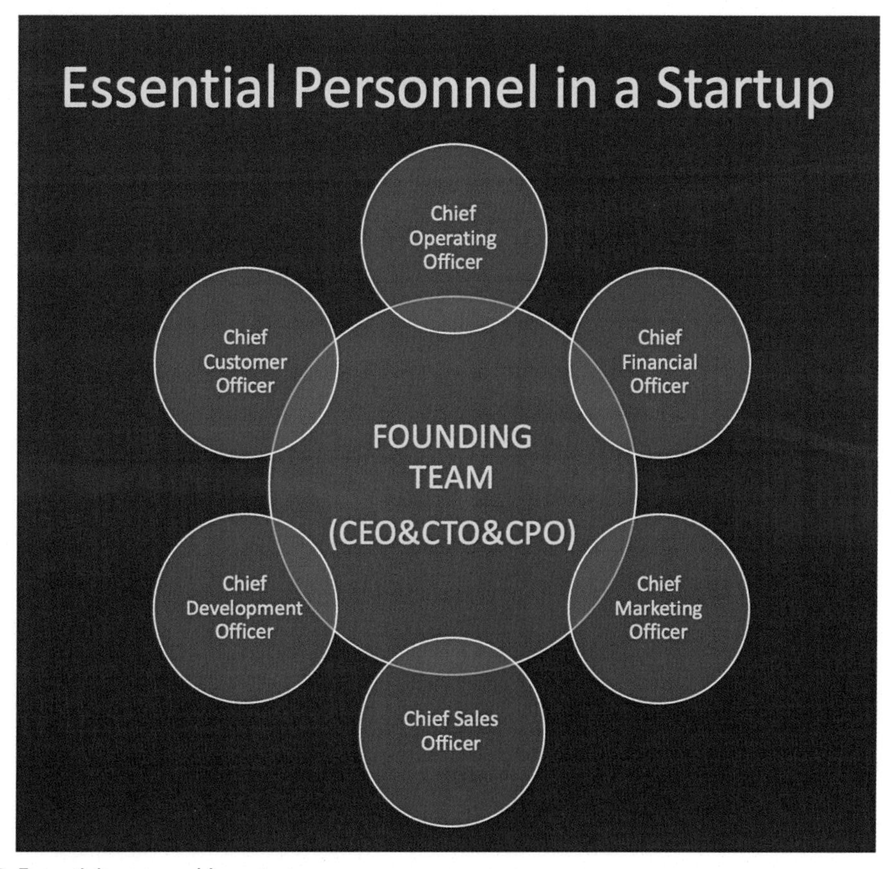

Figure 19.2 Essential personnel in a startup.

Chief Technology Officer

The CTO is the CEO's right-hand person who functions as the bridge between the startup's core team and the technology team and supports the CEO with the company's strategy and goals [1]. The CTO focuses on the technical "how" of the product and is responsible for using technology for product improvement to address market needs. [13,14].

Chief Product Officer

The CPO focuses on the nontechnical "what and why" of the product. This executive should ensure product-market fit, and develop nontechnical aspects of the product such as product strategy, vision, and development [1,6,13,14].

Chief Regulatory Officer/Chief Compliance Officer

The Chief Regulatory Officer/Chief Compliance Officer ensures that the startup's product and its business practices are in full compliance with rules and regulations. Especially in the healthcare space, having someone who can navigate the complex FDA approval process is crucial to the success of a startup. Ideally, this role should be held by an individual with a legal background with knowledge and experience in the FDA approval process.

Chief Operating Officer

The Chief Operating Officer's (COO's) role ranges from enhancing team productivity and making sure that day-to-day operations are being completed promptly to organizing systems and processes as the startup scales [1,15].

Chief Financial Officer

In addition to managing finances and monitoring the financial health of the startup, the Chief Financial Officer (CFO) is responsible for developing financial strategies for the startup's longer-term growth [1].

Chief Marketing Officer

The Chief Marketing Officer (CMO) is responsible for all product marketing communications and growing the startup's customer base to increase revenue and make profits [1,16].

Chief Sales Officer (also known as Vice President of Sales)

In addition to prospecting and closing sales and recruiting and managing a team of sales representatives, the Chief Sales Officer's primary role is to orchestrate sales strategy to maximize revenue [1].

Chief Development Officer

Not only does this person code and debug software, but this officer is also responsible for identifying new business opportunities [1].

Success factors of a startup team

While the skill set of each team member is important, the success of the startup depends on the complementary skills that each member provides. Research has shown the following to be the three success factors of a great startup team [17]: (1) Prior startup experience, (2) Product knowledge, and (3) Industry skills. However, excelling in only one is insufficient, as all aspects are required.

Get started

You have an idea with a market product that addresses a clinical or market need, a market of sufficient size with a confirmed interest, and a team to support the development from a number of directions. The next step is to identify a co-founder with complementary skills and then determine the core of your founding team. As your startup scales, you need to identify other key personnel described in this chapter to round out the required skill set for success.

Glossary

1. Solopreneur

   An entrepreneur who works alone without other team members [18].
2. Founder

   A person with the original idea then transforms it into a business or startup [19,20].

3. Co-founder

   A co-founder is a term used to give equal credit to multiple people who start a business together [19]. The two tests of whether someone is a co-founder or not are: "Do we have a company without them?" and "Can we find someone else just like them?" If the answer to both is no, then you have a co-founder [20].

4. Founding member

   A founding member is an early employee of the startup, and not a founder [19].

5. Founding team

   A founding team is a group of people who are going to build the company. It includes the founder and a few other co-founders who have skills that are complementary to those of the founder [20].

6. Chief Executive Officer (CEO)

The highest-ranking executive in a company makes high-level decisions, manages the overall operations, and acts as the main point of contact between the company and its board of directors [21].

References

[1] Richards R. 9 Make-or-Break Startup Roles (and Why They Are Important). https://masschallenge.org/article/important-startup-roles, 2020 (accessed 26.04.21).

[2] Taggar H. How to Find the Right Co-Founder. https://www.ycombinator.com/library/8h-how-to-find-the-right-co-founder, 2020 (accessed 26.04.21).

[3] Marmer M., Herrmann B.L., Dogrultan E., and Berman R. The Startup Genome Report. https://media.rbcdn.ru/media/reports/StartupGenomeReport1_Why_Startups_Succeed_v2.pdf, 2012 (accessed 26.04.21).

[4] Chan J. The Ultimate Guide To Creating The Perfect Founding Team. https://foundr.com/founding-team, 2020 (accessed 26.04.21).

[5] Dryka M., Warcholinski M. 5 Crucial Roles in Every Successful Tech Startup. https://brainhub.eu/library/5-crucial-roles-in-every-tech-startup/, 2021 (accessed 26.04.21).

[6] Schooley S. How to Hire for Your Startup: The First 8 People You Should Hire. https://www.businessnewsdaily.com/15186-first-startup-hires.html, 2019 (accessed 26.04.21).

[7] Vozza S. The Only 6 People You Need On Your Founding Startup Team. https://www.fastcompany.com/3032548/the-only-6-people-you-need-on-your-founding-startup-team, 2014 (accessed 26.04.21).

[8] Rowghani A. How to Lead. https://www.ycombinator.com/library/6s-how-to-lead, 2019 (accessed 26.04.21).

[9] Rowghani A. How Do You Measure Leadership? https://www.ycombinator.com/library/3j-hRoow-do-you-measure-leadership, 2021 (accessed 26.04.21).

[10] Cohen A. The 6 Most Important Roles of a Startup CEO. https://www.entrepreneur.com/article/244391, 2015 (accessed 26.04.21).

[11] Rowghani A. What's the Second Job of a Startup CEO? https://www.ycombinator.com/library/3k-what-s-the-second-job-of-a-startup-ceo, 2021 (accessed 26.04.21).
[12] Wasserman N. The Founder's Dilemma. https://hbr.org/2008/02/the-founders-dilemma, 2008 (accessed 26.04.21).
[13] De Haaff B. The Chief Product Officer vs. the Chief Technology Officer. https://www.aha.io/blog/the-chief-product-officer-vs-the-chief-technology-officer, 2018 (accessed 26.04.21).
[14] Harvey R. The Importance of CPO-CTO Collaboration. https://www.forbes.com/sites/forbestechcouncil/2020/10/16/the-importance-of-cpo-cto-collaboration/?sh = 1e0d2c465f05, 2020 (accessed 26.04.21).
[15] Bennett N., Miles S.A. Second in Command: The Misunderstood Role of the Chief Operating Officer. https://hbr.org/2006/05/second-in-command-the-misunderstood-role-of-the-chief-operating-officer, 2006 (accessed 26.04.21).
[16] Jordan J. Hire a CMO. https://a16z.com/2017/05/26/hiring-cmo-why-what/, 2017 (accessed 26.04.21).
[17] De Mol E. What Makes a Successful Startup Team. https://hbr.org/2019/03/what-makes-a-successful-startup-team, 2019 (accessed 26.04.21).
[18] Patel N. 12 Things That Are Awesome About Being a Solopreneur. https://www.entrepreneur.com/article/251738, 2015 (accessed 26.04.21).
[19] The Startups Team. Everything You Need to Know About Startup Founders and Co-Founders. https://www.startups.com/library/expert-advice/startup-founders-and-cofounders, 2019 (accessed 26.04.21).
[20] Blank S. Building Great Founding Teams. https://www.forbes.com/sites/steveblank/2013/07/29/building-great-founding-teams/?sh = 3ed81aa43505, 2013 (accessed 26.04.21).
[21] Hayes A. Chief Executive Officer (CEO). https://www.investopedia.com/terms/c/ceo.asp) 2020 (accessed 26.04.21).

20

Physician entrepreneurship—three steps to success

Reid Rubsamen[1,2]

[1]Department of Anesthesiology and Perioperative Medicine, University Hospitals Cleveland Medical Center, Cleveland, OH, United States [2]Case Western Reserve University School of Medicine, Cleveland, OH, United States

Chapter outline

Innovation in Anesthesiology. DOI: https://doi.org/10.1016/B978-0-12-818381-6.00010-3

Abstract

When an entrepreneur has developed a technology, they anticipate it will be valuable for clinical care and they may desire to turn their prototype into a product. The entrepreneur is required to determine if they want to build a company or partner early. Determining such goals informs the best path forward. Joint ventures (JV) are generally easier to establish than full partnerships. In addition, since design notebooks are no longer sufficient to protect one's invention, it is best advised to file for a patent before speaking to any companies about forming a JV to protect intellectual property. When a device receives regulatory clearance/approval decisions, the entrepreneur must decide the best method to deliver maximum benefits to patients. Waiting for the regulatory sanction can maximize the price paid by acquirers for an investment in production, distribution, and marketing. There are several marketing and sales options for technology, and it is valuable to consider the best fit for the innovation. Understanding the range of options and the different requirements for each can result in the best and most profitable option. Most companies are funded by investments until they become profitable. Financiers expect a return on their investment through the sale or licensing of the business to a strategic company or potentially an initial public offering (IPO). Selecting and executing the right exit strategy are complex activities that typically result in better outcomes by leveraging the services of an investment banker. The external economic environment is a critical factor. For example, the level of IPO activity in the public equity markets will typically dictate whether going public should be considered. The entrepreneur needs to consider their company's role posttransaction. Advanced planning in this regard is helpful in preparing for the future.

Keywords: Joint venture; JV; CDRH; PMA; USPTO; 510(K); HDE; SCI; NDA; FDA; OEM; B to B; partnering; pitch; IPO; S-1;

C-suite; road show; Sarbanes Oxley; private placement memorandum; data room; SEC; earn-out; capitalization table

Key points

- Determining your development goals
- Deciding to partner early or select a joint venture
- Securing a patent
- Postapproval is typically the best time to partner.
- Avoid reaching out to potential partners who are unlikely to be interested.
- Take financial modeling seriously and secure professional help.
- The external economic environment can change rapidly and heavily influence exit strategy.
- Investment banker involvement is critical for any significant exit transaction.
- Founders should be realistic about their role in the company after an exit.

Glossary terms and definitions

Patent Prosecution The process of getting a patent application from filing to grant
JV Joint Venture
CDRH Center for Devices and Radiologic Health—, the branch of the FDA responsible for medical devices
510(k) clearance Authorization to sell a medical device based on substantial equivalence to an already marketed device
PMA Premarket approval, required when no predicate device is available which would allow a 510(k) filing
HDE The FDA's Humanitarian Device Exemption, applicable if no comparable device exists, and without the HDE the device cannot be brought to market
USPTO United States Patent and Trademark Office
SCI Spinal Cord Injury
Pitch A presentation intended to sell a product or service
FEV1 Forced Expiratory Volume in One Second
NDA NonDisclosure Agreement
FDA US Food and Drug Administration
OEM Original Equipment Manufacturer
B to B Business to Business
CEO Chief Medical Officer
CFO Chief Financial Officer
USPTO United States Patent and Trademark Office
C-Suite Chief corporate officers e.g., CFO, CEO

Dilution Reduction in the percent of company ownership by investors after the subsequent sale of additional stock
IPO Initial Public Offering
S-1 Required SEC registration statement before a public offering
C-Suite Chief corporate officers, e.g., CFO, CEO
Road Show Series of rapid-succession meetings with potential investors before an IPO
Sarbanes Oxley Auditing and reporting requirements for public companies enacted in 2002
Data Room Collection of files placed online for potential investors with financial and technical company information
SEC Securities and Exchange Commission
Earn-out Payments made for achieving agreed-upon posttransaction milestones
Capitalization Table A list of shareholders indicating the number of shares held by each shareholder.

Step 1—Make a plan

If you have developed an advanced medical device, best practices dictate determining your desired trajectory will improve the probability that you will attain your objectives. Identifying joint venture (JV) partners with resources that could help has been proven valuable. Filing a patent before forming a JV will protect your invention.

Develop a concept

A concept that improves patient care is most often appreciated by colleagues also providing patient care.

Determine your goals

Reviewing one's objectives to either maximize financial profits or improve patient care with a reasonable economic reward is valuable as one initiates the development process. Is the objective to set up a company initially or seek a commercial partner early in the development? Does the inventor desire to work for the company full-time or continue working full-time in their clinical practice?

Valuable tools exist to help one assess their options. The book *The Goal,* by Eliyahu M. Goldratt [1] guides the reader through a process of understanding how critical it is to define the most important goals that drive enterprise activities. This chapter provides the reader with examples of the possible hurdles when business leadership fails to convey the desired development goals.

Determine the path forward

Since higher value is most often obtained once a concept has received FDA regulatory clearance/approval, delaying exit until this time generally provides a greater return. Conversely, a JV or partnership will decrease the amount of work required, although may also decrease potential returns.

A JV is not the same as a partnership. In a JV, two entities join with each other to contribute to an outlined task they both desire to accomplish. Both parties essentially operate as separate business entities, each responsible for their obligations, coming together to accomplish a common goal. JVs can be easier to form than partnerships because they typically do not involve a significant blending of resources between the two different entities and avoid shared liability.

Initiate premeeting preparations

There are some important points to consider before discussing a new idea with potential partners. In 2013, The United States patent rules changed from "first-to-invent" to "first-to-file" [2]. This means an inventor, even with notebooks documenting the date on which their invention was conceived, will not be able to secure a patent after another inventor files a patent on the same idea. Securing patent protection before revealing the concept is critical. This should be done before delivering lectures or authoring papers about the technology, and before meeting with any potential commercial partners. This can be executed with a provisional patent application, allowing discussions with companies to initiate before the patent application issues. Competitive entities have been known to copy ideas that have been presented to them for potential acquisition. Even nondisclosure agreements (NDAs) may be insufficient to protect an idea from a potential competitor who would like to reengineer the concept.

Securing a patent application requires significant effort. The gap between issued and granted utility patent applications has been steadily increasing over the last two decades as shown in Fig. 20.1. The United States Patent Office recently issued patent number 11,000,000, which is the number of patents issued since 1836. The number of US patents issued over the past two centuries is less than the number of US peer-reviewed publications that appeared over the past two decades; it is much harder to get a patent to issue than to publish a paper. As can be seen from Fig. 20.1, the difference between patent applications and issued patents was relatively constant from 1970 through the early 1980s and then the

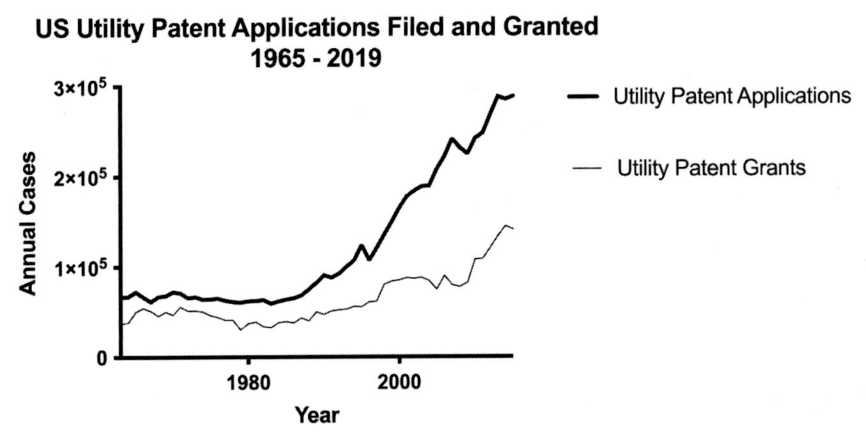

Figure 20.1 Annual patent applications are steadily increasing relative to patents granted. Credit: Plot made using USPTO government data.

number of applications began to climb away from the number of patents issued. One possible explanation for this increase in patent application filings is the switch from patent dispute jurisdiction from the United States Court of Customs and Patent Appeals (CCPA) to the United States Court of Appeals for the Federal Circuit that occurred in 1982. From the CCPA's inception in 1936, patents had to be defended or challenged on a state-by-state basis in front of the CCPA. Beginning in 1982, when all patent disputes were handed over to the Federal Circuit Court, the Court's decisions then had national standing which made patents more valuable [3]. Identifying a patent lawyer, with pertinent experience in medical device patent prosecution, will increase the odds of securing a patent with relevant protection more rapidly.

Potential pitfalls

If the partner and inventor have diverging goals, this can take time to manifest and ultimately result in enterprise failure. Outsourcing critical technology such as a customized sensor can be helpful to a startup. If, however, the JV partner picked for the critical technology contribution is unable to deliver, this can be very damaging unless a second source for the critical component is available. Even if the technology is successfully implemented and tested, if patents fail to be issued or are otherwise compromised, the business will have trouble with financing and attracting partners.

Real word examples

Make a solid plan for success

Dr. David Albert, founder of AliveCor, maker of the first iPhone-based ECG monitor, has publicly stated his formula for success. His inventions include a wrist-based pulse monitor, a Doppler hemodynamic monitor, and a high-resolution PC-based ECG machine which he turned into his first company, Corazonix. His second company, Data Critical, developed wireless ECG monitors and entered into multiple partnership arrangements with companies ranging from Hewlett-Packard to Nokia. After selling Data Critical to GE Healthcare, he eventually developed an iPhone ECG monitor prototype which went viral on YouTube in 2010 and later became AliveCor, which has now sold over three million devices worldwide. Dr. Albert has said that he has used this formula over and over again to achieve success for each of his companies [4]:

1. Invent
2. Protect the invention
3. Clinically prove the invention
4. Jump regulatory hurdles
5. Commercialize

Partner/inventor mismatched visions collide

Aradigm Corporation developed a microprocessor-controlled metered dose inhaler that could administer a dose of insulin to a patient in a single breath [5]. In large part because of the broad patent protection Aradigm was able to obtain covering its technology, Novo Nordisk paid Aradigm nearly $50 M to take the product through the final stages of development [6,7]. Pfizer obtained FDA approval for a competitive product that Novo believed infringed on Aradigm's patents, which at that point were assigned to Novo as part of their partnership agreement with Aradigm. Novo Nordisk attempted to obtain a preliminary injunction against Pfizer's product by arguing willful patent infringement, but this effort failed [8]. Novo subsequently decided to withdraw from its partnership with Aradigm. Aradigm's scientists were disappointed and felt that their product was superior to Pfizer's (which ultimately failed in the marketplace) but Novo made a business decision to discontinue trying to market against a very large company with a competitive product.

Joint venture to access critical technology

The pulse oximeter is arguably one of the most significant medical devices developed for anesthesia and critical care. The first commercial pulse oximeter was developed by Minolta and used a large fiber optic cable to send and receive light from the patient's finger [9]. The bulky cable, heat from the light source, and other factors limited the adoption of the Minolta device. Engineers from startup Nellcor determined that commercially available light-emitting diodes and a small photosensor could be used to make what would become the world's first disposable silicon biosensor, by eliminating the need to send hot light up and down fiberoptic cables extending to and from the patient's finger. Nellcor approached Japan-based Stanley Electric to supply them with the small circuits incorporated into commercial LED systems that Stanley manufactured and sold. Stanley responded by saying that medical applications were too low volume for them to be interested in making a custom LED product for Nellcor [10]. Nellcor management was at a loss to find an alternative source for this critical component. A young Nellcor engineer, David Goodman, wrote to the president of Stanley Electric explaining that a disposable sensor could be a high-volume product with significant benefits for patients. Stanley agreed to be a JV partner and supplied Nellcor with custom LED components for many years. Goodman went on to Harvard Medical School and later became a serial entrepreneur.

Joint venture with a university

General surgeon Ray Onders developed a prototype of a diaphragm pacing device to make it possible for spinal cord injury patients to breathe without a ventilator [11]. After extensive animal experiments, he implanted the first device in a human patient over two decades ago and, after a few adjustments, the patient was no longer ventilator-dependent. Other successes followed and Dr. Onders received requests from patients and physicians from all over the world for the device. Because these instruments were being made one at a time in Dr. Onders' lab, additional funding was essential for production to meet the need. Venture capital firms and potential industry partners were not interested in funding a product addressing a condition affecting so few patients. Dr. Onders was able to convince Case Western Reserve University and University Hospitals in Cleveland to form a JV with him after he entered into a technology licensing agreement and assigned his patent to the University [12]. This seed funding allowed Dr. Onders to team up with Tony Ignagni, who had relevant experience, and the two started Synapse Medical. They were able to ramp up production in a facility they built. With help from some venture capital firms betting on more patient populations who could benefit from diaphragm pacing, they were able to conduct clinical trials sufficient to earn a CE mark for sales in Europe. Given that there are only 300–500 new ventilator-dependent spinal cord injury patients in the US each year, the device was approved in 2008 under the FDA's Humanitarian Device Exemption, meaning the device could be sold for implantation into any SCI patient who needed diaphragm pacing under local IRB approval. Today, Dr. Onders proudly says that SCI and ALS patients around the world receive ten million breaths a day from his implanted devices. He is intimately associated with his device's success, implanting the devices every week, teaching the procedure, and giving lectures to surgeons all over the world. He has data showing that his device can be used temporarily to wean patients from ventilators, potentially reducing the 100,000 tracheostomies performed annually for ventilator-dependent patients in ICUs. This larger patient population could open the door for Dr. Onders to partner with large entities that could further extend the reach of his technology.

Step 2—Identify potential partners

When an entrepreneur has received regulatory approval for a technology it is important to determine how best to ensure that patients benefit from the advantages of the innovation. Delaying exit until regulatory clearance/approval has been achieved can increase the probability of improved financial gain from an investment from a strategic company. Understanding the different options from potential partners and the best ways to approach these partners will improve both the probability of acquisition and the level of payment.

Post approval—a good time to partner

Large medical device companies have significant field sales forces and often prefer to access approved products they can start selling right away rather than develop new products themselves. They often prefer to reduce technical and regulatory risk, making a company with an approved product potentially more interesting as a partner than a company with an earlier-stage concept requiring a full development effort. The result is typically higher valuations and better partnership deals. High valuations are important because they can lead to higher payments with less dilution. A potential partner may acquire stock in a company, license technology, pay for further development, or some combination of these. Many other partnership arrangements are possible. Identifying an attorney familiar with these types of deals is beneficial in achieving success.

Picking the best potential partners

Miller and Heiman's [13] classic book *Strategic Selling* makes an important observation about how to identify the best potential business-to-business deal partners. They divide potential business partners into three broad categories: companies that are in "Even Keel Mode," "Growth Mode," and "Trouble Mode." Companies in "Even Keel Mode," and its variant which Miller and Heiman call "Over-Confident Mode," believe that everything is progressing well, and they don't need to partner with anyone to improve any aspect of their business.

The point emphasized by Miller and Heiman is that a company in "Trouble Mode" is almost always the best of the three as a potential partner for business expansion.

How to pitch potential partners

A developer who is a respected member of the faculty at their academic institution may have delivered presentations to fellow faculty and residents and at various anesthesia conferences. Such forums are good opportunities to present the operation and workings of a new technology. A pitch to the business team at a potential partner company needs to be focused on the business opportunity more than on how the product works. The problem is that generating a model that describes the market opportunity is not a trivial task. Identifying and securing the assistance of a financial modeling expert with domain experience will assist with this process.

Potential pitfalls

The list of potential partners for an entrepreneur can be long, thus it is wise to consider all options seeking Growth

Mode and, preferably, Trouble Mode companies to approach. Technical success and FDA clearance/approval alone are insufficient since the business case for an invention must be sound and be presented effectively to potential partners. It is important to have both a signed NDA and at least a provisional patent application on file with the USPTO before approaching potential partners.

Real world examples

FDA approval/clearance alone is not always sufficient

Aradigm Corporation developed SmartMist: the first microprocessor-controlled metered dose inhaler accessory that monitored the patient's breathing and delivered a dose from the loaded metered dose inhaler at the optimal flow/volume point during the patient's inspiratory cycle. It also contained a spirometer that measured peak flow and FEV1. Dosing events and pulmonary function values were recorded and could be downloaded. The device had broad patent protection, peer-reviewed literature showing that it worked, and had received 510(k) clearance from the FDA [14–17]. Aradigm approached metered dose inhaler companies and medical device companies as potential marketing partners but was unable to generate interest because no insurance coverage was available resulting in patients having to cover the full cost of the device. Aradigm was subsequently able to develop its propellant-free, microprocessor-controlled metered dose inhaler technology capable of delivering partner company drugs including opiates in an inhaled PCA system, and insulin by inhalation to free patients from daily insulin injections. This platform technology was attractive enough to enable preapproval partnerships with Novo Nordisk, SmithKline Beecham, and Genentech [18].

Partner discussions can lead to problems

Even with an executed NDA and filed patent applications, partner discussions can lead to problems. Aradigm Corporation had multiple meetings with Eli Lilly to discuss partnering for the development of an inhaled insulin product delivered using Aradigm's microprocessor-controlled aerosol device. These discussions did not lead to a partnership agreement and Aradigm ultimately partnered with Lilly's competitor, Novo Nordisk [19].

Aradigm filed a patent claiming that monomeric insulin, such as Lilly's insulin analog Lispro, had certain advantages over insulin for inhalation delivery [20]. After that patent was issued, Lilly sued Aradigm claiming that two Lilly scientists should have been named as coinventors of Aradigm's patent based on discussions that took place during partnering meetings. The Indiana district court agreed and made one of the Lilly scientists a coinventor. This decision was later reversed by the Federal appeals court [21].

OEM partnership success stories—Nellcor and Aspect Medical

Many successful medical device businesses are based on the razor-razor blade model. These companies make disposable sensors that connect to durable devices. This creates a steady stream of disposable product sales across the installed base of devices. This is an interesting business model because it creates sales beyond the sale of the durable device. This model can be taken a step further by making the durable device a module that can be inserted into another company's device being sold to the same customer base. In the case of anesthesia monitors with disposable sensors, entering into OEM partnerships with companies that make stand-alone monitors and anesthesia machines has proved to be an effective strategy. Nellcor, the producer of the first widely available pulse oximeter, entered into OEM deals with multiple companies at a time when all of their sensors were disposable [22]. This income stream has been eroding over time as the use of nondisposable sensor probes has expanded. Aspect Medical similarly did many OEM partnerships for their Bispectral Index (BIS) monitor for measuring the depth of anesthesia [23]. Unlike the pulse oximeter, the probes used in Aspect's BIS monitor have always been disposable due to the nature of the sensor and where it is positioned.

Step 3—Research an exit strategy

Most companies are funded by investor dollars until they generate revenue. Financers expect a return on investment which can include the sale or licensing of the business to a strategic company or possibly an initial public offering (IPO). Selecting and executing the right exit strategy are complex activities that will typically reach better outcomes with the services of an investment banker. The external economic environment is a critical factor. For example, the level of IPO activity in the public equity markets will typically dictate whether going public should even be considered. The entrepreneur should consider their company's role posttransaction. Advanced planning is helpful in preparing for the future.

IPO or sell the company?

Most companies raise capital for product development from investors. The investor pool can include friends and family, angel investors, venture capital firms, or a combination of these as well as other potential sources. These investors expect a return on investment and generally desire to discuss possible exit strategies before agreeing to participate in a private stock offering. In a typical exit event, all preferred classes of stock are converted to common stock and join the founder's common in a pool sold for an agreed price per share. This sale can be in the form of an IPO or a part of a transaction to sell to another company that may include various other provisions such as retention of key employees for a contracted period and earn-out payments triggered by certain development or sales milestones.

However, going public comes with significant costs and risks. A successful IPO depends on market conditions with robust IPO deal flow which can reduce to almost zero during periods of economic contraction. What was once a great market environment at the initiation of investor pitches can deteriorate rapidly even during a brief "roadshow," potentially making an IPO transaction impossible.

Once public, a nimble technology-heavy company will have to start building more financial control infrastructure to address SEC reporting requirements, made more stringent by Sarbanes Oxley.

Selling to another company has significant advantages over an IPO, especially if multiple companies are bidding. Acquisition by a larger company generally results in significant management and technical infrastructure being established to secure the proposed product regulatory approval and successfully market the product to the intended target audience.

Hire an investment banker

An investment banker is an essential part of an IPO transaction and is highly recommended to assist with the company's private sale. Investment bankers initially help set up and market the deal by working with the company's business team and legal counsel to draft an S-1 (in the case of an IPO) or a Private

Placement Memorandum to support the private sale of the company. The banker can also identify potential purchasers of the company's stock for an IPO, or potential purchasers of the whole company for a private sale, and actively participate in pitches the company makes to potential suitors. Finally, the banker will assist in closing the transaction making sure that all the critical documents and various supporting materials are in order. This may sound like the role played by a real estate broker hired to help you sell your house. However, transactions involving the sale of a company are much more complex and are subject to state and federal regulations and the consequences of making a mistake can be far-reaching. Hiring an investment banker requires due diligence. It is important to speak with other companies who have engaged them for similar transactions before entering into a typically required exclusive banker-client engagement.

Potential pitfalls

Before contemplating a transaction, it is critical that all the corporate structure documents, required filings, financial documents, financial history, and all supporting documents (including a correct capitalization table), be complete and audited. Unless a lawyer with experience setting up startup companies created the founding documents and has been looking after the company since its inception, significant issues can surface when the company's status is being reviewed before a transaction. These issues can include disputes over the capitalization table; for example with investors or others claiming that they are owed equity for services not in the capitalization table, articles of incorporation specifying insufficient available shares to support the number of issued shares, missing board meeting minutes, missing key contracts and/or NDAs with employees or development partners and a host of other potential issues overlooked by technical founders who are busy in the lab and using part-time help to manage the business.

The external economic environment can complicate an exit

Fig. 20.2 shows the number of US IPO transactions with an offering price of at least five dollars per share by year from 1980 through 2021. Note the exceptionally high deal flow in 1986 at the dawn of the internet boom and the rapid fall during the

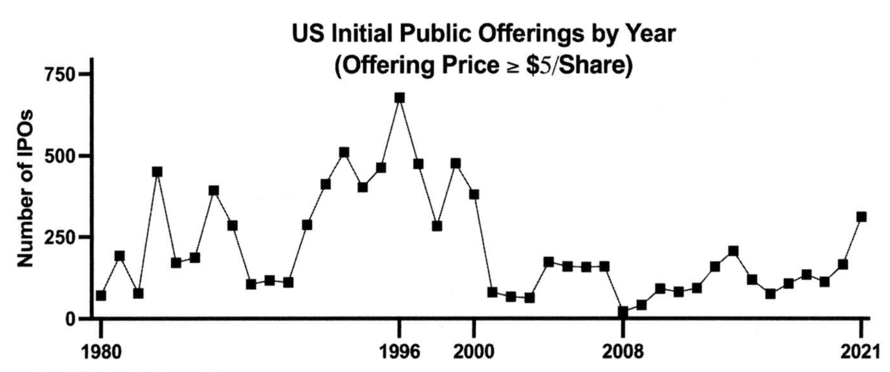

Figure 20.2 US IPO deal flow since 1980. *Credit:* Plot made using data from "Initial Public Offerings: Updated Statistics" Jay R. Ritter, Warrington College of Business, University of Florida with permission from the author.

internet bust beginning in 2000. We can also see the drop in IPOs occurring with the market crash in 2008. Companies planning IPOs during 2000 or 2007 would have seen a dramatic unfavorable change in market conditions when they tried to complete the transaction the following year.

Lack of planning may take a negative turn

An exit event, whether by IPO or a private sale transaction, can deliver big changes to the roles of the technical founders who now need to have regular interaction with a new or larger C-Suite to stay relevant. The posttransaction company will likely be more focused on sales and marketing rather than on product development.

When the technical founders consider their posttransaction role, they might be able to suggest ways in which they could add value while continuing their clinical practice. One example would be to continue to champion the device advantages to anesthesia providers.

Real world examples

HeartPort's magic ingredients for a successful IPO

The mid 1990s marked the beginning of the internet revolution with Netscape and Yahoo both going public with no earnings for what were then regarded as astronomical valuations. HeartPort, a medical device company with no earnings and no website, went public the same month as Yahoo with what analysts called the most successful medical IPO in recent memory [24].

Wes Sterman, the physician CEO of HeartPort, famously placed two photographs on the inside cover of the HeartPort S-1 referred to by analysts as "Sad Guy" and "Happy Guy." Sad Guy was a photograph of a middle-aged, sedated, and intubated patient in an ICU after a coronary artery bypass graft done via a very visible median sternotomy incision. Happy Guy was a thin, young, awake smiling patient with small band aids over the thoracic port access sites used to introduce the instrumentation needed to perform his minimally invasive CABG. I was on an IPO roadshow with Aradigm a few months after HeartPort's very successful offering. I had many conversations with Wall Street investment bankers who said that those photographs significantly motivated them to become believers in what was then a very new approach to cardiac surgery. This underscores the importance of being able to distill a complex technology story into a compelling description by using two captioned photographs.

Aradigm goes public early

Aradigm Corporation was founded to develop novel medical devices to optimize the delivery of inhaled pharmaceuticals. This began with the SmartMist microprocessor controlled metered dose inhaler accessory and continued with the AERX device, allowing precision delivery of systemically active drugs such as insulin by inhalation specially formulated and packaged by Aradigm for single breath delivery. The Aradigm management team expected to exit via the sale of the Company to big pharma after advancing into late-stage product development and achieving profitability through product sales and/or technology licensing. This strategy was based on the concept that the best way to add value to a company is to focus on achieving profitability and not rely on an early exit "off-ramp." The US IPO market in 1996 was very active with more companies going public than ever before. The year marked the beginning of internet public offerings with Netscape and Yahoo achieving record high valuations without being profitable companies. The early-stage minimally invasive cardiac surgery company HeartPort also had a very successful IPO in April 1996. The Aradigm board of directors decided to take advantage of this very strong external environment and took the company public on the NASDAQ over-the-counter market in July of that year [25]. This illustrates the importance of being able to respond quickly to changing market conditions which may require going public earlier than anticipated.

Successful acquisitions can happen post-IPO: Aspect and Nellcor

An “exit” by IPO does not preclude the possibility of a subsequent acquisition. Nassib Chamoun invented the BIS depth of anesthesia monitor while he was a graduate student at MIT. He founded Aspect Medical and established multiple partnerships with medical device companies willing to add BIS to their anesthesia monitoring devices. He successfully exited by IPO and later was able to broker the sale of Aspect to the very large medical device company Covidien [26]. Covidien had a much bigger marketing organization than Aspect and the resources to take the technology through further development to potentially benefit more patients.

Stanford anesthesiologist Bill New and engineer Jack Lloyd developed the first practical pulse oximeter and founded Nellcor, a name created from the first two letters of the founder’s last names. After a successful IPO, the company began to look for a potential acquisition target in the critical care space that could help them market pulse oximeters more widely. Nellcor acquired Puritan Bennett, recognizing that the two companies had nonoverlapping technologies addressing the same markets [27]. Nellcor was unable to increase its profitability after the deal, making the merger with Puritan Bennett more of a burden than an asset. Nellcor Puritan Bennett was purchased by Mallinckrodt two years later, taking the company private at a 36% premium to the publicly traded stock price [28]. This is a good example of a private sale valuing the company higher than the public markets, creating a good opportunity for Nellcor.

Patents can be worth more than the business

AliveCor has sold more than three million smartphone-based ECG monitors all over the world. That is an impressive number, but at an average sales price of less than $100 per unit, it does not represent a large amount of gross revenue when compared to Apple which is also able to monitor ECG starting with its Series 4 Apple Watch technology released in 2018. As of 2023, Apple had sold over $200 billion worth of Apple Watches with ECG monitor capability. AliveCor had demonstrated its patented technology to Apple executives and is now arguing in court that Apple incorporated patented AliveCor technology into its Series 4 Apple Watch infringing on AliveCor patents. If AliveCor prevails in its case against Apple, it could claim a portion of Apple’s past and future Apple Watch revenue, creating value that could significantly exceed AliveCor's valuation based solely on the sale of its devices. On February 6, 2024, a federal judge dismissed one of AliveCor's lawsuits against Apple [29]. AliveCor still has more lawsuits pending against Apple.

Medical Anesthesia Consultants sale to Sheridan

Investment bankers add value in exit transactions by helping position the company's story, helping to prepare the transaction documents, and finding parties interested in purchasing the company's shares either as part of an IPO or a private transaction. The sale of a medical group is another area where an investment banker can be helpful, especially in identifying potential purchasers. Medical Anesthesia Consultants (MAC), the largest anesthesia group in the San Francisco Bay Area, was sold to Sheridan in 2013 with the help of an investment bank [30]. The banking relationship was an important asset, sorting through multiple potential buyers to find the best fit for MAC, helping facilitate due diligence using a Data Room, and facilitating the closing of the transaction.

References

[1] Goldratt EM. The goal. <https://www.amazon.com/Goal-Process-Ongoing-Improvement/dp/0884271951/ref = sr_1_1?crid = 1Y1U54K99BQD0&keywords = the + goal + by + eliyahu + goldratt&qid = 1662392116&sprefix = the + goal%2Caps%2C84&sr = 8-1>.

[2] USPTO, First Inventor to File (FITF) resources <https://www.uspto.gov/patents/first-inventor-file-fitf-resources>.

[3] The number of patents has exploded since 1982, and one court is to blame <https://www.mercatus.org/research/data-visualizations/number-patents-has-exploded-1982-and-one-court-blame>.

[4] How Dave Albert (AliveCor) innovates in medical devices and wins. The 10x Medical Device Conference <https://www.youtube.com/watch?v = X1_4upWqe04>.

[5] USPTO Patent, Inhaled insulin dosage control delivery enhanced by controlling total inhaled volume. 5,884,620. <https://patents.google.com/patent/US5884620A/en?oq = US + 5%2c884%2c620>.

[6] Global News Wire <https://www.globenewswire.com/en/news-release/2001/10/25/1732607/0/en/Novo-Nordisk-invests-USD-20-million-in-Aradigm-Corporation.html>, October 25, 2001.

[7] Biospace <https://www.biospace.com/article/releases/aradigm-corporation-gets-27-5-million-in-insulin-deal-with-novo-nordisk-a-s-/>, July 5, 2006.

[8] Medical Marketing and Media <https://www.mmm-online.com/home/channel/sales/novo-nordisk-lawsuit-wont-delay-pfizers-launch-of-exubera/>, August 15, 2006.

[9] Anesthesia Patient Safety Foundation <https://www.apsf.org/article/the-development-of-pulse-oximeters-in-japan-good-competitors-nihon-kohden-and-minolta-camera/#:~:text = Pulse%20oximetry%20was%20first%20invented.for%20use%20in%20clinical%20settings>, June 10, 2021.
[10] Stanley Electric, Japan Web Page <https://www.stanley.co.jp/e/index.php>.
[11] Onders R. The diaphragm: how it affected my career and my life. The search for stability when the problem is instability. Am. J. Surg. <https://doi.org/10.1016/j.amjsurg.2014.12.003>, December 17, 2014.
[12] USPTO Database, Mapping probe system for neuromuscular electrical stimulation apparatus. US Patent 7,206,641 <https://patents.google.com/patent/US7206641B2/en?oq = 7%2c206%2c641>.
[13] Miller RB, Heiman SE. The new strategic selling <https://www.amazon.com/New-Strategic-Selling-Successful-Companies-ebook/dp/B001J8PQX8>.
[14] Cipolla. Personalizing aerosol medicine: development of delivery systems tailored to the individual". Therap. Deliv. 2010;1(5):667–82 https://www.researchgate.net/figure/MDIs-and-supplemental-technologies-including-clockwisefrom-upper-left-a-Standard_fig1_230571233.
[15] Goodman DE, Rubsamen RM. Delivery of aerosol medications for inspiration. US Patent 5,540,871 <https://patft.uspto.gov/netacgi/nph-Parser?Sect1 = PTO1&Sect2 = HITOFF&d = PALL&p = 1&u = %2Fnetahtml%2FPTO%2Fsrchnum.htm&r = 1&f = G&l = 50&s1 = 5,404,871.PN.&OS = PN/5,404,871&RS = PN/5,404,871>.
[16] Farr SJ. Aerosol deposition in the human lung following administration from a microprocessor controlled pressurised metered dose inhaler. Thorax 1995;50:639–44. Available from: https://thorax.bmj.com/content/50/6/639.short.
[17] FDA Database, SmartMist FDA 510(k) clearance <https://www.accessdata.fda.gov/cdrh_docs/pdf/K960593.pdf>.
[18] SEC Database, Aradigm SEC 10-K filing <https://sec.report/Document/0000891618-00-001215/>, March 2000.
[19] Medtech Insight, Novo Nordisk/Aradigm link on insulin device <https://medtech.pharmaintelligence.informa.com/MT081609/Novo-Nordisk-Aradigm-link-on-insulin-device>, June 1998.
[20] Gonda I, Rubsamen M, Farr SJ. Use of monomeric insulin as a means for improving the bioavailability of inhaled insulin. US Patent 5,888,477 <https://patft.uspto.gov/netacgi/nph-Parser?Sect1 = PTO1&Sect2 = HITOFF&d = PALL&p = 1&u = %2Fnetahtml%2FPTO%2Fsrchnum.htm&r = 1&f = G&l = 50&s1 = 5888477.PN.&OS = PN/5888477&RS = PN/5888477>.
[21] Nature, Inventor dispute over rapid-uptake insulin. Nat. Rev. Drug Discov. 3 (2004) 728 <https://www.nature.com/articles/nrd1505>.
[22] Pink Sheet, Nellcor 50% growth in oem oximetry module sales in fiscal 1994. <https://pink.pharmaintelligence.informa.com/MT002636/NELLCOR-50-GROWTH-IN-OEM-OXIMETRY-MODULE-SALES-IN-FISCAL-1994>, August 1, 1994.
[23] SEC Archives, Aspect medical SEC 10-K filing. <https://www.sec.gov/Archives/edgar/data/886235/000095013504001244/b48843ame10vk.htm>, December 2003.
[24] SfGate, A hot debut for heartport/medical device maker soars 67 percent in trading offer. https://www.sfgate.com/business/article/A-Hot-Debut-For-Heartport-Medical-device-maker-2983841.php, April 27, 1996.
[25] Pink Sheet, Aradigm raises $24.7 mil. in IPO; Proceeds to fund pulmonary drug delivery programs. <https://pink.pharmaintelligence.informa.com/

PS028441/Aradigm-raises-247-mil-in-IPO-proceeds-to-fund-pulmonary-drug-delivery-programs>, July 8, 1996
[26] Today's Medical Developments, Covidien to acquire Aspect Medical Systems Inc. <https://www.todaysmedicaldevelopments.com/article/covidien-to-acquire-aspect-medical-systems-inc-/>, September 9, 2009.
[27] SfGate, Nellcor adds on business/company to buy Puritan. <https://www.sfgate.com/business/article/Nellcor-Adds-On-Business-Company-to-buy-Puritan-3032717.php>.
[28] SfGate, Nellcor bought for $1.9 billion. <https://www.sfgate.com/business/article/Nellcor-bought-for-1-9-billion-3108438.php>, July 24, 1997.
[29] Reuters. Apple beats AliveCor lawsuit over heart-rate apps for Apple Watch. Available from: https://www.reuters.com/legal/apple-beats-alivecor-lawsuit-over-heart-rate-apps-apple-watch-2024-02-07/.
[30] Business Wire. Sheridan Announces Acquisition of Anesthesiology Practice in California. Available from: https://www.businesswire.com/news/home/20131118006519/en/Sheridan-Announces-Acquisition-of-Anesthesiology-Practice-in-California.

21

Choosing an entity structure for your business

Poorwa G. Bhaskar[1], **Ty L. Bullard**[2] **and David N. Flynn**[2]
[1]*Department of Health Care Management, The Wharton School, University of Pennsylvania, Philadelphia, PA, United States* [2]*Department of Anesthesiology, University of North Carolina School of Medicine, Chapel Hill, NC, United States*

Chapter outline

Abstract

This chapter reviews common business structures that entrepreneurs must understand when forming a business. Key features of sole proprietorships, partnerships, corporations, and limited liability companies are reviewed. The impact of each entity type on ownership, taxation, liability, and fundraising is discussed. Practical advice for entrepreneurs operating in the healthcare field is provided.

Keywords: Corporation; incorporate; sole proprietorship; partnership; limited liability company; business entity; taxation; liability

Innovation in Anesthesiology. DOI: https://doi.org/10.1016/B978-0-12-818381-6.00024-3

Abbreviations

LLC limited liability company
LLP limited liability partnership
PTE pass-through-entity
IRS Internal Revenue Service

Choosing an entity structure for your business

Every entrepreneur must decide how to structure his or her new business. The choice of organizational structure will impact startup costs, record keeping and reporting requirements, personal liability protection of owners and managers, taxation status, and ability to recruit investors.

Common types of business entities include sole proprietorships, partnerships, corporations (C and S), and limited liability companies (LLCs). Key attributes and tax implications of each business type are summarized in Figs. 21.1 and 21.2, respectively.

Sole proprietorships

Sole proprietorships are simple and inexpensive to establish, making them the most common form of business in the United States. With a sole proprietorship, the business and the owner are indistinguishable. Assets of the business and personal assets of the owner are one and the same. Owners have unlimited personal liability for debts or obligations associated with the business, meaning their personal assets can be seized in the event of a successful lawsuit against the business [1–3].

A sole proprietorship is a pass-through-entity (PTE), meaning income or losses from the business are passed directly to the business owner with no corporate income tax. The business's taxable income is included on the owner's personal tax return (Schedule C of Form 1040), and losses can be used to offset other income [1,2]. Additionally, sole proprietors are considered self-employed; thus they must pay self-employment (Medicare and Social Security) taxes [1].

Though simple and inexpensive to operate, sole proprietorships have disadvantages that make them inappropriate for most healthcare businesses. First, a sole proprietorship can have only one owner, which limits the business's ability to raise money from outside investors [1]. Second, the liability risk is prohibitive for many businesses operating in the medical field.

Business structure	Ownership model	Cost and administrative complexity	Liability
Sole proprietorship	1 owner	Very low	Unlimited
Partnership	2+ owners	Low - moderate	General partners have unlimited liability Limited partners have limited liability
Limited Liability Company (LLC)	1+ member	Moderate	Limited
S-corporation	1-100 shareholders	High	Limited
C-corporation	1+ shareholder, easily transferable	High	Limited

Figure 21.1 Business structures and key attributes.

Partnerships

A business with two or more owners can be operated as a partnership. Though not always required, a partnership agreement should be created to outline the terms of the partnership, including ownership interests, management responsibilities, and distribution of profits. Like sole proprietorships, partnerships are pass-through-entities, though a tax return must be filed with the Internal Revenue Service (IRS) to report profits and losses (Form 1065) and a breakdown of income allocation to each partner (Schedule K-1). Each partner is then responsible

Business structure	Taxation	Key Tax Forms
Sole proprietorship	PTE; owner includes business income or loss on personal tax return Owner responsible for self-employment tax	Schedule C of Form 1040
Partnership	PTE; business allocates income to partners who each report on their personal tax returns Partners pay self-employment tax on ordinary income	Form 1065, Schedule K-1
Limited Liability Company (LLC)	Default is PTE, can elect to be taxed as a corporation (C or S) Members responsible for self-employment tax	Form 1065 and Schedule K-1 if taxed as partnership Form 1120 if taxed as C-corporation Form 1120 and Schedule K-1 if taxed as S-corporation
S-corporation	PTE; owners include allocated income on personal tax returns	Form 1120, Schedule K-1
C-corporation	Entity pays corporate income tax	Form 1120

Figure 21.2 Business structures, tax implications, and key forms.

for reporting his or her personal share of the income on his or her individual return. Partners must also pay self-employment tax on ordinary income received from the partnership [1–3].

There are multiple types of partnerships, the most common of which are general partnerships, limited partnerships, and limited liability partnerships (LLPs). A major distinction between these forms of partnership is liability protection provided to partners. In a general partnership, all partners are fully liable for debts and obligations of the business, including for actions taken by other partners. With a limited partnership, there are two partnership classes: (1) general partners, who manage the business and retain full liability, and (2) limited partners, who invest in the business but are not involved in managing the business (they are also called "silent" partners since they are not involved in management). Serving as a limited partner provides liability protection, limiting potential losses to the amount invested. In an LLP, partners are protected from liability from negligent acts of other partners, but they remain fully liable for their own actions. They may or may not be protected from general debts or obligations of the partnership, depending on state laws [1–3].

Partnerships are typically formed by service professionals such as lawyers, accountants, or physicians. A partnership would likely be a poor structure for a business developing an innovative medical device or technology due to liability risk and inability to issue stock to outside investors or employees. Although investors can obtain some liability protection as silent partners, professional investors are unlikely to accept a role in which they cannot provide input on management decisions.

Corporations

A corporation is a legal entity that is distinct from its owners and managers. A corporation can enter into contracts, sue or be sued, and incur debts or acquire assets. A key advantage of this structure is liability protection. Owners and managers of the corporation enjoy limited liability, meaning that they are not personally liable for debts or actions of the corporation; rather, the corporation itself is liable for its own debts and actions [1–5].

Another advantage of corporations is the ability to issue stock to owners, known as shareholders. Stock provides a straightforward and convenient way to establish and transfer ownership of the company to founders, investors, employees, and the public. The ability to issue stock makes corporations the preferred entity form of many professional investors, such as venture capitalists [1–5].

A corporation is formed by filing articles of incorporation with the state where the business will operate, which can be distinct from where the owners live or headquarters are located. Many businesses choose to incorporate in Delaware, Nevada, or Wyoming, due to business-friendly laws, taxes, and court systems in those states [6,7]. The articles of incorporation outline many important aspects of business operations, including the company name, stock quantity and price, corporate officers, and shareholder rights [8].

Forming a corporation is more complicated and expensive than creating a sole proprietorship or partnership. Moreover, ongoing legal, accounting, and record-keeping requirements are more onerous, creating additional administrative burden and cost for owners. However, for businesses operating in fields with significant liability risk (such as healthcare) or that require outside investment, the benefits of forming a corporation may outweigh these downsides.

S-corporation and C-corporation

An entrepreneur electing to establish a corporation must decide whether to form a C-corporation or an S-corporation. C-corporations are not pass-through-entities. C-corporations must file income taxes with the IRS and pay corporate taxes on income. Shareholders also pay taxes on capital gains and dividends distributed to them by the C-corporation (this is often referred to as double-taxation) [4].

A corporation can avoid double-taxation by electing S-corporation tax status (technically, an S-corporation is a subtype of C-corporation, though they are often discussed as being distinct entity forms). An S-corporation is a pass-through-entity; the corporation must file a tax return with the IRS, but it does not pay corporate income tax. Rather, profits and losses are passed through to the shareholders who report them on their individual tax returns [5].

The advantaged tax status of S-corporations comes with important limitations. An S-corporation can only issue one class of stock, have no more than 100 shareholders, and shareholders must be individuals, estates, trusts, or charitable organizations. Corporations and partnerships may not own shares of an S-corporation, an important limitation for businesses that need to raise capital from professional investors [5].

In contrast to S-corporations, few limitations are placed on C-corporations. A C-corporation may issue multiple classes of stock and is not bound by restrictions on the number or type of its shareholders. This flexibility, combined with the liability protection afforded to shareholders, makes the C-corporation an attractive entity form for businesses that require significant investment from outsiders or operate in fields with significant liability risk [4].

Limited liability companies

A limited liability company (LLC) is a business structure that combines many of the benefits of alternative business structures, including:

1. Liability protection to its owners (called members): LLCs, like corporations, are legal entities distinct from their owners. Thus in the event of a lawsuit against the business, only the assets of the LLC are at risk.
2. Tax benefits of a partnership: By default, an LLC is taxed as a partnership, avoiding double-taxation. However, members can elect to be taxed as an S- or C-corporation.
3. Ownership flexibility of a C-corporation: LLCs do not issue stock; however, they may create multiple ownership classes, which outline each member's ownership interest and rights.

If taxed as a partnership, each owner will receive a Schedule K-1 with his or her income allocation to report on a personal tax return. Though taxes are passed through, LLC members are considered self-employed; thus they must also pay self-employment taxes [1–4].

An LLC is formed at the state level by filing articles of organization and paying a filing fee. Owners of an LLC should also create an operating agreement, outlining ownership interests, ownership classes and rights, management structure, and rules for membership changes [8].

Forming an LLC may be attractive for some healthcare businesses because LLCs are generally less expensive and administratively complex than corporations, yet they offer liability protection and PTE status. However, many professional investors will only invest in C-corporations. Thus, an LLC may not be an ideal structure for businesses that require substantial capital from outside investors.

Changing the structure of a business

Successful businesses change over time. Managers may find that the legal structure of the business needs to change to better align with the growth and evolution of the business. Generally, businesses change from simple to more complex entities to accommodate growth, change ownership, or add liability protection for those involved in the business. For example, an entrepreneur working on an innovation in anesthesia may initially operate as a sole proprietor to enjoy PTE status (deducting losses against his or her personal income from other sources) and low startup costs. After conducting research and creating a prototype, he or she may try to license the product to a larger company or distributor. At this point, the entrepreneur could form an LLC to provide liability protection, while maintaining PTE status and avoiding the expense of creating a corporation. Later, if the entrepreneur decides to expand his or her operations and needs to raise money from investors, then he or she may convert the LLC to a C-corporation to issue preferred stock to professional investors [9–11].

Although changing entity type is possible, it is not always a straightforward process. Depending on the change being made, it may impact the personal finances of the owners; require new business licenses, employer identification numbers, bank accounts, and insurance policies; and require consultation and agreement of other parties who share ownership in the business. Ideally, owners should anticipate such changes far in advance as part of their business strategy, ensuring that planned changes are possible and can be executed efficiently [9–11].

Key considerations for business founders

When choosing a business structure, there are multiple considerations, including:

1. How many owners/members/shareholders will the business require (in the short, medium, and long term)?

   A sole proprietorship can have only one owner, while partnerships must have more than one. An S-corporation can have no more than 100 owners. There are no restrictions on the number of owners for LLCs or C-corporations.

2. Is the business operating in a field or providing a service/product that could expose owners to significant liability risk?

   The nature of the industry and business offering impacts the liability risk to owners. The liability potential in healthcare is high. Structuring a business as an LLC or corporation provides owners with limited liability protection.

3. What are the financial implications of taxation status on the owners and the business?

   Businesses and their owners could be subject to multiple forms of taxation—personal income tax, self-employment tax, corporate income tax, and taxation on capital gains and dividends. Business owners must understand the impact of taxation on the overall finances of the business, as well as their personal finances. Early consultation with an experienced accountant is essential.

4. What are the fundraising needs of the business?
   Entrepreneurs who will require outside investment must structure their businesses accordingly. Many large, professional investors prefer C-corporations. Smaller investors, such as friends or family members, may be willing to invest in LLCs.

Summary

Entrepreneurs have multiple options when choosing a structure for their businesses. They must understand the attributes of each and choose the option that best matches the strategic goals of their business. Due to the complexity and potential costs involved, it is essential to seek the advice of a lawyer and accountant experienced with startups before selecting a business entity.

References

[1] Scholes MS, Wolfson MA, Erickson MM, Hanlon ML, Maydew EL, Shevlin T. Choosing the optimal organizational form. Taxes and business strategy: a planning approach. New Jersey: Pearson; 2015, p. 58–89.
[2] Fox R. The business entity. Tax strategies for the small business owner: reduce your taxes and fatten your profits. New York: Apress; 2013 (accessed 4.02.21).
[3] U.S. Small Business Administration, Choose a business structure. <https://www.sba.gov/business-guide/launch-your-business/choose-business-structure>, (accessed 04.01.21).
[4] Mintz, Choice of Business Entity: Pros and Cons of Corporations and LLCs. <https://www.mintz.com/insights-center/viewpoints/2166/2020-04-21-choice-business-entity-pros-and-cons-corporations-and>, 2021 (accessed 4.02.21).
[5] DLA Piper, S corporation basics. <https://www.dlapiperaccelerate.com/knowledge/2017/s-corporation-basics.html>, 2021 (accessed 4.01.21).
[6] Gregory, A. The balance small business, best states to incorporate a business <https://www.thebalancesmb.com/best-states-to-incorporate-a-business-4178799>, 2019 (accessed 4.15.21).
[7] Legal Nature, Top 3 Best States to Incorporate a Business. <https://www.legalnature.com/guides/top-states-for-incorporation-delaware-nevada-and-wyoming-explained>, (accessed 4.15.21).
[8] U.S. Small Business Administration, Register your business. <https://www.sba.gov/business-guide/launch-your-business/register-your-business>, (accessed 4.01.21).
[9] Accion Opportunity Fund, Small Business Growth: Changing Your Business Structure. <https://aofund.org/resource/small-business-growth-changing-your-business-structure/>, (accessed 4.03.21).
[10] Akalp, N. Small business trends, is it too late to change your business structure? <https://smallbiztrends.com/2014/11/change-your-business-structure.html>, 2014 accessed (4.15.21).
[11] Smith, PC. How to convert your company from an LLC to a C-Corp, <https://carta.com/blog/convert-llc-c-corp/>, 2020 accessed (4.15.21).

22

FDA regulation of anesthesia and critical care medical devices

Bahram Parvinian
Lighthouse Regulatory Consulting Group, Rockville, MD, United States

Chapter outline

Abstract

The focus of this chapter is the United States medical device regulatory landscape as related to the regulation of anesthesiology and critical care devices. Fundamental concepts such as, medical device definition, classification process, and various premarket regulatory pathways are discussed. The chapter then focuses on the currently evolving regulatory requirements related to digital health technologies, such as the regulatory framework for Software as a Medical Device, Clinical Decision Support Software, automation, and artificial intelligence in anesthesia and critical care medical devices. Through extensive review of FDA guidance documents and recent regulatory frameworks related to digital health medical devices, the chapter provides a comprehensive overview of FDA regulation applied to innovative anesthesia and critical care medical devices.

Innovation in Anesthesiology. DOI: https://doi.org/10.1016/B978-0-12-818381-6.00035-8

Keywords: Medical devices; premarket pathway; FDA; anesthesiology; critical care; 510(k); regulatory; De Novo; digital health; software; software as a medical device; clinical decision support; artificial intelligence; machine learning; anesthesia automation; closed-loop critical care

Introduction

The Center for Devices and Radiological Health (CDRH) within the Food and Drug Administration (FDA) is responsible for regulating the medical device industry in the United States. FDA's mission is to ensure that only safe and effective medical devices are marketed within the United States. The focus of this chapter is to provide an overview of the medical device premarket regulatory landscape in the United States, with special regard for anesthesiology and critical care devices.

The 1938 Federal Food, Drug, and Cosmetic Act (FD&C Act) authorized the FDA to regulate medical products [1]. Medical Device Amendments to the FD&C act were introduced in 1976 which established a risk-based classification system (Class I: low risk, Class II: moderate risk, Class III: high risk), premarket pathways, and postmarket requirements; all intended to provide reasonable assurance of safety and effectiveness of medical devices [1]. These Amendments created the foundation of the medical device regulatory framework of today. The 1990 Safe Medical Devices Act (SMDA) continued to build upon the previous regulations by improving postmarket surveillance requirements, modifying procedures, and by further defining premarket pathways (e.g., defining substantial equivalence for premarket notifications) and introducing requirements for investigational devices [1]. The 1997 Food and Drug Modernization Act (FDAMA) introduced the least burdensome provisions and established the De Novo premarket pathway which could be utilized for certain low to moderate risk devices.

The Medical Device User Fee and Modernization Act (MDUFMA) introduced in 2002 established device review performance goals for the FDA. These goals were focused on increasing the timeliness and predictability of the review process; with incrementally more rigorous performance goals being added each year. Medical Device User Fee Amendments (MDUFA) II concentrated on efficient decision making through the creation of performance goals with a distinct review time and performance level consisting of a certain percentage of submissions reviewed on time [2]. MDUFA III included additional and more stringent review performance goals. It also created shared outcome goals between FDA and industry to bring medical devices to market [2].

MDUFA IV introduced new shared outcome goals and performance goals were for most premarket applications including presubmissions and De Novo applications. For example, the FDA committed to issuing a MDUFA decision within 150 FDA days of receipt of the De Novo submissions for 70% of De Novo requests received in FY 2022 [3].

Medical devices and combination products

The FDA defines a medical device as “an instrument, apparatus, implement, machine, contrivance, implant, in vitro reagent, or other similar or related article, including a component part, or accessory which is intended to diagnose, cure, mitigate, treat, or prevent disease” [4]. A medical device is “intended to affect the structure or function” of the body and does not achieve its primary intended purpose through chemical action or metabolization [4].

Anesthesia and respiratory medical devices are regulated under 21 CFR §868 and as part of FDA’s anesthesiology devices review panel. Some anesthesia and critical care devices such as pulse oximeters and blood pressure monitors are regulated under the cardiovascular devices panel (21 CFR §870) and few utilizing electroencephalogram (EEG) technology with anesthesia indications fall under neurological device regulation (21 CFR §882).

A medical device may be combined with other regulated products such as drugs and biologics to form a single product. These “combination products” are regulated by the Office of Combination Products (OCP) at FDA formally formed in 2002 as part of MDUFMA amendments. OCP coordinates the review of the combination product between lead review centers such as CDRH or Center for Drug Evaluation and Research (CDER). A combination product may contain a combination drug, device, and/or biological product which are physically, chemically, or otherwise integrated to create a single product [5]. Combination products can be two or more separate products which are packaged together in a single package or a separately packaged product (drug, device, or biological product) which is intended to be used with an approved drug, device, or biological product to achieve its intended use [5]. Combination products can also apply to investigational drugs, devices, or biological products which are packaged separately but intended for use with another individually packaged investigational product [5]. Examples of anesthesia and critical care devices with potential combination product designation include metered dose inhalers, drug-coated endotracheal tubes, and closed-loop anesthesia drug delivery systems.

For some products with complex scientific technology and primary mode of action (PMOA), it can be difficult to determine the device's designation or its agency jurisdiction. A Request for Designation (RFD) application may be submitted for a combination product when clarity is needed for designating the lead review center [6]. In response to the RFD, FDA will issue a letter with regulatory designation of the product as a drug, device, biological product, or combination product. If the device is a combination product, the designation letter will identify the agency center which will be primarily responsible for reviewing and regulating the device [6]. The agency center (e.g., CDER) is determined by identifying which component of the combination product serves as the PMOA [5]. For example, an anesthesia system intended for delivery of an already approved drug consistent with the approved drug labeling is likely be deemed a combination product with the CDRH as the lead review center. Currently, the FDA believes that a single premarket application reflective of the device's PMOA may be sufficient as long as there is adequate information to evaluate the safety and effectiveness of the device's constituent parts [5].

Device classification

The FDA adopts a risk-based approach for classifying medical devices based on the device's intended use, indications for use, and technological characteristics [7]. To obtain a formal classification determination, companies may file a 513(g) application, with the FDA [8]. This application allows the agency to provide formal feedback regarding the generic type of device and the class of devices within that type. Each generic device type is assigned a specific product code which corresponds to a regulation number. The product code identifies the device class, premarket review panel, submission type, and associated consensus standards for a device type. For new approval and authorizations, FDA generates product codes, as these are effective tools in tracking medical device postmarket performance [9]. It is worthy to note that the 513(g) application is not intended to facilitate discussions regarding specific types of testing or any information surrounding safety, effectiveness, nor substantial equivalency of a device [8]. Although not intended for formal classification determinations, the presubmission process may be used to understand FDA's tendencies towards particular device classification [10,11].

The FDA classifies medical devices, including in vitro diagnostic (IVD) products, into three classes (I-III), based on the device's level of risk. Class I devices have the lowest risk, while class III devices have the highest risk [7]. Most class I devices are exempt from premarket clearance but must meet all General Controls including (1) the registration of establishment with the FDA, (2) medical device listing, (3) general labeling requirements, and (4) compliance with the Quality System Regulation (QSR), except for design controls [12]. Class II devices generally pose a moderate risk and must adhere to both General and Special Controls such as FDA guidance documents, device-specific consensus standards, labeling requirements, and postmarket surveillance [7]. Lastly, a class III device is usually a high-risk, life-sustaining/life supporting device with novel intended/indications for use and/or technological characteristics for which Special Controls are inadequate to establish safety and effectiveness. Therefore, in addition to meeting both General and Special Controls, class III devices require extensive testing including clinical studies to demonstrate reasonable assurance of safety and effectiveness [13]. Table 22.1 lists different FDA medical device classifications, the associated regulation pathway, and examples of anesthesia medical devices for each classification.

Premarket pathways

Classification of the medical device in turn determines the premarket pathway. Most Class I devices do not require FDA clearance prior to marketing. The 510(k) notification pathway is predominantly used for moderate risk devices (both non-exempt class I and II devices) and requires the subject device to be substantially equivalent to a previously marketed predicate device in terms of safety and effectiveness [14]. There are three types of 510(k) submissions: A traditional 510(k) is commonly used for a device seeking an initial marketing clearance [14]. For devices which are legally marketed but have undergone modifications by the manufacturer, a special 510(k) may be submitted to clear the changes as long as it is a change to the manufacturer's own predicate device, the methods to evaluate the changes are well-established, and if the FDA can review the results in a summary or risk analysis format [15]. An abbreviated 510(k) submission can be utilized when a device

Table 22.1 Summary of regulatory classifications, pathways, and requirements.

Regulatory classification	Level of risk	Regulatory pathway[a]	Applicable premarket regulation and controls	Regulation number	Examples of anesthesia devices
Class I	Low	510(k) exemption	General Controls per applicable regulation: • Listing and registration requirements • Labeling requirements • Medical Device Reporting • Quality Systems Regulation/Good Manufacturing Practices (GMPs)	FD&C Act, sections 501, 502, 510, 516, 518, 519, and 520 21 CFR 801, 803, 807, 820	• Tracheal Tube Fixation Device (CBH) • Gas Calibration Flowmeter (BXY) • Anesthetic Gas Mask (BSJ) • Laryngoscope (CAL, CCW) • Tracheobronchial Suction Catheter (OYI)
Class II	Moderate	510(K)	510(k): General and existing Special Controls Special controls can include • Performance standards • Postmarket surveillance • Patient registries • Special labeling requirements • FDA guidance documents	21 CFR § 807(e)	• Spirometer (BZG, BZK, BWF) • Nebulizer (CAF) • Critical Care ventilator (CBK) • Positive Airway Pressure System (QBY) • Gas Machine (BSZ) • Tracheal Tube (OQU, LNZ)
	Moderate	De Novo	De Novo: General, existing Special Controls, and new Special Controls drafted to address new risks. Eligibility: A 510(k) deemed NSE or direct submission of a De Novo application	FD&C Act, 513(f)(2)	• Ventilatory electrical impedance tomograph (QEB) • Retrograde Intubation Device (QCX) • Positive Airway pressure delivery system (QBY) • High flow humidified oxygen delivery device (QAV)

(*Continued*)

Table 22.1 (Continued)

Regulatory classification	Level of risk	Regulatory pathway[a]	Applicable premarket regulation and controls	Regulation number	Examples of anesthesia devices
Class III	High	PMA	General and Special Controls and additional testing for determination of reasonable assurance of safety and effectiveness	21 CFR § 814 FD&C Act, 513(a)(1)(C), 515	• Adjunctive Predictive Cardiovascular Indicator (QAQ) • Adjunctive Cardiovascular Status Indicator (PPW) • End-tidal anesthetic gas control (QSF) • Computer assisted personalized sedation system (PDR) • Upper Airway Stimulator (MNQ) • High Frequency Ventilator (LSZ)

[a]Not all class I devices may be 510(k) exempt. Similarly, some class II devices may be 510(k) exempt.

complies with an FDA guidance document, adheres to device-specific special controls, or follows voluntary consensus standards [16]. Abbreviated 510(k) applications are composed of "summary reports" which explain how guidance documents, special controls, or consensus standards were satisfied in the design and development of the device [16]. Abbreviated 510(k) applications may be advantageous for both parties as submissions are more efficient to review for the FDA and more succinct to compile for the manufacturer.

Irrespective of the type of 510(k) notification, determination of substantial equivalence (SE) between a predicate and subject device requires that the subject device have the same intended use as the predicate and any differences in the indications for use or technological characteristics do not result in new questions of safety and effectiveness [17]. If aspects of the technological characteristics differ between the two devices, those differences should not raise different questions of safety and effectiveness [17]. Determination of SE is usually supported by performance testing such as bench, animal, usability, and sometimes clinical testing [14]. As part of 510(k) review, the FDA also compares the benefit-risk profiles of the subject and predicate devices to ensure that the subject device is as safe and effective as its predicate. Otherwise, if there is an increase in risk or decrease in benefit, the device may be deemed not substantially equivalent (NSE) [17].

The De Novo pathway was first established in the 1997 Food and Drug Administration Modernization Act (FDAMA) as a marketing pathway for moderate-risk devices which submitted a 510(k) application and were deemed NSE [10]. The 2012 Food and Drug Administration Safety and Innovation Act (FDASIA) updated this eligibility criteria to allow manufacturers to directly apply for a De Novo classification [1]. As a result, the number of De Novo applications granted by the agency has increased substantially and several De Novo authorizations have been granted for devices with anesthesia indications (see Table 22.1) [18]. The De Novo pathway is for moderate risk devices with novel technological characteristics that raise new questions of safety and effectiveness (hence not appropriate for 510(k) pathway) for which special controls can be drafted to effectively mitigate the new risks [10]. If a De Novo request is granted, the FDA will establish a new classification regulation, new product code, and special controls will be established for the new device type. Upon a De Novo request being granted, the device can now be used as a predicate for a future 510(k) submission [10].

A premarket approval (PMA) is required for novel high risk, life-sustaining/life supporting devices where an extensive pivotal clinical study is typically necessary to obtain evidence of safety and effectiveness. Prior to initiation, the clinical study protocol requires FDA approval under an Investigational Device Exemption (IDE) and the applicable regulations for IDEs which are reserved for medical devices with significant risk [19]. To support the PMA marketing application, the IDE study must demonstrate that the possible benefits of the device outweigh any potential risks and the device must benefit the majority of its intended target population [20]. A traditional PMA application allows an "all-in-one" submission of reports to the FDA once all clinical testing has been completed [21]. Alternatively, a modular PMA permits modules of the PMA to be submitted as they are completed [21]. Within the PMA pathway, the Product Development Protocol (PDP) method, used for high-risk, well-established devices, provides both the FDA and the sponsor the opportunity to agree upon the details of design and development activities which will demonstrate safety and effectiveness of the new device [21]. Through this method, sponsors can save time and money associated with clinical studies by receiving FDA feedback as important information is generated at specific development milestones.

Regulation of software and medical devices with digital health technology

Anesthesia and critical care devices have been at the forefront of the fast-evolving field of digital health. Many anesthesiologists provide patient care in pre/perioperative, intraoperative, and postoperative environments where advanced sensor-enabled monitoring software, data sharing, medical device interoperability, and algorithms with decision support or direct physiological control capabilities have the potential to improve quality of care and patient outcomes. FDA's regulation of Digital Health Technology (DHT) directly affects innovation in anesthesia and critical care. Therefore it is highly relevant to the mission of this chapter to discuss FDA's current perspective on regulation of digital health products. In this section we discuss recent developments in regulation of medical devices with evolving digital health attributes.

In 2020 FDA established a center for digital health in charge of modernizing the agency's approach towards regulating digital health products [22]. Some of the recent efforts towards clarification of policy and facilitating review of devices with DHT technology include a number of guidance documents and discussion papers on: FDA's interpretation of legislature introduced as part of the 2016 Cures Act [23], regulation of software function intended for general wellness and mobile medical applications [24,25], interoperability and cybersecurity requirements [26,27], guidance document on clinical decision support (CDS) software [28], and proposed framework on regulation of medical device software with Artificial Intelligence (AI) and Machine Learning (ML) features [29–31] including, most recently, a draft guidance document on Predetermined Change Control Plan(PCCP) intended for continuously learning machine learning algorithms [32]. Furthermore, FDA has published a draft guidance on technical considerations applicable to automated anesthesia and critical care devices [33].

It is important to note that FDA's regulatory perspective and expectations in some of these areas may still be evolving. Table 22.2 summarizes recent milestones in evolving medical

Table 22.2 Digital health regulatory updates.[a]

Regulatory updates	Summary	Stage of FDA's thinking	Potential anesthesia and critical care examples
21st century cures act	Excluded some categories of medical software as regulated medical devices. The FDA has clarified its interpretation of the act and its extent of regulatory purview regarding various types of digital health.	Final guidance [23]	• Administrative support of a healthcare facility • Electronic patient records • Software that provides the clinician with a list of medications approved for a disease or condition
Software as a Medical Device (SaMD) or software in a medical device	Risk based classification of software based on the significance of information provided by the SaMD and the state of the patient's healthcare condition. Clinical evaluation of a SaMD includes generating data to demonstrate **1.** a valid clinical association, **2.** analytical validation of SaMD outputs, and **3.** clinical validation of the SaMD.	Final guidance [34–36]	• A machine learning software that uses EEG and heart rate signals to provide information on the depth of sedation. • A software algorithm that analyzes ventilatory waveforms and estimates time to weaning • Machine learning algorithm used for a reducing number of false alarms in a multiparameter physiological monitor
Clinical decision support (CDS) software	Distinguishes device and nondevice CDS Risk-based policy for regulation of CDS functions based on the intended use, intended user, inputs used to generate a recommendation, and the basis for making a recommendation. Regulatory oversight is focused on higher risk CDS functions used by healthcare providers, patients, and caregivers to inform clinical management for serious and critical healthcare conditions.	Final guidance [28]	• Software that analyzes breathing patterns from a sleep apnea monitor to provide therapy decisions • Machine learning algorithm that is trained on ECG and blood pressure data to detect onset of sepsis and recommends a vasopressor dose

(*Continued*)

Table 22.2 (Continued)

Regulatory updates	Summary	Stage of FDA's thinking	Potential anesthesia and critical care examples
Automated critical care devices guidance	Highlights key technical requirements including in silico and usability testing for closed-loop medical devices	Draft guidance [33]	• An algorithm that adjusts anesthetic concentration-based end tidal anesthetic and CO_2 concentrations • An algorithm that maintains blood pressure by titrating vasopressor based on feedback from blood pressure monitors
Proposed regulatory framework for modifications to artificial intelligence/machine learning-based software as a medical device Good machine learning principles Predetermined change control plan	Introduced the concept of Predetermined Change Control Plan (PCCP), SaMD Pre Specifications (SPS) and Algorithm Change Protocol (ACP) and evaluation of continuously learning algorithm based on a total product life cycle approach and based on Good Machine Learning Practice Provided 10 guiding principles based on which GMLP can be developed. Intended for Machine Learning enabled Device Software Functions (ML-DSFs). The PCCP is to be developed based on planned and anticipated modification, method of implementation, and an impact assessment.	White paper [31] White paper [30] Draft guidance [32]	SaMD and CDS examples apply

[a]For a complete list of regulatory developments in digital health products refer to https://www.fda.gov/medical-devices/digital-health-center-excellence.

device digital health regulations. It is beyond the scope of this chapter to discuss every facet of DHT regulation. Therefore, important developments more relevant to anesthesiology and critical care devices are discussed in the following sections.

The 21st century cures act

FDA defines "function" as the distinct purpose of the product which may be part of the intended use of the product [37]. The Cures Act Software Provisions defines software functions which are not devices as those that are intended for administrative support of a healthcare facility, encouraging a healthy lifestyle, serving as electronic patient records, and those which transfer, store, convert formats, or display data and results [23]. The act also provides provisions for multiple software functions including cases where a product includes both a device and a nondevice function [37].

Software in and as a medical device

Medical Devices with DHT may be categorized based on whether the software component is embedded with device's hardware to enable it is intended use (e.g., automated drug delivery) or the software functions independent of device hardware. The latter is termed Software as a Medical Device (SaMD) while the former is termed Software in a Medical Device (SiMD). SaMD is software which is "intended to be used for one or more medical purposes that perform these purposes without being part of a hardware medical device" [34]. FDA has also initiated close collaboration with international organizations such as the International Medical Device Regulators Forum (IMDRF) and has adopted the same potential framework for risk categorization of software as a medical device (SaMD) [35]. According to this framework, SaMD risks are categorized based on the software's level of impact to the patient with category I SaMD having the lowest impact in nonserious situations and category IV having the highest impact in critical situations [34].

SaMD evaluation prior to marketing focuses on determining the device's "accuracy, specificity, sensitivity, reliability, limitations, and scope of use in the intended environment" through three levels of clinical evaluation [36]. First, a valid clinical association determines if there is a solid association between the SaMD output and the targeted clinical condition. Second, analytical validation evaluates the accuracy to which the SaMD processes input data and generates reliable outputs [36]. Lastly, the SaMD undergoes clinical validation to ensure that the accurate output data achieves its intended purpose in the target population using it for clinical care [36]. It is expected that many of the regulatory and technical considerations for SaMDs will be applicable to SiMDs.

Clinical decision support software

SaMD and SiMDs may serve as clinical decision support systems in anesthesia and critical care. Examples of such systems include a device that analyzes streams of data from physiological monitors, such as electrocardiogram, blood pressure, and pulse oximeters to provide recommendations for drug-specific dose delivery. The extent of FDA regulatory oversight on CDS and whether they are deemed as medical devices subject to FDA regulation may not always be clear. For example, a software solely intended for notifying the clinician of drug-drug interaction based on established drug libraries may not be considered a medical device. FDA has published their interpretation of the definition of device CDS vs non-device CDS functions based on four criteria which were introduced in section 520(o)(1)(E) of the FD&C Act [28]. As part of this guidance the FDA clarifies how factors such as type of input used in the CDS, frequency at which medical information is provided to the clinician, time criticality of the clinical decision, and whether the clinician can independently review the basis for the recommendation as factors that may differentiate a device CDS from a non-device CDS. FDA is expected to continue to follow a risk-based approach and elect not to enforce certain premarket clearance requirements for some low risk device CDS products [28].

Automation in anesthesia

An anesthesia CDS which provides recommendations for titration of anesthetics and/or analgesics, in its next iteration, may be embedded within device hardware (hence a SiMD) to enable direct command to an infusion pump or anesthesia gas machine to deliver dose or manage the platform's alarms automatically. Future iterations may involve addition of other physiological sensors to automatically manage oxygenation and hemodynamic stability under clinical supervision.

The concept of automated anesthesia is not new. It was first introduced in the 1950s [38]. With ever increasing rate of data streams from physiological monitors and clinician fatigue in ICU and surgical environments, there is need for intelligent anesthesia to improve safety and enhance patient care. This is critically important for ensuring patient safety in low resource/high demand emergency settings and environments such as battlefield, disaster struck regions, and in the event of a pandemic [39].

From the FDA perspective, automated anesthesia and critical care delivery (e.g., automated mechanical ventilation) if enabled by physiological sensors, is part of a broader category of medical devices known as Physiological Closed-Loop Controlled (PCLC) systems [33,39]. PCLCs may be regarded as an advanced digital health medical device with interoperable subsystems. Critical aspects of automation safety, which will be evaluated by FDA as part of any marketing application include reliability and accuracy of feedback sensors, control algorithm robustness, presence of adequate fail-safe mechanisms, and usability/ human-machine interface design considerations for automated system [39].

In the context of introducing automation in anesthesia and critical care, defining a framework that delineates the roles of clinician and the PCLC medical device as automation progresses to higher levels, may be helpful in characterizing benefits and risks and evidence needed for regulatory approval. The concept of levels of automation (LOA) has been used successfully in other industry domains with safety-critical systems such as aviation and automobile, with regulatory bodies adopting the framework in order to allow a gradual introduction of automation [40–42]. This concept was discussed in a 2015 FDA held workshop focusing on automated critical care systems [39] and holds promise to serve as a framework with potential to facilitate regulatory approval of automated critical care systems. It is likely that commercialization of CDS devices as a first step will simplify the regulatory evaluation of anesthesia medical devices at higher levels of automation and with more advanced autonomous decision-making capabilities.

While establishing safety and proving risks arising from automation have been mitigated down to an acceptable level is necessary for regulatory approval, it may not be sufficient. Reasonable expectation of improvements in patient care and outcomes combined with community consensus particularly on physiological targets may facilitate FDA approval.

FDA regulation of AI/ML technology

AI technology may be leveraged to design, evaluate, and improve the performance of CDS and PCLC devices (i.e., at different LOAs) with anesthesia or critical care indications. Data-driven methods such as machine learning may be used to enhance performance of physiological monitors and controllers [43–45]. They may further be combined with alternative methods such as computational modeling and simulation (e.g., in silico testing) to evaluate and generate safety evidence required for regulatory approval [33,39].

FDA has published a number of white papers and proposed a regulatory framework for regulating AI/ML enabled medical devices which mainly focus on "locked" algorithms [29–31]. Under FDA's current regulatory requirements, algorithmic changes impacting safety and effectiveness of previously approved or cleared medical devices such as a device CDS or PCLC would generally require submission of a new marketing application or supplements to inform FDA of the change and reassessment of the safety and effectiveness of the device [46,47]. The advent of continuously learning algorithms capable of updating device output by processing and analyzing new or better-quality input data requires a new paradigm in regulation of software algorithms. Most recently FDA released a draft guidance document to propose a new framework known as Predetermined Change Control (PCCP) to be reported in marketing application for ML-enabled software function. By submitting a PCCP the manufacturer may seek approval for planned modifications, their method of implementation, and assessment. If authorized, a new submission for each modification described in accordance with the PCCP may not be necessary [33].

References

[1] Center for Devices and Radiological Health, A history of medical device regulation & oversight in the United States, 24 June 2019. https://www.fda.gov/medical-devices/overview-device-regulation/history-medical-device-regulation-oversight-united-states.
[2] U.S. Food and Drug Administration, MDUFA performance reports, 1 July 2023. https://www.fda.gov/about-fda/user-fee-performance-reports/mdufa-performance-reports.
[3] U.S. Food and Drug Administration, FY 2018 MDUFA performance report, 2018. https://www.fda.gov/industry/medical-device-user-fee-amendments-mdufa/medical-device-user-fee-amendments-2017-mdufa-iv.
[4] U.S. Food and Drug Administration, How to determine if your product is a medical device. <https://www.fda.gov/medical-devices/classify-your-medical-device/how-determine-if-your-product-medical-device> [accessed 07.07.23].
[5] U.S. Food and Drug Administration, Principles of premarket pathways for combination products, 2022. <https://www.fda.gov/regulatory-information/se.arch-fda-guidance-documents/principles-premarket-pathways-combination-products> [accessed 07.07.23].
[6] U.S. Food and Drug Administration, How to write a request for designation. <https://www.fda.gov/regulatory-information/search-fda-guidance-documents/how-write-request-designation-rfd> [accessed 07.07.23].

[7] Center for Devices and Radiological Health, Classify your medical device. <https://www.fda.gov/medical-devices/overview-device-regulation/classify-your-medical-device>. 2020 [accessed 07.07.23].

[8] U.S. Food and Drug Administration, FDA and Industry procedures for Section 513(g) requests for information under the Federal Food. Drug, and Cosmetic Act 2019. <https://www.fda.gov/regulatory-information/search-fda-guidance-documents/fda-and-industry-procedures-section-513g-requests-information-under-federal-food-drug-and-cosmetic> [accessed 07.07.23].

[9] U.S. Food and Drug Administration, Medical device classification product codes, 2013. https://www.fda.gov/regulatory-information/search-fda-guidance-documents/medical-device-classification-product-codes-guidance-industry-and-food-and-drug-administration-staff.

[10] U.S. Food and Drug Administration, De Novo classification process (Evaluation of automatic class III designation). <https://www.fda.gov/regulatory-information/search-fda-guidance-documents/de-novo-classification-process-evaluation-automatic-class-iii-designation>; 2021 [accessed 07.07.23].

[11] U.S. Food and Drug Administration, Requests for feedback and meetings for medical device submissions: the Q-submission program, 2019. https://www.fda.gov/regulatory-information/search-fda-guidance-documents/requests-feedback-and-meetings-medical-device-submissions-q-submission-program.

[12] Center for Devices and Radiological Health, General controls for medical devices, 22 March 2018. https://www.fda.gov/medical-devices/regulatory-controls/general-controls-medical-devices#establishment_registration_requirements.

[13] U.S Food and Drug Administration. Premarket Approval (PMA). <https://www.fda.gov/medical-devices/premarket-submissions-selecting-and-preparing-correct-submission/premarket-approval-pma>; 2019 [accessed 07.07.23].

[14] U.S. Food and Drug Administration, The 510(k) program: evaluating substantial equivalence in premarket notifications [510(k)], 2014. https://www.fda.gov/regulatory-information/search-fda-guidance-documents/510k-program-evaluating-substantial-equivalence-premarket-notifications-510k.

[15] U.S. Food and Drug Administration, The special 510(k) program. <https://www.fda.gov/regulatory-information/search-fda-guidance-documents/special-510k-program>; 2019 [accessed 07.07.23].

[16] U.S. Food and Drug Administration, The abbreviated 510(k) program, Silver Spring: https://www.fda.gov/regulatory-information/search-fda-guidance-documents/abbreviated-510k-program, 2019.

[17] U.S. Food and Drug Administration, Benefit-Risk Factors to Consider when Determining Substantial Equivalence in Premarket Notifications (510(k)) with Different Technological Characteristics, 2018. [Online]. https://www.fda.gov/regulatory-information/search-fda-guidance-documents/benefit-risk-factors-consider-when-determining-substantial-equivalence-premarket-notifications-510k.

[18] U.S. Food and Drug Administration, De Novo Classification request, 20 November 2019. [Online]. https://www.accessdata.fda.gov/scripts/cdrh/cfdocs/cfpmn/denovo.cfm.

[19] Center for Devices and Radiological Health, Investigational Device Exemption (IDE), 13 December 2019. [Online]. https://www.fda.gov/medical-devices/how-study-and-market-your-device/investigational-device-exemption-ide.

[20] U.S. Food and Drug Administration, Step 3: Pathway to Approval, 9 February 2018. https://www.fda.gov/patients/device-development-process/step-3-pathway-approval.

[21] U.S. Food and Drug Administration, PMA Application Methods, 27 September 2018. [Online]. https://www.fda.gov/medical-devices/premarket-approval-pma/pma-application-methods#:~:text = The%20PDP%20allows%20a%20sponsor,effectiveness%20of%20a%20new%20device.&text = It%20establishes%20reporting%20milestones%20that,to%20in%20a%20timely%20manner.

[22] U.S. Food and Drug Administration, Digital health center of excellence, 22 September 2020. https://www.fda.gov/medical-devices/digital-health-center-excellence.

[23] U.S. Food and Drug Administration. Changes to existing medical software policies resulting from section 3060 of the 21st century cures act. Silver Spring; 2019.

[24] U.S. Food and Drug Administration, General Wellness: Policy for Low-Risk Devices, 2019. https://www.fda.gov/regulatory-information/search-fda-guidance-documents/general-wellness-policy-low-risk-devices.

[25] U.S Food and Drug Administration Policy for Device Software Functions and Mobile Medical Applications, 2022 <https://www.fda.gov/regulatory-information/search-fda-guidance-documents/policy-device-software-functions-and-mobile-medical-applications> [accessed 07.07.23].

[26] U.S. Food and Drug Administration, Design considerations and pre-market submission recommendations for interoperable medical devices, Silver Spring, 2017.

[27] U.S. Food and Drug Administration. Content of premarket submissions for management of cybersecurity in medical devices. Silver Spring; 2018.

[28] U.S. Food and Drug Administration, Clinical decision support software, 2019. https://www.fda.gov/regulatory-information/search-fda-guidance-documents/clinical-decision-support-software.

[29] U.S Food and Drug Administration, Artificial intelligence/machine learning (AI/ML)-based software as a medical device (SaMD) action plan, 2021. <https://www.fda.gov/media/145022/download> [accessed 07.07.23].

[30] U.S Food and Drug Administration, Good machine learning practice for medical device development. [Online]. 2021; <https://www.fda.gov/media/153486/download> [accessed 07.07.23].

[31] U.S. Food and Drug Administration, Proposed regulatory framework for modifications to artificial intelligence/machine learning (AI/ML)- based software as a medical device (SaMD), Silver Spring, 2020.

[32] U.S Food and Drug Administration, Marketing submission recommendations for a predetermined change control plan for artificial intelligence/machine learning (AI/ML)-enabled device software functions, 2023. <https://www.fda.gov/regulatory-information/search-fda-guidance-documents/marketing-submission-recommendations-predetermined-change-control-plan-artificial> [accessed 07.07.23].

[33] U.S. Food and Drug Administration, Technical considerations for medical devices with physiologic closed-loop control technology. <https://www.fda.gov/regulatory-information/search-fda-guidance-documents/technical-considerations-medical-devices-physiologic-closed-loop-control-technology> [accessed 07.07.23].

[34] International Medical Regulators Forum, Software as a medical device: possible framework for risk categorization and corresponding considerations, 2014. https://www.fda.gov/medical-devices/software-medical-device-samd/global-approach-software-medical-device. [accessed 07.07.23].

[35] International Medical Regulators Forum, Software as a medical device: possible framework for risk categorization and corresponding considerations, 2014.
[36] U.S. Food and Drug Administration, Software as a medical device (SAMD): clinical evaluation, Silver Spring, 2017.
[37] U.S. Food and Drug Administration, Multiple function device products: policy and considerations. <https://www.fda.gov/regulatory-information/search-fda-guidance-documents/multiple-function-device-products-policy-and-considerations.2020> [accessed 07.07.23].
[38] J.C. Alexander and G.P. Joshi, Anesthesiology, automation, and artificial intelligence, In *Baylor University Medical Center Proceedings*, Dallas, 2018.
[39] Parvinian B, Scully C, Hanniebey W, Kumar A, Weininger S. Regulatory considerations for physiological closed-loop controlled medical devices used for automated critical care: food and drug administration workshop discussion topics. Anesthes Analg 2017;126(6):1916–25.
[40] M. Cummings, Automation bias in intelligent time critical decision support systems, in *AIAA intelligent systems conference*, Chicago, 2004.
[41] Federal Aviation Administration, Safety Alert for Operators SAFO #13002, U.S. Department of Transportation, Washington, D.C., 2013.
[42] Frohm J, Lindstrom V, Stahre J, Winroth M. Levels of automation in manufacturing. Int J Ergonomics Hum Factors 2008;30(3):1–28.
[43] Wingert T, Lee C, Maxime C. Machine learning, deep learning, and closed loop devices-anesthesia delivery. Anesthesiol Clin 2021;39(3):565–81.
[44] Pathmanathan P, Daluwatte C, Yaghoouby F, A Gray R, Weininger S, M Morrison T, et al. Credibility evidence for computational patient models used in the development of physiological closed-loop controlled devices for critical care medicine. Front Physiol 2019;10:220. p. Mar 26.
[45] Jian Z, Buddi S, Lee C, Settels J, Sibert K, Rinehart J, et al. Machine-learning algorithm to predict hypotension based on high-fidelity arterial pressure waveform analysis. Anesthesiology 2018;29(4):663–74.
[46] U.S. Food and Drug Administration, Deciding When to Submit a 510(k) for a Software Change to an Existing Device. <https://www.fda.gov/regulatory-information/search-fda-guidance-documents/deciding-when-submit-510k-software-change-existing-device> [accessed 07.07.23].
[47] U.S. Food and Drug Administration, Deciding when to submit a 510(k) for a change to an existing device. <https://www.fda.gov/regulatory-information/search-fda-guidance-documents/deciding-when-submit-510k-change-existing-device> [accessed 07.07.23].

23

Step-by-step guide for Food and Drug Administration listing, 510(k), premarket approval

Junichi Naganuma
Department of Anesthesiology, University of California, San Diego, CA, United States

Chapter outline

Innovation in Anesthesiology. DOI: https://doi.org/10.1016/B978-0-12-818381-6.00014-0

Abstract

Bringing a medical device to market is time-consuming and costly. The U.S. Food and Drug Administration (FDA) is the regulatory gatekeeper of bringing new medical devices to market in the United States and makes sure that the devices are safe and effective. While it takes significantly less time than bringing a new drug to market, it takes three to seven years to bring a medical device from idea to market, and costs between $10 million and $20 million even in the preclinical phase of development. Moreover, the average cost from idea to FDA approval is $31 million via the 510(k) pathway and $94 million via the PMA pathway. Unlike innovation in the nonmedical space, obtaining FDA approval adds significant time, cost, and complexity to the device's pre- and postmarketing process. This chapter aims to demystify the FDA's premarketing regulatory process and provide a step-by-step guide on how to obtain FDA approval.

Keywords: The U.S. Food and Drug Administration (FDA); the Federal Food, Drug, and Cosmetic Act (FD&C Act); premarket notification 510(k); substantial equivalence; predicate device; premarket approval (PMA); investigational device exemption (IDE); De Novo pathway; humanitarian use device (HUD); humanitarian device exemption (HDE)

Key points

- Bringing a medical device to market in the United States is a time-consuming and a costly process involving the FDA.
- The three-tiered risk classification system (Classes I, II, and III) determines which premarketing approval pathway is required for a medical device.
- While the FDA's medical device premarketing process may appear daunting at first, a thorough understanding of material available on the FDA website is critical in obtaining FDA approval for a medical device.

Glossary

The U.S. Food and Drug Administration A federal agency within the U.S. Department of Health and Human Services which is "responsible for protecting public health by regulating human drugs and biologics, animal drugs, medical devices, tobacco products, food (including animal food), cosmetics, and electronic products that emit radiation" [26].

The Federal Food, Drug, and Cosmetic Act (FD&C Act) This Act signed in 1938 authorized the FDA's regulation and oversight of medical products [4].

Premarket notification 510(k) "A premarket submission made to the FDA to demonstrate that the device to be marketed is as safe and effective, that is, substantially equivalent, to a legally marketed device" [12].

Predicate device A legally marketed device to which a "new" device is compared to demonstrate substantial equivalence [12].

Substantial equivalence "Substantial equivalence" means that the new premarket device "is as safe and effective as the predicate" [12].

Premarket approval The FDA's regulatory pathway which ensures safety and effectiveness of Class III medical devices [27].

Investigational device exemption (IDE) An exemption which allows a premarket device under consideration for a PMA to be used in a clinical study [18].

De Novo pathway The FDA's regulatory pathway for novel medical devices without predicate devices but which provide reasonable assurance of safety and effectiveness [20].

Humanitarian use device A medical device "intended to benefit patients in the treatment or diagnosis of a disease or condition that affects not more than 8,000 individuals in the United States per year" [22].

Humanitarian device exemption A premarketing application for an HUD which is exempt from the effectiveness requirements and subject to certain profit and use restrictions [22].

Introduction/why it matters?

Bringing a medical device to market in the United States is a time-consuming and a costly process. The U.S. Food and Drug Administration (FDA) is the regulatory gatekeeper of bringing new medical devices to market in the United States and makes sure that the devices are safe and effective. While it takes significantly less time than bringing a new drug to market, studies show that it takes three to seven years to bring a medical device from idea to market, and costs between $10 million and $20 million even in the preclinical phase of device development [1,2]. Moreover, the average cost to bring a device from concept to FDA clearance is $31 million (of which $24 million is spent on FDA-related activities) via the 510(k) pathway and $94 million (of which $75 million is spent on FDA-related activities) via the PMA pathway [3]. Unlike innovation in the nonmedical space, obtaining regulatory approval from the FDA adds significant time, cost, and complexity to the device's pre- and postmarketing process. This chapter aims to demystify the FDA's premarketing regulatory process and provide a step-by-step guide on how to obtain FDA approval for a medical device.

Key milestones in the U.S. Food and Drug Administration history and its oversight of medical devices

To better understand the FDA's regulatory role in medical device development, several key milestones in its history are noteworthy [4]. In 1938, the Federal Food, Drug, and Cosmetic Act (FD&C Act) authorized the FDA's oversight of medical devices. In 1976, the Medical Device Amendments to the FD&C Act created the three-tiered risk classification system (Classes I, II, and III) for medical devices, and established the premarket notification (510(k)) and premarket approval (PMA) pathways. In 2002, the Medical Device User Fee and Modernization Act (MDUFMA) was enacted to allow the FDA to collect fees for premarket applications.

The FDA's definition of a "device"

Before considering the FDA's premarketing approval process for a medical product, first determine whether it meets the FDA's definition of a medical device. Section 201(h) of the Federal Food, Drug and Cosmetic Act (FD&C Act) [5] defines a medical device as: an instrument, apparatus, implement, machine, contrivance, implant, in vitro reagent, including a component part, or accessory which is:

1. recognized in the official National Formulary, or the United States Pharmacopoeia,
2. intended for use in the diagnosis of disease, or in the cure, mitigation, treatment, or prevention of disease, or
3. intended to affect the structure or function of the body, and which does not achieve its primary intended purposes through chemical action or metabolism within or on the body.

Next, determine if a product classification exists for your product by using one of the following search functions on the FDA website [5]:

1. Search the FDA product classification database (https://www.accessdata.fda.gov/scripts/cdrh/cfdocs/cfpcd/classification.cfm).
2. Search for similar devices in the following databases: PMA (https://www.accessdata.fda.gov/scripts/cdrh/cfdocs/cfPMA/pma.cfm), 510(k) (https://www.accessdata.fda.gov/scripts/cdrh/cfdocs/cfPMN/pmn.cfm), De Novo (https://www.accessdata.fda.gov/scripts/cdrh/cfdocs/cfPMN/denovo.cfm), and humanitarian device exemption (HDE) (https://www.accessdata.fda.gov/scripts/cdrh/cfdocs/cfHDE/hde.cfm).
3. Search for similar devices by device listing via the FDA's establishment registration and device listing database (https://www.accessdata.fda.gov/scripts/cdrh/cfdocs/cfrl/textsearch.cfm).

Device classification

Classify your device into one of the following three categories (Classes I, II, or III); this will determine the required FDA premarket approval pathway. The two key determinants for device classification are: "intended use (general purpose of the device)" and "indications for use (disease or condition the device will diagnose or treat)" [5,6].

1. Class I (low risk of patient harm)

   Most devices in this category, for example, tongue depressors, are 510(k) exempt.
2. Class II (moderate risk of patient harm)

   Some devices in this category, for example, sutures, are 510(k) exempt, but those that are not require 510(k) submission.
3. Class III (high risk of patient harm)

Devices in this category, for example, pacemakers, require premarket approval (PMA) submission.

If you are unsure whether your device is a "device" per the FDA's definition, or which classification the device falls into, the FDA can be contacted for a feedback via the 513(g) request with a fee [7].

Regardless of which approval pathway your device requires, the FDA encourages the use of free presubmission meetings called "Q-sub" or "presub" early in the approval process. As the name implies, these meetings are intended to occur before application submission. Early feedback from the FDA may mean significant time and cost savings to the applicant [8].

FDA listing

Medical device establishments in the United States must register annually with the FDA. You must first pay the annual registration user fee via the FDA's Device Facility User Fee (DFUF) website (https://userfees.fda.gov/OA_HTML/furls.jsp), then complete the registration and listing via the FDA's Unified Registration and Listing System (FURLS)/Device Registration and Listing Module (DRLM) website (https://www.access.fda.gov/oaa/). If your device requires 510(k) or PMA submission, you can only register your establishment and list your device after the premarket submissions are approved as described below [9–11].

Premarket notification 510(k)

The purpose of the 510(k) submission is to demonstrate that your device is as safe and effective, or "substantially equivalent," to an already marketed device called a predicate. Unless the device is 510(k) exempt, all Classes I, II, and III devices not requiring the most stringent Premarket Approval (PMA) will require 510(k) submission [12].

Required contents of 510(k) application

Before starting the 510(k) application, you should have the following information available:

1. Classification of your device
2. Predicate device
3. Final draft labeling
4. Specifications, for example, drawings and photos
5. Performance data, for example, nonclinical or clinical trial results
6. Sterilization information
7. Guidance documents, that is, documents regarding FDA-regulated products or approval submissions [13].

The following is a list of required contents of the 510(k) application [14]:

1. General Information
   a. Medical device user fee cover sheet (Form FDA 3601)
      The first page of a 510(k) application should be the medical device user fee cover sheet. See https://userfees.fda.gov/OA_HTML/mdufmaCAcdLogin.jsp for detailed instructions.
   b. CDRH premarket review submission cover sheet (Form FDA 3514).

A cover letter and/or the CDRH premarket review submission cover sheet should follow the user fee cover sheet above. Although you are not required to use the CDRH Cover Sheet, it may help expedite the processing of the 510(k) application.

c. Cover letter

A cover letter can be used to provide information covered on the CDRH premarket review submission cover sheet.

The Cover Letter must include the following items:

1. Submission date (month/day/year) and labeled as "510(k) Submission."
2. Submitter's contact information (name, address, phone, fax and e-mail address).
3. Establishment registration number (or state that you will register following FDA clearance).
4. The common name of the device, for example, syringe.
5. The trade name of the device.
6. The classification name for the device and the class in which the device has been placed.
7. The reason for the 510(k).
8. Identification of the predicate device.
9. If applicable, the registration number, name and address of manufacturing facility of the finished device.
10. Compliance with any special controls [section 513 (b) of the Food, Drug, and Cosmetic Act (FD&C Act)], FDA mandatory performance standards (section 514 of the FD&C Act), standards under the Radiation Control for Health and Safety Act (RCHSA), or voluntary consensus standards.

2. Table of contents

List each required item with page numbers, including a list of attachments and appendices.

3. 510(k) acceptance checklist

The 510(k) acceptance checklist (see https://www.fda.gov/medical-devices/premarket-notification-510k/acceptance-checklists-510ks) is used to determine whether the 510(k) application meets the minimum requirement and should be accepted for substantive review. Attach this checklist after the table of contents and include page numbers where each of the required elements of the 510(k) application is located.

4. Statement of indications for use
   Attach the indications for use form (see https://www.fda.gov/medical-devices/premarket-notification-510k/510k-forms). The Statement of Indications for Use should include specific indications, clinical settings, target population, etc.
5. 510(k) summary or statement
   Prepare either a 510(k) Summary or a 510(k) Statement for all 510(k) applications.
   a. 510(k) Summary: a summary of information which support that your product is substantially equivalent to a predicate.
   b. 510(k) Statement: a certification that the 510(k) owner will provide safety and effectiveness information to any person within 30 days of a written request.
6. Truthful and accurate statement
   Include a statement certifying that all information submitted in the 510(k) is truthful and accurate. Attach this statement as a separate page if the CDRH premarket review submission cover sheet is used.
7. Proposed labeling
   Include copies of all final draft labels, for example, package inserts, service manuals, and instructions for use.
8. Specifications
   Include both a narrative description, for example, indications for use, principles of operation, and composition, and a physical or technical description, for example, diagrams, photographs or pictures, and engineering drawings, of the device.
9. Substantial equivalence comparison
   Include a side-by-side comparison table of your device compared to the predicate device, for example, intended use, indications for use, and target population. Include whether the predicate device is a preamendment device (device legally marketed before May 8, 1976) or a device which has already been 510(k) cleared.
10. Performance
    Provide data to support that your device is substantially equivalent to a predicate.
11. Additional requirements
    Additional information may be required, for example, if your device contains software or a color additive, or emits electronic radiation.

510(k) application review timeline

510(k) applications are reviewed by the Office of Product Evaluation and Quality (OPEQ) within the FDA's Center for

Devices and Radiological Health (CDRH) [15]. When CDRH's Document Control Center (DCC) receives the 510(k) application (one electronic copy and the other hard copy), it assigns a "510(k) number," or "K number."

The 510(k) number begins with the letter "K" followed by 6 digits: the first two digits represent the calendar year of submission and the last four represent the submission number for that year starting with 0001 (Fig. 23.1).

Once the DCC confirms that the user fee was paid and a valid eCopy of the 510(k) application was submitted, it will email the submitter an Acknowledgment Letter which includes the date of 510(k) submission receipt and the 510(k) number. The application will then be assigned to a Lead Reviewer in the appropriate department and will undergo the Acceptance Review.

Within 15 calendar days of the receipt, the submitter will receive an email with the contact information of the Lead Reviewer and one of the following three application statuses: (1) accepted for Substantive Review, (2) not accepted (or refused to accept or RTA), or (3) under Substantive Review because the FDA did not complete the Acceptance Review within 15 calendar days. If placed on RTA hold, the applicant has 180 calendar days to amend the application.

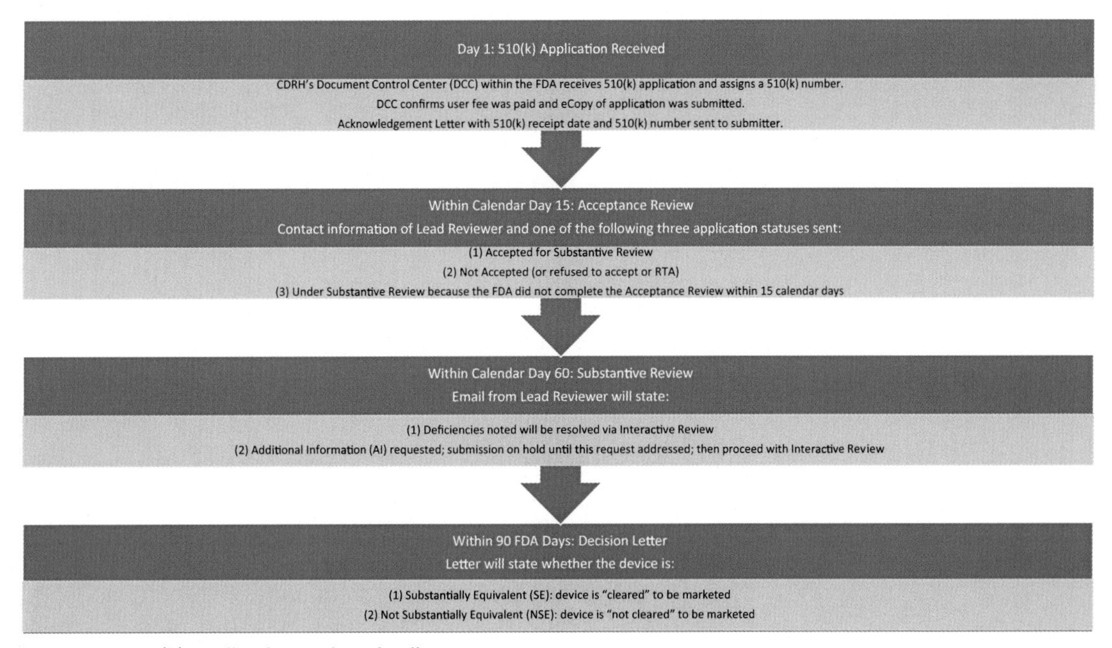

Figure 23.1 510(k) application review timeline.

The 510(k) application then undergoes the Substantive Review. Within 60 calendar days of 510(k) submission, the Lead Reviewer will send the submitter an email stating that the FDA will proceed to either (1) resolve any outstanding deficiencies via the subsequent Interactive Review or (2) place an Additional Information (AI) request, putting the submission on hold, and proceed with an Interactive Review once this request is addressed.

Within 90 FDA days from 510(k) submission, the Decision Letter indicating whether the device is "substantially equivalent (SE)" or "not substantially equivalent (NSE)" is sent. "FDA Days" are the number of calendar days between the date the 510(k) was received and the date of a Medical Device User Fee Amendments (MDUFA) decision, excluding the days the submission was on hold. A 510(k) application that receives an SE decision is "cleared" to proceed with marketing the device. If the FDA does not reach decision within 100 FDA days, the submitter will receive a "Missed MDUFA Communication" indicating reasons for the delay.

There are three types of 510(k) submissions: traditional, special, and abbreviated. Some 510(k) submissions may qualify for the Special and Abbreviated submission programs introduced in 1998 to facilitate the review process. The user fee is the same for all three types of 510(k) submission [16].

Premarket approval

All Class III devices and some Class II devices require premarket approval (PMA), the most stringent FDA review process which involves clinical trials. While there is no PMA form to complete, the following is a list of the required items to be included in a PMA application [17]:

1. Name and address of the applicant.
2. Table of contents.
3. Summary section which includes: indications for use, device description, alternative practices and procedures, marketing history, summary of studies, and conclusions drawn from the studies.
4. Complete description of: the device, the components, the device properties, the device operation principles, and the methods, facilities, and controls used in manufacturing, processing, packing, and storage.
5. Reference to any performance standard or voluntary standard.
6. Technical sections which include nonclinical and clinical trial results.
7. Bibliography.

8. Samples of the device.
9. Copies of all proposed labeling.
10. Environmental assessment.
11. Financial certification or disclosure.

Completing the optional premarket submission coversheet (Form FDA 3514) may help expedite the processing of PMA application. Furthermore, a sample Cover Letter with suggested format and items to be included is available on the FDA website [17].

Investigational device exemption

Before conducting clinical trials using your device required for PMA applications, the device must have already been approved as an Investigational Device Exemption (IDE) [18].

PMA application review timeline

The FDA's goal in the PMA review process is to reach a decision within 180 days from the application filing date [19] (Fig. 23.2).

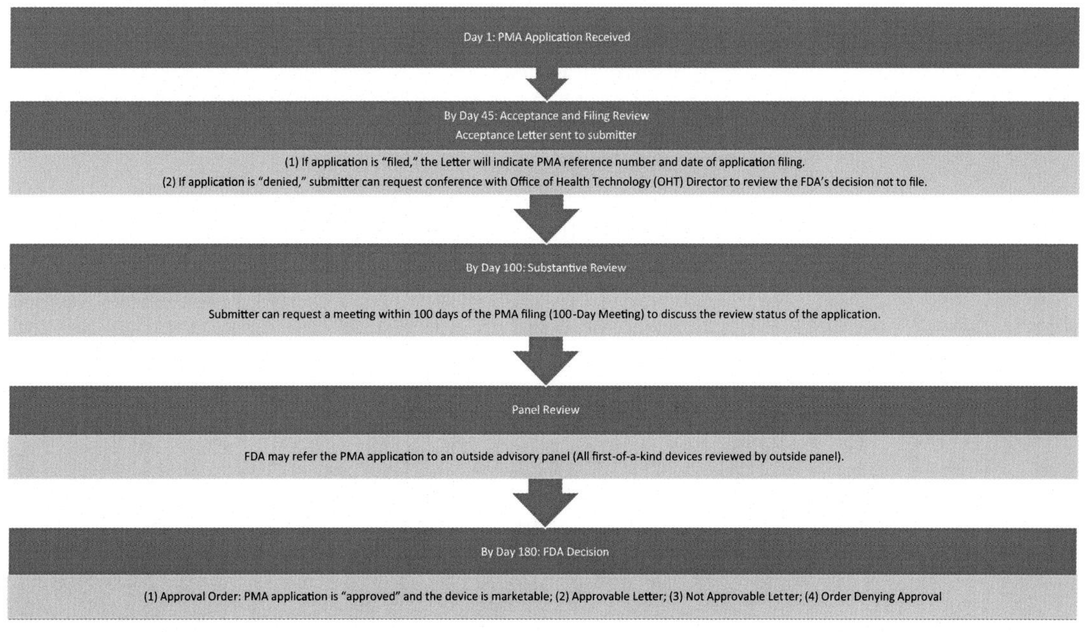

Figure 23.2 PMA application review timeline.

The PMA review process can be divided into the following four stages:

1. Acceptance and filing reviews

   Within 45 days after a PMA application is received, the submitter is notified whether the application has been filed, with the date the application was filed and the PMA reference number. If the application is denied, the submitter will be notified with the reasons. The submitter may submit a request for a conference with the Office of Health Technology (OHT) Director to review the FDA's decision not to file the PMA.

2. Substantive review

   If the application is filed, the FDA may ask for additional information necessary to complete the review. The submitter can request a meeting within 100 days of the PMA filing ("100-Day Meeting") to discuss the status of the application, but the request must be submitted before 70 days from the filing date to allow time for scheduling.

3. Panel review

   The FDA may refer the application to an outside panel of experts called the advisory committee or the advisory panel, especially for first-of-a-kind devices.

4. The decision

   The FDA will issue one of the following decisions within 180 days of PMA filing:

   a. Approval order

      The PMA application is "approved," and the device can be marketed.

   b. Approvable letter

      The PMA application will likely be "approved," pending additional information requested.

   c. Not approvable letter

      The PMA application may not be approved for the reasons included in the letter. The applicant has 180 days from the date of the letter to address the reasons cited.

   d. Order denying approval

      The PMA application does not meet the requirements and is denied.

De Novo classification request

As the name implies, the De Novo pathway is for devices with no predicates but provide reasonable assurance of safety and effectiveness [20]. The De Novo application can be submitted for the following reasons: (1) There is no predicate device, or (2) 510(k) submission resulted in "not substantially equivalent (NSE)" decision. Before submitting a De Novo request, however, consider obtaining feedback via presubmission.

Required contents of De Novo request

1. Cover sheet indicating the request as a "Request for Evaluation of Automatic Class III Designation".
2. Administrative information, for example, the device's intended use.
3. Device description, for example, technology, proposed conditions of use, accessory, and components.
4. Classification information and supporting data

De Novo request review timeline

1. Acceptance review

   Once a De Novo request is received, an Acceptance Review is conducted to assess the completeness of the application (Fig. 23.3).

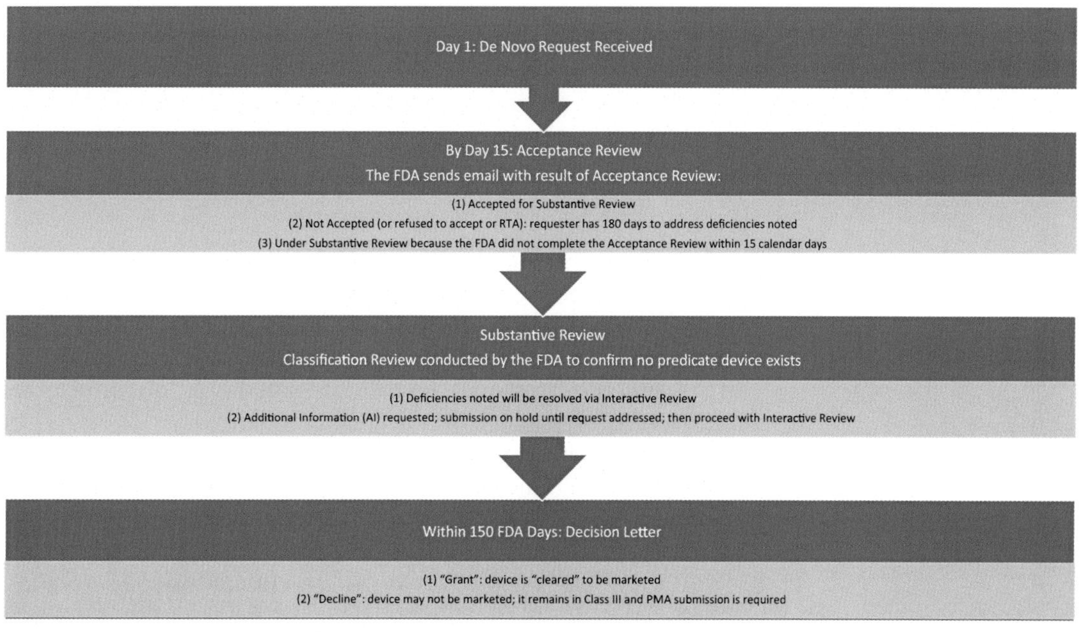

Day 1: De Novo Request Received
By Day 15: Acceptance Review
The FDA sends email with result of Acceptance Review:
(1) Accepted for Substantive Review
(2) Not Accepted (or refused to accept or RTA): requester has 180 days to address deficiencies noted
(3) Under Substantive Review because the FDA did not complete the Acceptance Review within 15 calendar days
Substantive Review
Classification Review conducted by the FDA to confirm no predicate device exists
(1) Deficiencies noted will be resolved via Interactive Review
(2) Additional Information (AI) requested; submission on hold until request addressed; then proceed with Interactive Review
Within 150 FDA Days: Decision Letter
(1) "Grant": device is "cleared" to be marketed
(2) "Decline": device may not be marketed; it remains in Class III and PMA submission is required

Figure 23.3 De Novo request review timeline.

Within 15 calendar days of the Document Control Center (DCC) receiving the De Novo request, the FDA will send the requester an email of the acceptance review result:

a. accepted for Substantive Review

b. not accepted for Substantive Review, that is, considered refused to accept or RTA; the requester has 180 calendar days to address issues cited.

c. under Substantive Review because the FDA did not complete the Acceptance Review within 15 calendar days.

2. Substantive review

During the Substantive Review, the FDA conducts a classification review of predicate device types and analyzes whether a predicate device of the same type exists. After the classification review is complete, the Lead Reviewer may identify deficiencies that can be addressed through the Interactive Review, or an Additional Information letter will be sent to the requester who has 180 calendar days to amend the request.

3. De Novo request decision

The FDA's goal is to reach a decision about a De Novo request within 150 review days. Review days are the number of calendar days between the date the De Novo request was received and the date of the FDA's decision, excluding the days the request was on hold.

The FDA may either grant or decline a De Novo request:

a. Grant: the device is authorized to be marketed.

b. Decline: the device remains in Class III and a PMA application is needed.

Humanitarian device exemption

This PMA-exempt pathway is for a device intended to treat disease or condition that affects not more than 8000 individuals in the United States per year. The 21st Century Cures Act enacted in 2016 increased the patient population size limit from the previous "fewer than 4000" to "not more than 8000" [4,21].

Getting such a device to market requires the applicant to first acquire Humanitarian Use Device (HUD) designation for the device and then submit a Humanitarian Device Exemption (HDE) application. There is no user fee for an HDE application. When the CDRH receives an HDE application, it assigns a unique HDE number which begins with "H" followed by six digits. Similar to the 510(k)-numbering system, the first two digits represent the calendar year of the application receipt, and the last four digits represent the submission number for that year starting with 0001 [22].

The FDA will send an Acknowledgment Letter with an HDE number within 7 days of the application receipt. Unlike PMA-requiring devices, HDE devices are only required to demonstrate probable benefit to patients but has restrictions on its use and profit [22].

Required contents of HDE application

1. A complete description of the HUD.
2. A copy of the HUD designation letter.
3. An explanation of why the device would not be available unless an HDE were granted.
4. A statement that no predicate device is available for the intended use.
5. An explanation of why the probable benefit to health from the use of the device outweighs the risk of harm.
6. Technical information including nonclinical and clinical trial data.
7. Quality Systems information.

HDE application review timeline

The FDA targets 75 FDA days to reach a decision on an HDE application and will notify the submitter within 30 days of receipt of the application as to whether the HDE was filed based on a filing review [23] (Fig. 23.4).

Once the application is filed, the FDA conducts the Substantive Review and will issue one of the following decisions, similar to those for PMA submissions: Approval Order, Approvable Letter, Major Deficiency Letter, or Not Approvable Letter.

Medical device user fees (FY 2021)

The following is an abridged version of the medical device user fee schedule for fiscal year 2021 (October 1, 2020 through September 30, 2021) [24]:

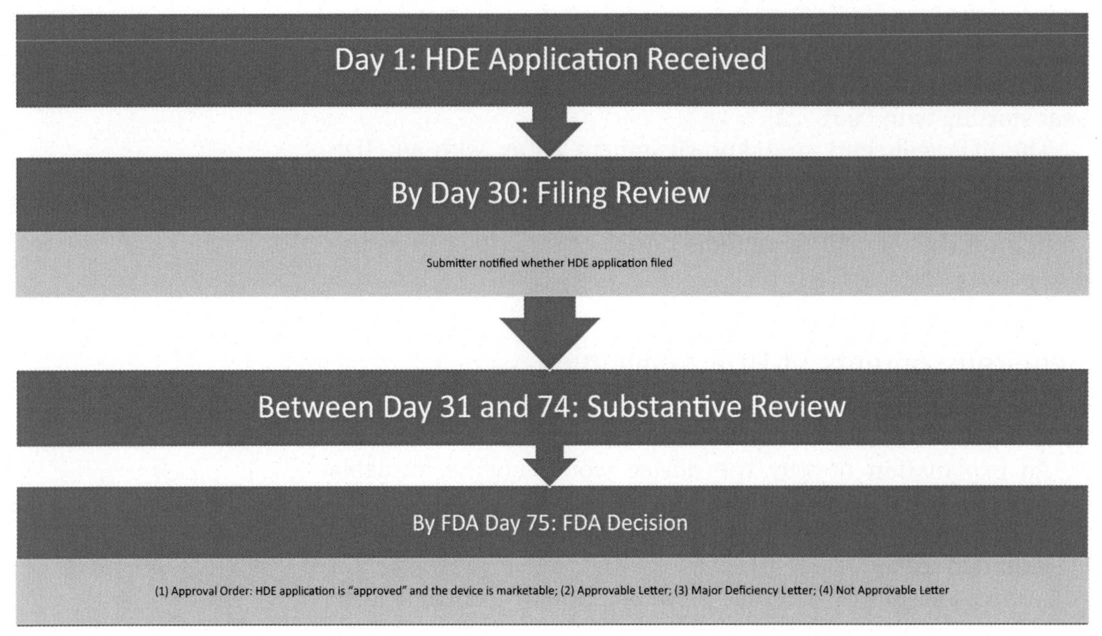

Figure 23.4 HDE application review timeline.

Annual establishment registration fee: $5,546*
Other fees:

Application type	Standard fee	Small business reduced fee†
513(g)	$4936	$2468
510(k)‡	$12,432	$3108
PMA	$365,657	$91,414
De Novo classification request	$109,697	$27,424

*There are no reduced fees for small businesses.
†Small Business Fee: For small businesses with an approved Small Business Determination (SBD).
‡510(k) Fees: Fees for all three types of 510(k)s (Traditional, Abbreviated, and Special) are the same.

Small business determination program

The small business determination (SBD) Program by the Center for Devices and Radiological Health (CDRH) within the FDA determines whether a business qualifies as a "small business" and is eligible for a reduced fee for some of CDRH submissions [25]. A qualifying small business may not have gross receipts and sales of more than $100 million for the most recent tax year. Additionally, a small business is eligible for a "first premarket application/report" fee waiver, if its gross receipts or sales are no more than $30 million. The FDA will complete its review of the small business certification request within 60 calendar days of receipt. There is no fee for this small business certification request submission.

Get started

Readers of this chapter should now have a better and clearer understanding of the FDA's clearance and approval processes for medical devices. While navigating FDA's complex premarketing pathways may seem daunting at first, appropriate knowledge and planning is critical in successfully and efficiently obtaining premarketing clearance and approval.

References

[1] Van Norman GA. Drugs, devices, and the FDA: part 2: an overview of approval processes: FDA approval of medical devices. JACC Basic Transl Sci 2016;1(4):277–87 Jun 27.

[2] Kaplan AV, Baim DS, Smith JJ, Feigal DA, Simons M, Jefferys D, et al. Medical device development: from prototype to regulatory approval. Circulation. 2004;109(25):3068–72 Jun 29.

[3] Makower, J. FDA impact on U.S. medical technology innovation: a survey of over 200 medical technology companies. https://cdn.ymaws.com/http://www.medicaldevices.org/resource/resmgr/docs/FDA_impact_on_US_med_tech_in.pdf, 2010 (accessed 15.05.21).

[4] U.S. Food and Drug Administration. A history of medical device regulation & oversight in the United States. https://www.fda.gov/medical-devices/overview-device-regulation/history-medical-device-regulation-oversight-united-states, 2019. (accessed 15.05.21).

[5] U.S. Food and Drug Administration. How to determine if your product is a medical device. https://www.fda.gov/medical-devices/classify-your-medical-device/how-determine-if-your-product-medical-device, 2019. (accessed 15.05.21).

[6] U.S. Food and Drug Administration. Class I / II exemptions. https://www.fda.gov/medical-devices/classify-your-medical-device/class-i-ii-exemptions, 2019. (accessed 15.05.21).

[7] U.S. Food and Drug Administration. Classify your medical device. https://www.fda.gov/medical-devices/overview-device-regulation/classify-your-medical-device, 2020. (accessed 15.05.21).

[8] Van Batavia J, Goldenberg SJ. Strategic planning and costs of FDA regulation. Academic Entrepreneurship Med Health Scientists 2019;1(2):8.

[9] U.S. Food and Drug Administration. How to register and list. https://www.fda.gov/medical-devices/device-registration-and-listing/how-register-and-list, 2018. (accessed 15.05.21).

[10] U.S. Food and Drug Administration. Device registration and listing. https://www.fda.gov/medical-devices/how-study-and-market-your-device/device-registration-and-listing, 2021. (accessed 15.05.21).

[11] U.S. Food and Drug Administration. When to register and list. https://www.fda.gov/medical-devices/device-registration-and-listing/when-register-and-list, 2017. (accessed 15.05.21).
[12] U.S. Food and Drug Administration. Premarket notification 510(k). https://www.fda.gov/medical-devices/premarket-submissions/premarket-notification-510k, 2020. (accessed 15.05.21).
[13] U.S. Food and Drug Administration. Guidance documents (medical devices and radiation-emitting products). https://www.fda.gov/medical-devices/device-advice-comprehensive-regulatory-assistance/guidance-documents-medical-devices-and-radiation-emitting-products, 2021. (accessed 15.05.21).
[14] U.S. Food and Drug Administration. Content of a 510(k). https://www.fda.gov/medical-devices/premarket-notification-510k/content-510k, 2019. (accessed 15.05.21).
[15] U.S. Food and Drug Administration. 510(k) Submission process. https://www.fda.gov/medical-devices/premarket-notification-510k/510k-submission-process, 2020. (accessed 15.05.21).
[16] U.S. Food and Drug Administration. 510(k) Submission programs. https://www.fda.gov/medical-devices/premarket-notification-510k/510k-submission-programs, 2019. (accessed 15.05.21).
[17] U.S. Food and Drug Administration. PMA application contents. https://www.fda.gov/medical-devices/premarket-approval-pma/pma-application-contents, 2020. (accessed 15.05.21).
[18] U.S. Food and Drug Administration. Investigational device exemption (IDE). https://www.fda.gov/medical-devices/how-study-and-market-your-device/investigational-device-exemption-ide, 2019. (accessed 15.05.21).
[19] U.S. Food and Drug Administration. PMA review process. https://www.fda.gov/medical-devices/premarket-approval-pma/pma-review-process, 2020. (accessed 15.05.21).
[20] U.S. Food and Drug Administration. De Novo classification request. https://www.fda.gov/medical-devices/premarket-submissions/de-novo-classification-request, 2019. (accessed 15.05.21).
[21] U.S. Food and Drug Administration. PMA application methods. https://www.fda.gov/medical-devices/premarket-approval-pma/pma-application-methods, 2018. (accessed 15.05.21).
[22] U.S. Food and Drug Administration. Humanitarian device exemption. https://www.fda.gov/medical-devices/premarket-submissions/humanitarian-device-exemption, 2019. (accessed 15.05.21).
[23] U.S. Food and Drug Administration. Getting a humanitarian use device to market. https://www.fda.gov/medical-devices/humanitarian-device-exemption/getting-humanitarian-use-device-market, 2019. (accessed 15.05.21).
[24] U.S. Food and Drug Administration. Medical device user fee amendments (MDUFA). https://www.fda.gov/industry/fda-user-fee-programs/medical-device-user-fee-amendments-mdufa, 2021. (accessed 15.05.21).
[25] U.S. Food and Drug Administration. Reduced medical device user fees: small business determination (SBD) program. https://www.fda.gov/medical-devices/premarket-submissions/reduced-medical-device-user-fees-small-business-determination-sbd-program, 2019. (accessed 15.05.21).
[26] U.S. Food and Drug Administration. Is it really 'FDA approved?' https://www.fda.gov/consumers/consumer-updates/it-really-fda-approved, 2017. (accessed 15.05.21).
[27] U.S. Food and Drug Administration. Premarket approval (PMA). https://www.fda.gov/medical-devices/premarket-submissions/premarket-approval-pma, 2019. (accessed 15.05.21).

24

Anesthesia reimbursement

Michael P. Harlander-Locke, Zhuo Sun, Xiaodong Hua and Anterpreet Dua

Department of Anesthesiology and Perioperative Medicine, Medical College of Georgia, Augusta University, Augusta, GA, United States

Chapter outline

Abstract

The purpose of this chapter is to provide an overview of basic concepts surrounding billing for anesthesia services. While there are numerous sources available on the topic of reimbursement, most include significant detail, which can be overwhelming. The authors aim to organize and present the material in a straightforward format even for the reader with no prior exposure to the subject. To help facilitate this the authors have minimized detailed

Innovation in Anesthesiology. DOI: https://doi.org/10.1016/B978-0-12-818381-6.00022-X

technical language and organized the chapter so that information builds in a stepwise manner from section to section. Key concepts have been extracted from the literature and supplementary sources and rearranged into focused and easily digestible sections.

Keywords: Anesthesia; billing; reimbursement; modifiers; maximum allowable fee; conversion factors

Introduction: why does billing matter?

Billing is a crucial part of any successful healthcare practice to ensure that anesthesia providers are properly compensated for their time and expertise. Standards for billing have largely aligned with requirements set forth by the Center for Medicare & Medicaid Services (CMS) to claim payment [1,2]. When distilled, the essential aspects required to bill are the 5 Ws:

Who *is performing the service?*	Is it the anesthesiologist, CRNA, or AA
What *services is the patient receiving?*	Open femur fracture repair, laparoscopic lower abdominal hernia repair, etc.
When *(how long) is the service being provided?*	Is it 22 min, 30 min, . . .
Where *did the services take place?*	Geographic location: Zip code
Why *were the interventions performed?*	Diagnosis codes (ICD-10)

In addition to these factors, other details such as the patient history and medical decision-making that led to the intervention, patient comorbidities, the length of care, and the personnel involved all play a role in determining the level of reimbursement. Insurance companies allocate significant resources to identify submitted claims deficiencies forming the basis for claim denial—an avoidable situation with proper document completion.

Getting started: maximum allowable fee

The following calculation is used by CMS and private insurance carriers to determine the maximum allowable fee (MAF) that can be billed for anesthesia services [1]:

$$\text{Maximum allowable fee} = (\text{time units} + \text{base units}) \times \text{conversion factor}$$

The MAF refers to the maximum amount of money that an anesthesia provider can bill CMS or insurance carriers for their services. Each component of this equation will be discussed throughout the remainder of the chapter.

Time units are a measure of time spent providing services to the patient. Time units are equal to total anesthesia time (in minutes) divided by 15, and rounded to the nearest whole number. [2] Therefore, 1 time unit = 15 minutes. Example: Time spent with the patient during Colonoscopy was 40 minutes, (40/15) = 2.66, time units for this procedure = 3 when rounded to the nearest whole number. It is important to note that if the total time spent with the patient is less than 15 minutes, the time unit is automatically rounded up to 1. Service start and stop times are measured to the nearest minute and no rounding for these times should be done (e.g., 10:12 am start time should be reported as such, and not be rounded to 10:10 a.m. or 10:15 a.m.). Taking all this information into consideration, it can be seen that when required to round to the nearest whole number, it can impact billable time units in a significant way (Table 24.1).

Base units relate to procedure complexity, and their values are preassigned and published by CMS [1]. These procedure base units are independent of procedure time and remain the same whether the case is completed in 50 minutes or 6 hours. This follows to reason, that services provided for emergent intracranial tumor resection are not equal to services provided for an inguinal hernia repair. The trends show that more invasive and complex procedures correspond to higher numbers of assigned base units. Examples:

Anesthesia for Intracranial procedure in sitting position: CPT 00218; Base units = 13

Anesthesia for lower abdomen hernia repair: CPT 00830; Base units = 4

Table 24.1 Anesthesia time in minutes and corresponding time units.

Anesthesia time (min)	Time units (rounded to the nearest whole number)
1[a]	1
22	22/15 = 1.46 → 1
23	23/15 = 1.53 → 2
37	37/15 = 2.46 → 2
38	38/15 = 2.53 → 3
52	52/15 = 3.46 → 3

[a]Service time <15 min automatically rounded up to 1.

A full list of CPT codes and their corresponding base units can be accessed at the US Dept. of Labor, Office of Workers Compensation Programs (OWCP) website [1].

Conversion factor (CF) relates to geographic location (zip code) where services are provided and is the rate that can be billed per unit (time or base unit). This conversion factor is the same for all qualified anesthesia providers in that zip code.

CF for services in Miami, FL as of 7/1/2022: 59.65

CF for services in Piedmont, SD as of 7/1/2022: 51.70

A full list of zip codes and their corresponding CF can be accessed at the US Dept. of Labor, Office of Workers Compensation Programs (OWCP) website—OWCP Medical Fee Schedule.[1]

***Putting The Pieces Together*...**

An anesthesiologist located in Miami, Florida (zip code 33090), and another in Piedmont South Dakota (zip code 57769), each provide anesthesia services for a 78 years-old patient who undergoes open repair of a femur neck fracture (CPT 01214) after suffering a ground level fall. The time spent with the patient was 3 hours and 20 minutes. What is the MAF amount that the anesthesiologist may bill for in each of these locations?

Maximum allowable fee = (time units + base units) × conversion factor

	Miami, FL (zip code 33090)	**Piedmont, SD (zip code 57769)**
Time units	(200 min/15 min) = 13.33 Rounded: 13	(200 min/15 min) = 13.33 Rounded: 13
Base units (CPT 01230)	6	6
Conversion factor	59.65	51.70
Maximum allowable fee	(13 + 6) × 59.65 $1133.35	(13 + 6) × 51.70 $982.30

The examples above help to illustrate the impact of the conversion factor (specific to the geographic location of services rendered) and its effect on MAF. An anesthesiologist who provides services for the same procedure of equal duration will earn $151.05 less for services in Piedmont, SD compared to Miami, FL. National trends show that differences in conversion factors typically correspond to differences in cost of living (South Dakota vs Florida) and costs associated with providing healthcare delivery (Alaska vs Oregon).

Details matter: modifiers

Prior content and examples discussed thus far have considered only a single scenario: An anesthesiologist provides services to one patient at a single time who is healthy otherwise. While this helps keep things simple, everyday practice typically involves more complex patients and scenarios that require the addition of specific coding and the use of modifiers.

Modifiers

In short, modifiers are additional information provided to payers as the basis for requesting higher payment. Example: Services provided for emergent ureteral stent placement in a profoundly septic patient are not the same as those for a patient undergoing routine outpatient cystoscopy and ureteral stent placement. When calculating the MAF (discussed in prior sections), there are situations in which additional base units are added to the procedural base units depending on the patient's physical status (corresponds with ASA Physical Status Classification, discussed next), and special qualifying circumstances. It is important to note that Medicare does not recognize physical status modifiers or qualifying circumstance modifiers, although many private insurance companies do cover these risks and conditions.

With the addition of modifiers, the MAF calculation is expanded:

$$\text{Maximum allowable fee} = (\text{time units} + \text{base units} + \text{modifier units}) \times \text{conversion factor}$$

Physical status modifier

The American Society of Anesthesiologists physical status classification system, utilized to characterize every patient in daily practice, serves to objectively separate patients based on their comorbidities and current condition (Table 24.2). This classification system is necessary as the basis for comparing similar patients and comparing patient groups. To compensate for the acuity differences between patients (e.g., ASA 1 vs ASA 4), additional base units can be added to the MAF calculation. Based on physical status modifiers, providing services to ASA 3–5 patients can increase the total base units by 1–3 units (Table 24.3).

Table 24.2 The American Society of Anesthesiology (ASA) physical status classification of the patient's present physical condition on a scale from 1–6 as it appears on the anesthesia record.

Classification	Definition
ASA 1	Normal healthy patient
ASA 2	Patient with mild systemic disease
ASA 3	Patient with severe systemic disease
ASA 4	Patient with severe systemic disease that is a constant threat to life
ASA 5	Moribund patient who is not expected to survive without the operation
ASA 6	Declared brain-dead patient whose organs are being removed for donor purposes

Table 24.3 Relationship between the American Society of Anesthesiologists physical status classification system and the physical status nomenclature utilized by CMS.

ASA physical status classification	CMS physical status classification	Additional base units
ASA 1	P1	0 Unit
ASA 2	P2	0 Unit
ASA 3	P3	1 Unit
ASA 4	P4	2 Units
ASA 5	P5	3 Units
ASA 6	P6	0 Unit

ASA, American Society of Anesthesiologists.

Qualifying circumstances

Special circumstances warrant special consideration—this modifying factor aims to add additional base units when services are provided to particularly vulnerable populations and in special situations. Although these modifiers correspond to a reported CPT code, their value in units is added to the "base units" portion of the maximal allowable fee equation and not separately (Table 24.4).

To review the use of modifiers let's go back to the original example calculating MAF for services administered in Miami, Florida.

An anesthesiologist located in Miami, Florida (zip code 33090) provides anesthesia services for a 78-year-old patient

Table 24.4 Qualifying circumstances modifiers and their corresponding base unit value.

CPT	Qualifying circumstances	Additional base units
99100	Anesthesia for patients of extreme age (age < 1 years or age > 70 years)	1 unit
99116	Anesthesia complicated by utilization of total body hypothermia	5 units
99135	Anesthesia complicated by utilization of controlled hypotension	5 units
99140	Anesthesia complicated by emergency conditions (specify)	2 Units

List as an additional line item in addition to code for primary anesthesia procedure.

with multiple comorbidities including uncontrolled hypertension, who undergoes open repair of a femur neck fracture (CPT 01214) after suffering a ground-level fall. **The patient's physical status is determined to be ASA 3 and the procedure is deemed emergent**. The time spent with the patient was 3 hours and 20 minutes. What is the MAF amount that the anesthesiologist may bill for?

	Miami, FL (zip code 33090)
Time units	(200 min/15 min) = 13.33 Rounded: 13 units
Base units: CPT 01214	6 units
Conversion factor:	59.65
Physical status modifier: ASA 3 = P3	1 units
Qualifying circumstance modifier: CPT 99140	2 units
Maximum allowable dee:	(13 + >6 + 1 + 2) × 59.65 $1312.30

This detailed example illustrates the benefits of including modifiers, when applicable, to increase compensation for services rendered. The anesthesiologists in the above example increased their MAF by $178.95 with the addition of modifiers, a 15.7% increase.

Completing the puzzle: provider coding

When anesthesia services are provided for a patient, it must be documented who is providing the service and under what circumstances. This point becomes important when submitting payment claims, as a single Anesthesiologist can cover multiple ORs at once with the help of qualified providers (Certified Registered

Table 24.5 Anesthesia-specific codes from The Healthcare Common Procedure Coding System (HCPCS).

Service provider and degree of involvement	Percent reimbursed	Modifier code
Personally performed by an anesthesiologist	100%	AA
Direction of one qualified provider by an anesthesiologist	50%	QY
Direction (2–4 concurrent procedures) of qualified providers by an anesthesiologist	50%	QK
Supervision (> 4 concurrent procedures) of qualified providers by an anesthesiologist	(3 or 4 units)[a]	AD

Qualified providers include CRNA and CAAs.
[a]A fourth unit can be billed if present for induction.

Nurse Anesthetists, CRNA, or Certified Anesthesiologist Assistants, CAA). For Anesthesiologists, there are four recognized possible scenarios, each associated with its code, and for qualified providers, there are only two codes. The different scenarios and associated rates (percentage of total MAF) that the Anesthesiologist can bill for those services can be seen in Table 24.5. These modifiers are attached as a suffix to the end of the surgical procedure CPT codes.

Example: CPT 01230 which corresponds to open repair of proximal 2/3 femur. An anesthesiologist providing the services themselves with no other concurrent services would report the procedure with: CPT 01230—AA.

The terms medical direction and medical supervision are terms used to describe the involvement of the Anesthesiologist in the services provided. An anesthesiologist is said to be medically directing when working with 1–4 qualified providers. While medically directing, the Anesthesiologist must complete seven tasks:

1. Complete pre-op exam and evaluation,
2. Develop an anesthetic plan,
3. Personally participate in the most demanding aspects of the procedure including induction and emergence when applicable,
4. Ensure that procedures included in the plan are performed,
5. Monitor anesthetic administration at frequent intervals,
6. Remain physically present for diagnosis and management of any problems that may arise during anesthetic administration, and
7. provide appropriate post-op care

When working with >4 qualified providers the term changes from medical direction to a less involved state referred to as medical supervision. This distinction is important because when overseeing >4 qualified providers, the anesthesiologist is not always involved in all aspects of the case including induction or being in the immediate vicinity to provide expertise and management if needed urgently.

At the same time as an anesthesiologist bills for their service (Table 24.6), the qualified provider (CRNA or AA) also bills for their services using distinct modifiers. As qualified providers, they can either bill for 50% of the MAF or bill at 100% if they are personally performing all services and not working with an anesthesiologist (based on state-specific credentialing laws).

Modifier: AA—services personally performed by an Anesthesiologist

This code is used when an Anesthesiologist personally performs all the anesthesia services rendered throughout the entire duration of the procedure. Two exceptions to this scenario are when an anesthesiologist is working with resident(s) or involved in only a single case and with a student CRNA (see the section on special considerations). For anesthesiologists not working in an academic setting, these exception scenarios will rarely, if ever, arise. Usage of the AA modifier allows the anesthesiologist to bill for 100% of the MAF.

Modifier: QY—direction of one qualified provider by an Anesthesiologist

When an Anesthesiologist oversees one qualified provider, they continue to be responsible for ensuring that all seven tasks (listed

Table 24.6 Anesthesia-specific codes for qualified providers from The Healthcare Common Procedure Coding System (HCPCS).

Service provider and degree of involvement	Percent reimbursed	Modifier
Performed by a qualified provider under the medical direction of an anesthesiologist	50%	QX
Personally performed by a CRNA	100%	QZ

Note that CAAs are not permitted to provide anesthesia services while not under the medical direction or supervision of an anesthesiologist.

above) from pre-op eval to post-op care are completed and documented. Under this modifier, the Anesthesiologist can bill for 50% of the MAF, while the qualified provider can bill for the remaining 50% of the MAF (using modifier -QX, seen in Table 24.4).

Modifier: QK—direction of qualified providers (2–4 concurrent procedures) by an Anesthesiologist

The Anesthesiologist is required to ensure that all seven tasks from pre-op eval (listed above) to post-op care are documented. They must remain in the immediate area of the anesthesia service locations and be able to respond to changes and challenges as they arise. There are exceptions to these constraints which include:

1. Attending to an emergency of short duration in the immediate area,
2. Administering caudal or epidural anesthesia for a laboring patient to ease labor pain,
3. Periodic monitoring of laboring patients,
4. Receiving patient in pre-op area for next case,
5. Attending to, or discharging patient(s) from the post-op recovery area.

Modifier: AD—supervision of qualified providers (>4 concurrent procedures) by an Anesthesiologist

When an Anesthesiologist concurrently overseers >4 procedures, they are said to be medically supervising. This may be unintentional—attending to an emergency of a short duration can turn into a situation that requires the attention of the anesthesiologist

Table 24.7 MAF, Maximum allowable fee; qualified provider: CRNA or CAA.

Service provider and degree of involvement	Anesthesiologist MAF	Qualified provider MAF
Personally performed by an anesthesiologist	100%	0%
Direction of one qualified provider by an anesthesiologist	50%	50%
Direction (2–4 concurrent procedures) of qualified providers by an anesthesiologist	50%	50%
Supervision (>4 concurrent procedures) of qualified providers by an anesthesiologist	3 or 4 base units[a]	50%

[a]4 units If present for induction.

longer than intended. Under this modifier, the Anesthesiologist may bill for 3 base units per supervised location with an additional unit billed if present for induction of anesthesia.

When included, these modifiers impact the MAF for both anesthesiologists and qualified providers (Table 24.7):

Modifier QX: qualified provider operating under Anesthesiologist direction or supervision

When medically directed or supervised the qualified provider can bill for 50% of the maximum allowed fee.

Modifier QZ: personally performed by CRNA

This modifier is used by CRNAs to bill for anesthesia services not directed or supervised by an Anesthesiologist. The ability of CRNAs to perform this duty differs by location in accordance with state-specific laws.

When revisiting the last example of the anesthesiologist practicing in Miami, FL, it can be further expanded to real-world examples:

An anesthesiologist located in Miami, Florida (zip code 33090) provides anesthesia services at a local community hospital. He medically directs two qualified providers.

The first case in OR 1 is with a 78-year-old patient with multiple comorbidities including controlled hypertension, who undergoes open repair of a femur neck fracture (CPT 01214) after suffering a ground-level fall. The patient's physical status is determined to be ASA 2 and the procedure is deemed emergent. The time spent with the patient was 3 hours and 2 minutes.

The first case in OR 2 is with a 43-year-old patient with mild asthma who undergoes a total should replacement (CPT 01638). This is an electively scheduled case. The patient's physical status is determined to be ASA 1. The time spent with the patient was 1 hour and 40 minutes.

What is the MAF that the anesthesiologist can bill for by medically directing these two qualified providers?

Case #1.

Miami, FL (Zip Code 33090)	Anesthesiologist	Qualified provider
Time units	(185 min/15 min) = 12.13	

(Continued)

(Continued)

Miami, FL (Zip Code 33090)	Anesthesiologist	Qualified provider
	Rounded: 12 units	
Base units: CPT 01214	6 units	
Conversion factor	59.65	
Physical status modifier: ASA 2 = P2	0 units	
Qualifying circumstance modifier: CPT 99140	2 units	
Maximum allowable fee: total	(12 + 6 + 0 + 2) × 59.65 $1193.00	
Provider modifier	AA (50%)	QX (50%)
Maximum allowable fee:	$596.50	$596.50

Case #2.

Miami, FL (zip code 33090)	Anesthesiologist	Qualified provider
Time units	(100 min/15 min) = 6.67 Rounded: 7 units	
Base units: CPT 01638	10 units	
Conversion factor	59.65	
Physical status modifier: ASA 2 = P2	0 units	
Qualifying circumstance modifier: none	0 units	
Maximum allowable fee: total	(10 + 7 + 0 + 0) × 59.65 $1014.05	
Provider modifier	AA (50%)	QX (50%)
Maximum allowable fee	$507.03	$507.03

For this scenario, it would be helpful to understand (even with several brief statements) the reasons WHY the anesthesiologist would select to perform multiple cases directing qualified providers if the revenue would be less.

Based on the above example, the Anesthesiologist can bill a total of $1103.53 ($596.50 + 507.03) while medically directing the two qualified providers. This amount is less than what the Anesthesiologist would have collected had they done Case #1 themselves (-AA modifier, 100% maximum reimbursement fee). Although certain cases can be more financially rewarding, the MAF can vary significantly based on all the factors discussed here thus far—working with qualified providers can help reduce variance through billing diversification. This can be illustrated by considering what the anesthesiologists' total MAF would be with the addition of medically directing a third qualified provider concurrently:

An anesthesiologist medically directs a qualified provider with anesthesia for a 44-year-old patient who is undergoing an elective scheduled left adrenalectomy (CPT 00866) for an enlarging benign adenoma. His comorbidities include HTN, GERD, and moderate-severe COPD. The procedure takes 2 hours and 37 minutes.

Case #3

Miami, FL (zip code 33090)	Anesthesiologist	Qualified provider
Time units	(157 min/15 min) = 10.46 Rounded: 10 units	
Base units: CPT 00866	10 units	
Conversion factor	59.65	
Physical status modifier: ASA 3 = P3	1 unit	
Qualifying circumstance modifier: None	0 units	
Maximum allowable fee: total	(10 + 10 + 1 + 0) × 59.65 $1252.65	
Provider modifier	AA (50%)	QX (50%)
Maximum allowable fee	$626.33	$626.33

When the anesthesiologist medically directs the third room concurrent with Case 1 and Case 2, the anesthesiologists' total MAF during this period is $1729.86 ($596.50 + 507.03 + 626.33). The remaining 50% of the MAF is submitted by the qualified provider (-QX).

Special situations

Up until this point we have considered routine situations and examples; however, there are instances where things are not always clean and straightforward.

Complex cases/high acuity

There are times when the services of the anesthesiologist (while not medically directing or supervising any other concurrent rooms) are not enough to manage a procedure or situation and when the full services of a qualified provider are also needed. In this case, the Anesthesiologist can bill at 100% of the MAF (modifier—AA) and the qualified provider can also bill at

100% of the MAF (modifier—QZ). This necessitates additional documentation submitted by both the anesthesiologist and qualified provider to support this necessity.

Supervising residents (teaching)

An anesthesiologist may medically direct up to two resident physicians concurrently, or one resident and another qualified provider, and use the modifier -AA GC. This allows the anesthesiologist to submit claims for 100% of the MAF for both cases. The addition of the GC to the -AA modifier implies the Anesthesiologist meets the criteria for providing teaching services and that they (or another member of the teaching group) are present during all critical portions of the involved procedure.

Case cancellation

If the case is canceled before induction, the anesthesia provider may bill for the preoperative as an evaluation and management service. If the case is canceled for any reason after induction of anesthesia, the services can be billed as other procedures would be, including the full number of base units (and associated procedure time). Documentation is required in the anesthesia record stating when the case was canceled and the reason. When completing the billing, the addition of modifier -53 should be added to the CPT code. Example: After induction of anesthesia for a vitrectomy, the case is canceled by the surgeon. The anesthesiologist would list the CPT code on their payment claim as CPT 00145–53.

Special events during medical direction

While providing medical direction to up to four qualified providers concurrently, the anesthesiologist cannot ordinarily provide services to any other patients except in the following extenuating circumstances:

1. Emergencies of short duration and in the immediate area,
2. Administering an epidural or caudal anesthetic for labor pain,
3. Intermittently monitoring a laboring patient. If the Anesthesiologist is unable to return on time to re-establish direct patient contact in the setting while overseeing qualified providers then it is no longer appropriate to submit

claims with the medical direction modifier, and services would instead be submitted with the medical supervision modifier (–AD).

Several technological innovations can potentially increase revenue and patient safety for anesthesia practice. Implementing an anesthesia information management system can streamline documentation, billing, and coding processes. Embracing telemedicine and remote monitoring technology can expand the anesthesia practice's reach, thus generating additional income. Increased patient throughput and revenue generation can be achieved through investment in enhanced patient monitoring technology. Value-based care models and bundled payments can incentivize quality outcomes and cost savings, resulting in increased revenue for the anesthesia practice. Diversifying the range of services can help tap into additional revenue streams. Shared staffing models and increased patient volume can generate higher revenue by maximizing resource utilization. Marketing through social media, online advertising, and search engine optimization can attract more patients and referrals. Artificial intelligence (AI) can be harnessed to improve patient scheduling, enhance billing and coding accuracy, provide predictive analysis for cost reduction, and accelerate the research and development process. AI-based Intelligent Decision support systems can assist anesthesiologists in making informed decisions during surgeries. Focusing on patient satisfaction, providing excellent care, and managing the practice's reputation can result in positive referrals and increased patient volume, ultimately leading to higher revenue for the anesthesia practice. With innovative techniques and technology expertise, anesthesiologists may be better positioned to negotiate favorable contracts with hospitals, surgical centers, or healthcare networks.

In summary, the value of innovative technology in anesthesia care lies in improved patient safety, efficiency, professional development, collaboration, reduced liability risks, job satisfaction, and potential cost savings. These advancements can ultimately benefit both the anesthesiologists and the patients they serve.

References

[1] U.S. Department of Labor, Office of Workers Compensation Programs (OWCP). July 1st, 2022. Anesthesia Service and Reimbursement Policy. Retrieved from: https://www.dol.gov/agencies/owcp/regs/feeschedule/fee.

[2] Centers for Medicare & Medicaid Services. Medicare Claims Processing Manual: Chapter 12 – Physician/Nonphysician practitioners. March 4, 2022. Retrieved from: https://www.cms.gov/Regulations-and-Guidance/Guidance/Manuals/Downloads/clm104c12.pdf.

25

Business models

Jungmin On, Anterpreet Dua and Zhuo Sun
Department of Anesthesiology and Perioperative Medicine, Medical College of Georgia, Augusta University, Augusta, GA, United States

Chapter outline

Abstract

The business of anesthesia is a continually evolving entity, as the role of the anesthesiologist has consistently expanded in both inpatient and outpatient management over time. In an effort to maintain their utility in the dynamic field that is medicine, anesthesiologists have developed new skills and adjusted their practice models to cater recent changes. In this chapter, we outline the different variety of anesthesia groups in today's practice, business entities, and types of contracts. We also explore the recent trend of anesthesia practice consolidation and its implications on the independent and academic anesthesia practice.

Keywords: Anesthesia; business; contracts; models; company; academic

Anesthesia practices continue to evolve and adapt to the rapidly changing economic environment, roles of anesthesia service, and the growing interdependence of our healthcare system. Some may disagree, but we must acknowledge that medicine is a big business. Centers for Medicare and Medicaid Services (CMS) projected that by 2028, health care spending would reach $6.19 trillion, and would account for 19.7% of (Gross Domestic Product) GDP of the United States, up from 17.7% in 2018. While no single model is appropriate for every

Innovation in Anesthesiology. DOI: https://doi.org/10.1016/B978-0-12-818381-6.00012-7

practice setting, in order for continued success toward improving efficiency and value, constant reassessment of the healthcare environment, the original mission, and its capacity to succeed in that mission is essential in business management. We also recognize that in addition to achieving business success and financial competence, the anesthesia practices and staffing models should be optimized to deliver high quality and safe care to patients. The expanding role of anesthesiologists in the perioperative setting and the healthcare system reform provides new opportunities and requires different approaches to practice management.

Types of anesthesia groups

- Local groups → These groups are typically provider-owned but may also be owned by nonclinicians. They generally have about 1–2 facility contracts in a smaller geographical area. Owners are very involved in the day-to-day management. If owners are clinicians, they are usually responsible for a great deal of the clinical load and are highly involved in the oversight of the group. While this grants the benefit of efficient decision-making, there is a danger of low patient volume leading to financial instability.
- Regional/midsized groups → These types of groups are typically owned by clinicians or nonclinical business owners. They tend to operate in a larger geographic area than local groups (state-wide or across states), commonly providing services to 10–50 facilities through 500 or more providers. Owners are usually less involved in day-to-day management, instead emphasizing strategic vision. They have greater financial stability and more staffing resources than local groups, and generally have more robust clinical protocols and quality/practice improvement programs. One drawback is that leadership may lose touch with employees compared to local groups due to the larger scale.
- National/large groups → These types of groups may be owned by clinicians, non-clinicians, private equity firms, or be publicly traded companies. They tend to operate across a much larger geographical area than local or regional groups, often spanning several states or the entire nation. It is typical for these groups to serve well over 100 facilities through 1000 or more providers. There has been a recent trend of market consolidation in the field of anesthesia due to increased involvement of private equity firms, which has driven a number of

recent acquisitions. These groups have the benefit of strong funding and well-established clinical guidelines and protocols. There is, however, an even higher degree of disconnect between owners and clinical providers than regional groups.

Types of business entities

- Sole proprietorship → Single person business that is not registered with the state and requires no paperwork. This entity is inseparable from its owner and so the owner must report all incomes and losses on their personal income tax return. The owner is personally responsible for all debt, liabilities, and court judgments.
- Limited liability company (LLC) → This type of entity requires registration with the Secretary of State and must create an Employee Identification Number (EIN) through the IRS, which takes about thirty minutes to complete. It also requires Articles of Organization and Operating Agreement forms. The benefit of this entity is that it limits the owner's personal liability, which includes debt, court judgments, etc. The LLC must file federal tax returns but does not pay any federal taxes ("Pass through Entity"). All taxes must be paid by the owner's share of income on their personal tax return.
- Corporation (Inc.) → Like an LLC, this entity requires registration with the Secretary of State and creation of an EIN through the IR. The process of creating a corporation is more arduous, as it requires Articles of Incorporation and Corporate Bylaws forms. It must also issue stock certificates to owners and shareholders of the business. Once created, annual shareholder and director meetings must be held with minutes kept. Corporations must file and pay taxes on any profits that are left in the business after paying out all salaries, bonuses, expenses, and other overheads. US Corporate tax rates start at 15% for the first $50,000 and tops out around 35% for $10 M and above.

Types of contracts

- Fee for service → Services are unbundled and paid for separately, which gives incentive to anesthesiologists to increase surgical volume because payment is dependent on number of patients rather than the quality of care provided. These types of contracts raise overall cost and discourages efficient and integrated healthcare.

- Fee for service with subsidy → Similar to FFS but the facility makes up for any potential losses to anesthesia groups through subsidy payment. Subsidy is typically either *floating* or *fixed*. Floating stipends vary based on the net collection, while fixed stipends subsidize the department with regular payments at a fixed rate.
- Pay-for-performance → Defined as "the use of incentives to encourage and reinforce the delivery of evidence-based practices and health care system transformation that promote better outcomes as efficiently as possible." There are three basic categories of performance measures commonly found in anesthesia contracts or departmental policies: (1) clinical quality, (2) efficiency, and (3) customer satisfaction.
- Management contract → The anesthesia group is contracted to manage an anesthesia service. The management company tends to support the group through billing, recruiting, scheduling, etc., in exchange for a management fee. There is a recent trend towards clinician-owned centers to view anesthesia as another service line that they can own and profit from but do not want to manage. These contracts are more common in ambulatory surgery and endoscopy centers.

Business models

- Physician only practices → In this business model, anesthesiologists traditionally provide care in the operating room setting. With the expansion of their role in recent years, it is common for anesthesiologists to work in ambulatory surgery centers, endoscopy suites, pain clinics, and other non-OR environments. Anesthesia is personally administered by the anesthesiologist, and pre-/postoperative care is provided by the same individual or a clinician colleague within the same anesthesia group. With the variety of subspecialties available to the field of anesthesia today, many departments now employee physicians with specialized training in pain medicine, critical care, and several others. Some anesthesia practices recruit physicians from outside specialties to strengthen clinical capabilities and improve coordination of care (e.g., hospitalists with training in perioperative care, internists for preoperative care, etc).
- Anesthesia care team models of practice → These practices employ anesthesiologists, Certified Registered Nurse Anesthetists (CRNAs), and Anesthesia Assistants (AAs). The group is responsible for the management of all business

practices and the recruitment of employees. Approximately half of US states allow for independent practice of CRNAs, which does not require a written collaborative agreement, physician supervision, or conditions for practice. In states that do not allow independent practice of CRNAs, the anesthesiologist is responsible for all care provided to patients, including the preoperative evaluation, perioperative/periprocedural period, critical portions of each case, and postoperative management. Anesthesiologists may supervise up to 4 CRNAs at a time in most cases based on guidelines set forth by the US Centers for Medicare and Medicaid Services. As budgets are squeezed tighter and tighter, every entity is looking at the safest and most efficient option to deliver care. The CRNA to anesthesiologist staffing ratios would influence the cost of care.

- Certified registered nurse anesthetists → A nurse that has trained specifically in the administration of anesthesia. This type of nursing requires a 4-year undergraduate degree, 2 years of relevant clinical experience (often ICU), and a 2-year CRNA master's program. Some CRNA programs have begun to offer a doctorate-level degree, referred to as a doctorate of nurse anesthesia practice (DNAP). These specialized nurses are able to perform preoperative evaluations and administer anesthesia to patients under supervision by an anesthesiologist, unless the state in which they practice allows for independent practice.
- Anesthesia assistant → These individuals work under the direction of anesthesiologists in the anesthesia care team model. AAs must complete a 4-year undergraduate degree followed by a 2-year master's degree. AAs can collect medical history, perform preoperative physical examinations, administer anesthetics, establish noninvasive and invasive monitors, and perform general and regional anesthetic techniques with the supervision of an anesthesiologist. AAs are currently unable to practice independently in any state in which they are licensed.

- Health system-employed physician models → These practices are managed by the health system instead of an independent anesthesia group. Anesthesiologists are salaried in this model and the health system or foundation manages all aspects of business including billing, coding, contracting, and collections. Many health systems and hospitals provide financial support for physician practices, both in the academic and community settings. The growth of this model in recent years has led to the diminished ability of independent physicians to compete, as a greater number of physicians are becoming employees of health systems or foundations. Due to this reason, there has been increased consolidation of anesthesia practices into both regional and national groups, which have greater size and infrastructure. This allows for more effective management of practices, and data documentation needed to negotiate with the health systems and payers. Such a trend poses a clear threat to independent and individual hospital-based practices.

- Company model → In this model, the referring physicians create a separate "Anesthesia Company" to profit from anesthesia revenues. These clinicians generally own the facility at which the procedures are performed. This separate "company" employees anesthesiologists, collect billings on their behalf, and share the profit made with corporate shareholders. The fees are often not based on the management cost of billing and collections, but as management fees retained by the corporation.
  Many questions have been raised by this model of practice, since the arrangement has serious financial and legal consequences for anesthesiologists. On June 1, 2012, the Department of the Health and Human Services Office of Inspector General (HHS-OIG) issued an Advisory Opinion (No. 12–06) on a "company model" arrangement and expressed the view that it could violate the Federal Anti-Kickback Statute (prohibits exchange of anything of value, in an effort to induce/reward the referral of business reimbursable by federal health care programs). In a proposed arrangement reviewed in OIG Advisory Opinion, the OIG concluded that the arrangement had all of the hallmarks of a referral fee. In another review, the proposed relationship was determined to pose more than a minimal risk of fraud and abuse. Recently, the Florida Society of Anesthesia (FSA) has alleged unlawful "company model" schemes by several practice groups, mainly gastroenterologists, and filed a federal false claims act complaint. Another recent agreement was settled in December 2018 between two physician-owners of a Florida pain management practice and the federal government in which the physicians each pain in excess of $1.7 million to settle claims brought by DOJ under the AKS and the federal False Claims Act. Surgical centers and their owners are encouraged that they must provide a fair market value, and ensure compliance to antikickback laws.
- Academic practices → In this business model, the anesthesiologist works within an academic facility as an employee of the

school of medicine or the faculty practice plan (typically funded by the medical center). In many models, the anesthesia group is a part of the faculty practice and shares business practices with other departments within the faculty practice plan. Each business model varies significantly, but in most cases, the faculty anesthesiologist is employed by the university, health system, or the faculty practice. Contracting and other business services are often provided by the practice plan, and compensation is based on clinical efforts, as well other responsibilities such as teaching, research, and administration.

- Academic medical centers (AMCs) are characterized by the combination of patient care, research, and education. Traditionally, AMCs have benefited from a combination of revenue streams, including Medicare, Medicaid, private insurance reimbursements, copayments, combined with research grants and contracts. Although only 6% of the hospital in United States are AMCs, they play an essential role in the healthcare system, which include providing 20% of the hospital care, receiving more than 40% of patients transferred from community hospitals for higher level of care, managing 61% of level 1 trauma centers, 62% of pediatric intensive care units and 75% of burn care units.
- Academic practices have multiple missions, which complicates the business models and creates potential and real conflicts between clinical and academic needs (academic physician who is an outstanding clinician, educator, and researcher perceived to be less efficient and less committed to patient care). In response, many academic departments have recruited faculty with a primary commitment to clinical care. At the same time, research funding has decreased and many departments are finding it difficult to maintain scholarship while also addressing the increasing clinical demands.
- Another challenge is the changing role of residents. In addition to providing high-quality, efficient, and safe care, the academic departments are responsible for training the next generation of anesthesiologists and for providing broad-based clinical experiences and didactic educational programs. These programs are accredited by the ACGME, and the oversight/accreditation process has become more rigorous and challenging for the academic departments to meet. Work hour restrictions, increased supervisory expectations, and implementation of simulation/other models of education limit clinical availability on both

ends. Because of these changes, the residents are no longer the primary providers of clinical care under faculty supervision; faculty members are providing personal care or supervising other nonphysician providers more commonly in academic environments. In some cases, entire academic departments have been replaced by large national anesthesia groups to manage the department and deliver care, putting into question how we can ensure the ongoing training of anesthesiologists.

- As health care is not immune to market pressures, academic medical centers' model of high-acuity and high cost care is at risk in the current landscape of healthcare reform. One solution to enhance their competitive advantage and achieve economies of scale is through mergers or partnerships with community health systems. Although academic-community mergers were increased steadily in past years, the majority have been unsuccessful. Culture is cited as a key factor in the failure of mergers, both in health care and other industries.

Competitive strategy

- Understand the market → Understanding and addressing market forces is essential to developing a competitive strategy. The threat of backward integration from proceduralists (perform sedation by themselves or through nurses) is increasing with the advent of many new procedures characterized with non/miniinvasive nature, enhanced monitoring, and drug administration devices. As drugs, monitoring, and skills among providers who are nonanesthesiologists improve, the barriers to entry in the non-OR environment will likely decrease. As a consequence, nonanesthesia personnel may enter the market to meet the need of service to non-OR procedures. Furthermore, commercial applications of artificial intelligence and machine learning have made remarkable progress recently. Technological innovations in combination with drug administration and monitoring devices may potentially replace human providers of sedation or anesthesia in certain areas in the near future.
- Strategy → A successful business model of anesthesia management should maintain a dynamic and profitable market presence, while providing high-quality and safe care to patients. The changes in the healthcare environment, financing and delivery of anesthesia care in both OR and non-OR

settings will continue to put our practice under challenge. While challenging, this may also create new opportunities to extend our strength in perioperative care, quality control, safety management, and cost efficiencies to clinical practice.

Further reading

Cohen NH, Eriksson LI. Anesthesia business models. In: Gropper M, editor. Miller's anesthesia. Philadelphia, Pennsylvania: Saunders Publishing; 2014.
French KF, Guzman AB, Rubio AC, Frenzel JC, Feeley TW. Valued based care and bundle payments: anesthesia care costs for outpatient oncology surgery using time-driven activity-based costing. Healthcare (Amst) 2016;4(3):173–80.
Gross WL, Gold B. Anesthesiology and competitive strategy. Anesthesiol Clin 2009;27:167–74.
Porter M. What is strategy? Harv Bus Rev 1996;(Nov-Dec).
Regev A, Mahajan A. Integrating academic and private practices: challenges and opportunities. Anesthesiol Clin 2018;36:321–32.
Young, T.P., The business of anesthesia. MANA, 2014: http://www.mana.us/wp-content/uploads/2014/10/Tracy-Young-Handout-Business-of-Anesthesia.pdf.

26

Marketing, sales, and distribution

Che Antonio Solla
University of Tennessee Medical Center, Knoxville, TN, United States

Chapter outline

Abstract

Creating a superior product does not guarantee its success. Gaining market acceptance of your product involves consideration of several interconnected factors. An appropriate marketing strategy is essential to success. This strategy must contain an appropriate mix of branding, price, distribution, and promotion as well as an understanding of the stakeholders involved. These factors will vary based on the product and market environment as well as the chosen business model.

Keywords: Distribution channel; lease; license; market constituent; sales cycle

Innovation in Anesthesiology. DOI: https://doi.org/10.1016/B978-0-12-818381-6.00031-0

Key points

- Identification and understanding of market constituents are essential to creating a successful marketing strategy.
- The ability to appropriately scale a new technology is essential to a product's success as it is a necessary component of generating demand.
- Working with competitors is often a good way to build and meet demand while preventing the creation of imitator products.

"Why it matters?"

Even well-thought-out and developed products can have difficulty gaining market acceptance without a sound marketing strategy. An inability to convey a product's full benefits may allow imitators to gain market share. Additionally, alternative products may exist that have a valid claim to performing the same function. Persuading consumers of the superiority of a new product can be difficult and, in the fast-paced atmosphere of consumer marketing, such superiority can dissipate quickly [1]. Creating a market able to withstand these factors requires the involvement of various market constituents. To mobilize these market constituents, an analysis of a new product's business model as well as the identification of actors influencing its development at every stage must be undertaken.

The type of business model chosen will determine the degree of direct involvement a product inventor will have in several areas. Involvement can range from completely do-it-yourself to working with a product manager. A common misconception holds that a choice must be made between either making and selling your product or selling off your invention to a third party with the means to market, manufacture, and sell it. In reality, business models exist along a spectrum of options that differ in terms of the extent to which the technology is shared in relation to the control maintained over its exploitation. This spectrum ranges from maximizing retention of whatever value is created through making and selling the final product to transferring virtually all the value added to partners or whoever is interested in the technology, to create a larger market for the technology overall. Some of the factors influencing the choice of business model include long-term ambition versus the immediate economics of contracting, the promotional needs of the technology, the need to mobilize certain market constituents to deliver the technology effectively, and the price of the technology itself versus the profitability of the end-product [2].

Once a business approach is determined, it is important to get an understanding of the marketing environment for a particular new technology. To do this, it is necessary to appreciate that a product's marketing will influence a wide range of market constituents. These stakeholders have an array of roles including making available needed components for the technology as well as influencing demand and helping with the cost-effective delivery of the product. Therefore the identification of key stakeholders and the development of a strategy to co-opt them in the delivery of the technology and the creation of its demand is necessary. Traditionally, three market constituents play major roles: companies that can engage in the delivery of the technology, advocates, and arbitrators that affect market acceptance, and companies commercializing the incumbent technology [2] (Fig. 26.1). Companies that engage in the delivery of the technology include manufacturing partners, distributors, intermediate adopters, and independent suppliers of complementary products or services needed to benefit fully from the product [2].

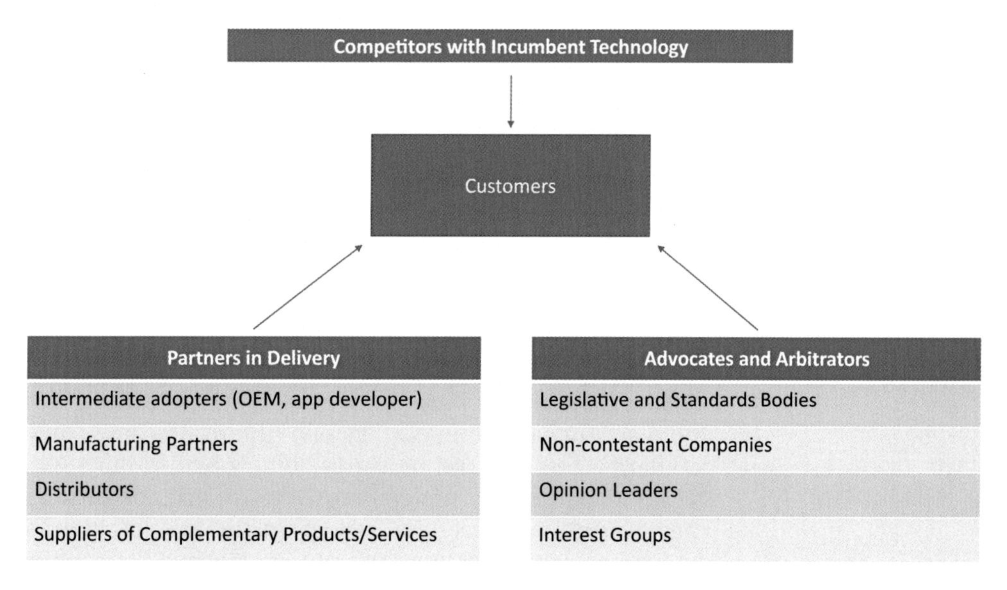

Figure 26.1 Market constituents for a new product. From [1].

Abbott Laboratories versus Edwards Lifesciences: transcatheter mitral valve devices

Mitral regurgitation is the most common type of valvular disease and impacts about 2% of the global population. Since its development in the early 2000s, transcatheter mitral repair device companies like Abbott Laboratories and Edwards Lifesciences have secured leading positions in the market. However, Abbott gained a significant global market share when its MitraClip device received Food and Drug Administration (FDA) approval in 2013. It was the only transcatheter mitral repair device approved for use in the United States until September 2022 when Edwards' competing transcatheter mitral product Pascal finally received FDA approval [3]. Pascal's primary market up to that point was in Europe, with the product available in over 10 countries. The delay in FDA approval for Edwards' products means that the company has a lot of ground to cover to obtain market share in the mitral valve repair space.

One way in which Edwards may be able to catch up comes from its relationship with surgeons built through its large presence in the transcatheter aortic valve replacement (TAVR) space where its device was the first to obtain FDA approval. Even with solid results, the Edwards timeline may make it difficult to catch Abbott in the near term. If the mitral market hits the 2025 projection, about $2 billion will be made up by Abbott and $500 million by Edwards [3]. Another benefit of Abbott's head start may come from the boon in revenue that may result from the Centers for Medicare and Medicaid Services' (CMS) recent reimbursement expansion for secondary MR that could double or triple MitraClip's U.S. patient base [3].

Market constituent: advocacy groups and arbitrators

Advocates and arbitrators play a role in market acceptance and can include opinion leaders, interest groups, industrial standards bodies, and governmental organizations [2]. This market constituency underscores the importance of waging an effective promotional and public relations campaign. In the medical device world, a strong presence at medical conferences and in the medical literature serves as some of the most effective means of swaying advocacy groups like opinion leaders. Additionally, lobbying legislators at the local, state, and federal levels is often essential in the market expansion of your product. This is illustrated in the Baxter example listed below. Another type of arbitrator is termed a noncontestant. These arbitrators have no vested interest in the solution but desire for quick action to be taken [2].

Government entities constitute perhaps the most important group of arbitrators as they can apply rules through fiat. Furthermore, legislation or government initiatives may impede or advance the viability of a particular technology. The Abbott versus Edwards example highlights the roles of government arbitrators. As Edwards did in the aortic valve market, Abbott was able to come to market in the United States with its mitral valve repair device first through FDA approval 9 years sooner than Edwards' device approval. This translated into Abbott's dominance over the U.S. market. The recent approval by another government arbitrator, CMS, to increase coverage to include more forms of mitral regurgitation further illustrates the importance of these stakeholders.

Outreach aimed at neutral arbitrators and advocates before campaigning to mobilize other partners may lay more favorable groundwork for the rollout of a product. In the previously mentioned example, Edwards' relationships with surgeons through its dominance in the aortic valve repair space may help to close the gap in its mitral valve repair market share position in the future.

Baxter spectrum IQ infusion pumps

Baxter International is the producer of one of the most used infusion pump devices in U.S. hospitals. They employ about 60,000 people worldwide and approximately 19,000 employees in the United States as of December 31, 2022. Their Spectrum IQ infusion system, built in Medina, New York requires approximately 70 chips per pump, however, recent interruptions to the supply chain of microprocessors resulting from the after-effects of the Covid-19 pandemic have resulted in a sharp decline in Baxter's revenue. This supply chain interruption affected the production of several devices and resulted in Baxter implementing a cost-cutting program that included a reduction in its workforce by 5%, or about 3000 staff, after a decline in fourth-quarter profit in 2022 [4]. Through lobbying efforts led by Senator Chuck Schumer, who represents the state where Baxter infusion pumps are produced, an alternative supply chain was established with Texas Instrument to acquire the needed microchips [5].

Market constituent: suppliers of complementary products

Products may sometimes require the simultaneous availability of complementary products. Without these products, the technology may become restricted to a narrower market and face greater financial challenges. This was illustrated in the Baxter example where supply chain issues largely outside their control affected the production of their devices.

Oftentimes, suppliers of complementary products may need to be wooed. In the Baxter example, this involved lobbying efforts on government arbitrators. However, getting these suppliers to back a new system can sometimes require selling them on the business potential of your product. This can be achieved by outlining a vision and strategy where each contributor recognizes a profitable role for itself. As mentioned in the role of business models in marketing, the degree of ownership can also play a role in gaining this group's support. Sharing a technology liberally with those prepared to help with its realization has the potential to lead major players to align themselves with the new technology. This can work to standardize the technology, facilitate consumer acceptance, and mobilize various market constituents. In this scenario, only value-enhancing features that consumers determine would enjoy a premium in the market [2].

Nevro HF10 spinal cord stimulators

Spinal cord stimulators (SCS) have been in clinical use for over 50 years. These devices are placed in the epidural space overlying the dorsal column of the spinal cord and provide pain relief by interrupting pain signal transmission from the periphery to the brain. In the last decade, the main driver in the SCS market has been the differing waveform therapies offered among competitors.

Traditional SCS relies on the generation of a paresthesia at the site of pain, but in 2006 Nevro introduced HF10, a 10 kHz high-frequency therapy that is entirely paresthesia-free [6]. This paresthesia-free therapy provides numerous benefits to physicians and patients. For physicians, paresthesia mapping is not required during placement as leads are placed anatomically. As a result, the patient does not need to be awakened from sedation to provide paresthesia coverage feedback. Patient benefits include the absence of paresthesia at rest, which can be perceived as uncomfortable. This proprietary waveform therapy has been closely guarded by Nevro but this has not stopped their competitors from producing similar paresthesia-free therapies.

Over the years, Nevro continued its primary marketing focus on high-frequency therapy while its competitors continued to market a more varied array of waveform therapies. It was not until 2019 with the release of their next-generation platform Omnia that Nevro attempted to fully embrace traditional low-frequency stimulation [6]. Nevro's late entrance to the traditional SCS market may be part of the reason that their SCS market share remains far behind competitors like Medtronic and Boston Scientific.

Market constituent: competitors with incumbent technologies

The final major market constituent is competitors offering the incumbent technology. Several adaptations by competitors can be expected in response to the opposition posed by a new product including improvement in price-performance characteristics and strengthened bonds with stakeholders [2]. Overcoming competition can occur in many forms. However, confrontation with competitors is usually undesirable and oftentimes unnecessary. New technologies can often create opportunities for all to gain. Indeed, guarding an innovation too strongly can impair its marketability as a result of creating unfulfilled demand and encouraging imitators into the market sooner and more aggressively than would occur otherwise [2]. The Nevro example above highlights this fact.

While confrontation should never be the initial plan, a strategy to address obstructing competitors is crucial. Two potential options for this include sidestepping the competition and their associated captive infrastructure or coopting some of them by offering a share of the business on favorable terms [2]. Bypassing competition is often possible when a new technology includes innovation that does not require infrastructure controlled by competitors.

An alternative solution to bypassing competitors involves coopting them. This can have the effect of aligning other market constituents in its favor. Antitrust laws often dictate the degree to which competitors can be marshaled to a technology. In general, behavior is considered acceptable if cooperation occurs mainly at the technology level and does not impede competition [2]. One potentially appealing strategy that maintains local market share while increasing growth and revenue is to target licensing on markets not served by your existing operation. By doing this, competitors are not only prevented from pursuing the development of alternative technologies, but this can also advance your technology [2].

Licensing and leasing

Many companies fail because they attempt to do too much with their technology. Therefore it is important to approach commercialization with an appreciation for all the components that make up a product. Sometimes a more lucrative option than the production of the entire device is to concentrate on the most important facet and contract for the rest of the business with partners. Contracting with partners can come in many forms. An investor could make and sell the component that leverages the technology most or combine the licensing of some components with the making and selling of others.

It is first important to understand the distinction between leasing and licensing. A lease is an agreement that conveys an exclusive possessory leasehold interest in an intellectual property (IP), while a license is a privilege to act on another's IP that does not confer a possessory interest in the property [7]. A license is generally revocable at any time by the licensor. The ultimate distinguishing characteristic of a lease is that the lessor surrenders exclusive possession of the intellectual property to the lessee for a specific term or period [7].

Leasing can offer similar benefits to licensing in terms of the ability to control the use of technology. While reducing the financial burden on buyers may be the main objective served by leasing, it allows the lessor to gain a range of collateral benefits, especially if the component concerned is covered by strong intellectual property rights [7]. Also, as opposed to sales transactions, leasing entails the retention of ownership by the lessor. This allows the lessor to charge a rental for the use of the product while controlling how it is used in practice, including issues such as operation, maintenance, and the use of spare parts [7].

Manufacturing and distribution

The ability to appropriately scale a new technology is essential to a product's success as it is a necessary component of generating demand. Even for companies looking to manufacture on their own, early involvement of manufacturing partners in technology commercialization can serve to "seed" the market quickly and provide the resources needed to pursue independence. Companies with adaptable manufacturing can create products at a minimized expense by limiting variability in production techniques [8]. This is achieved using an effective manufacturing flow management process. This process involves creating flexible agreements with suppliers and shippers so that unexpected demand surges are responded to appropriately. The goals of the manufacturing flow management process involve leveraging the capabilities of multiple supply chain members to improve manufacturing output, speed of delivery, and flexibility to maximize profit [8].

Launching and sustaining the commercialization of new technology is facilitated by access to capable and resource-rich distributors. Like manufacturing partners, distributing partners often commit significant resources and accept a share of the financial risk. Distribution decisions for the new product should be determined by the company's market segments and objectives. Distribution channels involve interdependent organizations within a business structure from production to consumption that facilitate the downstream physical movement of goods through the supply chain [8]. Fig. 26.2 gives a concise illustration of the types of distribution options available. The aim of distribution channels should identify the desired level of performance, the services needed, the control the company wishes to retain, and the financial support given to channel members [8].

With the advent of the internet, the ability to have more direct channels has become more possible than ever before, however, several factors must be considered to achieve the ideal

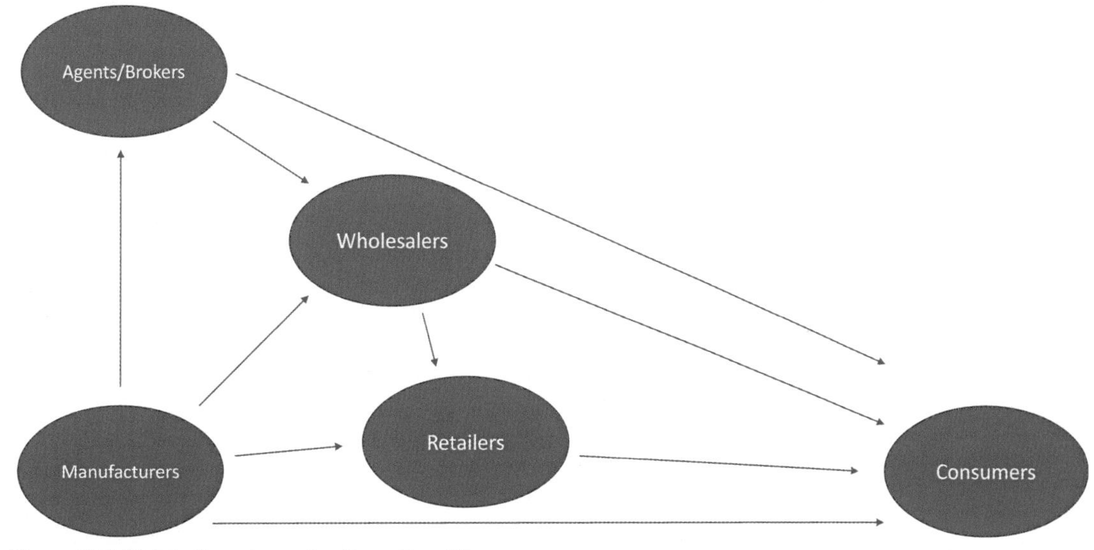

Figure 26.2 Distribution channel options. From [2].

distribution channel. The directness of the channel will be affected by the nature of the new product. For example, smaller companies will likely pay higher commissions to get channel members to store and deliver a new product whereas larger companies with greater infrastructure would be able to provide their storage and shipping, resulting in a more direct channel [8].

Certain channel patterns involve significant fiscal obligations like promotional support, inventory, credit, and transportation [8]. Other channels may need a considerable amount of time for training and development before they can perform efficiently whereas established channels would require less time but may provide little differentiation from competitors. Channel alternatives that place an unprofitable financial burden or unnecessary time constraints on the manufacturer should not be considered.

The ability of the manufacturer to control channel members is an important factor in the selection process. Channel members are typically most concerned with profit margins and the new product's demand potential [8]. Therefore high margins and strong promotional activity can give the manufacturer a strong bargaining position. Whatever decision is made on distribution channel patterns, it is important to recognize that large retailers and distributors are reluctant to add new products, especially from a company with an unknown track record. Large corporations with a good reputation will often have a better chance.

After implementing a channel system, it must be managed to ensure efficiency. This is achieved through operating policies and procedures as well as through motivating channel members. In the case of large manufacturers, leadership power results from financial strength and the ability to achieve economies of scale that create high margins [8]. Small companies on the other hand must build this control as the new product gains acceptance by consumers. Strong demand will allow the manufacturer to expand distribution and increase its control of the channel system. The larger the channel system selected the further the manufacturer is from the consumer. Therefore effective communication and information flow in both directions is crucial to ensure the manufacturer's objectives are satisfied. Appraisal of the system will be dependent on how well these objectives are met. Additionally, cost analysis, market share penetration, and consumer satisfaction can also be used to assess system efficacy [8].

Pricing

Pricing decisions should be based on several factors. First, competitor pricing of similar products should be taken into consideration. The percentage of sales devoted to research and development (R&D) can also help quantify the cost a license will need to cost to develop a technology [2]. Amortizing costs such as R&D investments, protection costs in the form of patents and trademarks, and the costs associated with transferring the technology to a licensee are other important factors [2].

Licensing on a royalty basis is another approach that involves sharing the value that the technology creates for a potential licensee. Compared to the other approaches to price determination, this method has the virtue of being based on the individual merits of a particular technology and the true benefits provided to the licensee. In general, the royalty charged to licensees tends to be between 10% and 40% of their profit, depending on the technology's contribution to the overall success of the business and the licensee's share of the risk [2]. The length of the contract, the guarantees given to the licensee, and its exclusivity all have an impact on the price. Royalties should be made to price the new product at competitive rates with existing products so that downstream benefits are achieved by the licensee and their customers. By doing this the royalty should be set to provide licensees with a good rate of return on their investment while also yielding a fair return to the licensor [2].

Promotion and sales

The final step in the commercialization of a new product is sales. Without this step, none of the preceding steps matter. The sales process is commonly referred to as the sales cycle (Fig. 26.3). This cycle has several steps and begins with the creation of leads. A lead is the contact and demographic information of a potential customer who shows interest in a specific product or service [9]. Several promotional methods exist for generating leads.

In today's market, a website is often a company's most powerful lead-generation tool and serves as the nexus of a company's online marketing efforts. Every marketing tactic employed will serve to send traffic to the website. It is then the task of the website to convert that traffic into leads, or for eCommerce sites, sales [9]. Branding is another important promotional tool that can help set your company apart in today's cluttered marketplace. Attracting the right demographic with the right message is essential to generating results. Brands should create identities that get to the core of the needs of their most common target customers. Fully developed brand identities can help you target

Figure 26.3 Sales cycle. From [3].

your messaging and ensure the message resonates and drives action from your prospects [9].

Sales management is one of marketing's most important specialties. Effective sales management begins with a determination of sales goals. These goals should be stated in precise and measurable terms with specific timelines for their completion. Salesforce goals are commonly stated in terms of dollar sales volume, market share, or profit level. Individual salespeople are also assigned goals in the form of quotas [9].

Once sales goals are identified, a determination of the sales force structure can be made. Table 26.1 outlines several common organizational options for sales force structure. In today's competitive selling environment, market-based and key account structures are gaining popularity. In the medical device industry specifically, organizing the sales force around specific customers can improve customer service, encourage collaboration, and unite salespeople in customer-focused sales teams. Assembling an effective sales force should take into consideration the type of traits desired in an applicant. Some ideal characteristics may include strong self-esteem that is resilient to rejection, competitive spirit, assertiveness, strong social skills, empathy, creativity, and an ability to understand complex concepts [9]. Appropriate compensation will afford the sales

Table 26.1 Types of sales force structures.

Types of sales force structures
Geographic region
Product line
Market function (account development or account maintenance)
Market or industry
Individual clients or account

Source: From [3].

force stability and quality employees. Once a sales force is assembled, an evaluation of their effectiveness is the final task of a sales manager. Monitoring sales performance metrics helps sales managers monitor individual salespeople's progress as well as identify where breakdowns may be occurring.

Get started

- Identify market constituents and devise an approach to address each one individually
- Consider lease and licensing options for your product
- Create flexible agreements with suppliers and shippers so that output may be scaled based on fluctuations in demand
- Determine the distribution channel pattern that works best for your product in its given market environment
- Determine pricing based on the distribution channel selected that is competitive and accounts for production costs
- Create sales goals and assemble a team to systematically achieve your stated sales metrics targets

Terms

1. Distribution channel—interdependent organizations within a business structure from production to consumption that facilitate the downstream physical movement of goods through the supply chain.
2. Lease—a contract that gives an exclusive possessory leasehold interest on an intellectual property.
3. License—a right to use another's intellectual property that does not confer a possessory interest in the intellectual property.
4. Market constituent—an entity within a product's market that has a significant effect on the manufacture, distribution, or sale of the product.
5. Sales cycle—a set of steps a salesperson takes to identify a customer and nurture a relationship to complete a sale.

6. Manufacturing flow management process—leveraging the capabilities of multiple supply chain members to improve manufacturing output, speed of delivery, and flexibility to maximize profit.
7. Royalty fee—remuneration to an owner of a product or patent for the use of that product or patent.

References

[1] Ekelman KB. New medical devices: invention, development, and use. National Academy Press; 1998.
[2] Jolly VK. Commercializing new technologies: getting from mind to market. Harvard Business School; 1997.
[3] Zipp, R. Abbott. Edwards focus on mitral valve as market projected to rival TAVR. Medtechdive. https://www.medtechdive.com/news/abbott-edwards-focus-on-mitral-valve-ahead-of-market-growth/593784/.
[4] Taylor, N. Baxter to cut about 3,000 jobs after earnings slump on supply chain woes. Medtechdive. https://www.medtechdive.com/news/baxter-BAX-4Q-Earnings-job-cuts/642481/.
[5] Office of U.S. Senate Majority Leader Charles Schumer. After launching push in Orleans County, Schumer successfully helps resolve Baxter international's chip shortage for western ny plant protecting 300 person workforce and ensuring company can keep pumping out lifesaving medical tech, January 2023. https://www.schumer.senate.gov/newsroom/press-releases/after-launching-push-in-orleans-county-schumer-successfully-helps-resolve-baxter-internationals-chip-shortage-for-western-ny-plant-protecting-300-person-workforce-and-ensuring-company-can-keep-pumping-out-lifesaving-medical-tech.
[6] Hagedorn JM, et al. Overview of HF10 spinal cord stimulation for the treatment of chronic pain and an introduction to the Senza Omnia™ system. Pain Management 2020;10(6):367–76.
[7] Rivkin B. Patenting and marketing your invention. Van Nostrand Reinhold; 1986.
[8] Hisrich RD, Perters MP. Marketing a new product: its planning, development, and control. Benjamin/Cummings Publishing Co; 1978.
[9] Lamb CW, Boivin M, Hair JF, Gaudet D, McDaniel C, & Snow, K. Mktg: Principles of marketing. Cengage, 2022.

Further reading

MacCracken CD. A handbook for inventors: how to protect, patent, finance, develop, manufacture, and market your ideas. Scribner; 1983.

27

Business strategy—valuing emerging technology companies

Kory Razaghi
Aptus Advisors Inc., Laguna Niguel, CA, United States

Chapter outline

Abstract

Determining the value of a medical device technology throughout the company's lifecycle is a key component of developing an operating plan, a sound funding strategy and ultimately a successful exit. Conducting a through valuation analysis allows the executive team to think through the entire trajectory of the business, which will be beneficial when approaching professional investors. Thus developing realistic financial projections that yield defensible financial valuation is a worthwhile exercise for firms developing novel technologies. The goal of this chapter is to provide the reader an understanding of basic valuation concepts and the logic used to derive enterprise value. Deducing a company's value requires a financial forecast combined with valuation analysis. The forecast is an assumption driven model that aims to capture the financial performance of the company as it matures throughout the forecast period, typically five to 10 years. Valuation analysis is a blend of scientific and subjective interpretation of risk data and applying these interpretations to the forecast model to triangulate a range of enterprise values. When combined, these two represent a risk-adjusted range of the company's present value based on executing a successful

Innovation in Anesthesiology. DOI: https://doi.org/10.1016/B978-0-12-818381-6.00019-X

business plan and reaching key development and financial milestones.

Keywords: Valuation; enterprise value; discounted cash flow; enterprise value multiple; return on invested capital; venture capital method; forecasting; profit and loss statement; income statement; cost of goods sold; gross profit; operating expenses; sales; General and Administrative; EBITDA; net income; market penetration rate; average sales price ("ASP"); design for manufacturing; weighted average cost of capital ("WACC"); cost of equity ("COS"); cost of debt ("COD"); risk free rate; beta; market risk premium; terminal value; internal rate of return ("IRR")

Introduction

Determining the value of a medical device technology throughout the company's lifecycle is a key component of developing an operating plan, a sound funding strategy and ultimately a successful exit. Conducting a thorough valuation analysis allows the executive team to think through the entire trajectory of the business, which will be beneficial when approaching professional investors. Thus, developing realistic financial projections that yield defensible financial valuation is a worthwhile exercise for firms developing novel technologies.

The goal of this chapter is to provide the reader an understanding of basic valuation concepts and the logic used to derive enterprise value. Deducing a company's value requires a financial forecast combined with valuation analysis. The forecast is an assumption driven model that aims to capture the financial performance of the company as it matures throughout the forecast period, typically five to 10 years. Valuation analysis is a blend of scientific and subjective interpretation of risk data and applying these interpretations to the forecast model to triangulate a range of enterprise values. When combined, these two represent a risk-adjusted range of the company's present value based on executing a successful business plan and reaching key development and financial milestones.

Reader should note that simplifying assumptions have been made to avoid extensive discussion about the nuances inherent in every valuation analysis while focusing on illustrating the crux of the valuation process.

Deriving a firm's valuation is not without its challenges. Especially for early-stage device companies with little more than a promise that the technology will achieve the necessary milestones, have a successful market launch, and gain sufficient market share to become a commercial success.

While segments of this discussion may seem complex to the new developer, the entrepreneur can gather the industry specific information, and ultimately seek assistance of a life science financial professional to guide them through the forecasting and valuation process to value their emerging technology.

Revenue forecasting and the anatomy of the profit and loss statement

Building a forecast model requires building a pro forma set of financial statements. In this discussion, the focus will be on the income statement, also referred to as the profit and loss statement, which is one of the key financial statements. The other two being balance sheet and cash flow statement.

Income statement summarizes a company's revenues, expenses and profit or loss within a specific period of time. The main categories of the income statement are listed in Table 27.1.

The income statement starts by reporting the company's revenues as the first line, also referred to as "top line." Revenue projections for medical device companies rely on the total number of units sold, which is determined by defining the addressable market segment, as previously discussed, and projecting penetration rates. This discussion focuses on sale of single product. Combining the number of units sold within a given year and the average sales price ("ASP") yields the revenues for that year. Medical innovations with a service component offer an additional revenue source, which should be added to capture total revenues.

First step is to project the target patient population throughout the model period. Combining population data and epidemiological rates provides the patient population forecast that will be the basis for the revenue projections. While most med tech start-ups do not generate revenues during the first years of its lifecycle, it is worthwhile to forecast the target patient population starting on the base year.

Once the base year for the target patient population has been determined, the next step is to identify factors that will change over the forecast period. There are obvious factors such as changes in overall population and within age groups. Then the patient specific drivers can be layered into this population

Table 27.1 Anatomy of a profit and loss statement.

Category	Definition
Revenues	Generated from sale of products or services.
Cost of Goods Sold (COGS)	Direct costs incurred in the production of a product or service, such as raw material costs and direct labor.
Gross profit	Defined as deducting COGS from Revenues.
Operating expenses	Ongoing expenses incurred in running daily operations. These expenses include SG&A, marketing & advertising, R&D, and regulatory & clinical.
Sales, General and Administrative	Nonproduction related expenses such as management salaries, legal, accounting, rent, utilities, and the like.
Marketing and advertising	Expenses related to marketing product or services including trade-show expenses, media and print advertising, etc.
Research and Development	Expenses related to developing new products.
Regulatory and clinical	Expenses related to receiving FDA approval and conducting clinical studies.
EBITDA	Earnings Before Interest, Tax, Depreciation, and Amortization. Revenues less COGS less operating expenses.
Interest	Interest paid within the period.
Tax	Taxes paid within the period.
Depreciation and amortization	Depreciation and amortization are noncash expenses that capture the cost of tangible and intangible assets over the asset's useful life, respectively.
Net income	Defined as gross profit less EBITDA, interest and tax expense.

data to project the target population through the forecast period. Examples of these factors include acuity of the disease state, patient improvement or declination with age, patient's response to treatment, recidivism or relapse rates, satisfaction with current treatment options, and the like.

Below is an example of a forecast model for end stage renal disease (ESRD) patients. The factors identified for this analysis included ESRD patients on dialysis, dialysis patients on hemodialysis, dialysis patients with fistulas, hemodialysis patient with occluded fistulas, dialysis patients with grafts, and hemodialysis patients with occluded grafts. The result yielded a market segment specific patient population forecast that was used as the basis for the financial model. A useful metric to include is the compound average growth rates (CAGR), which illustrates fluctuations within the target patient population within the forecast period. CAGR is determined by dividing the difference between the patient populations in two different years by the previous year's totals (Fig. 27.1).

End Stage Renal Disease (ESRD) Population US 2021 - 2031

Age	Prevelance Rate	2021	2022	2023	2024	2025	2026	2027	2028	2029	2030
45-64	0.36%	306,674	306,152	305,032	303,913	302,372	300,865	299,486	298,458	298,020	297,551
65-74	0.63%	189,300	195,129	202,213	209,928	217,593	221,116	225,751	230,400	235,432	239,811
75+	0.63%	133,218	138,195	142,425	146,621	150,658	159,281	166,656	173,737	180,884	188,029
	Total	**629,192**	**639,475**	**649,671**	**660,462**	**670,623**	**681,263**	**691,893**	**702,595**	**714,335**	**725,391**

	Target Population	**76,801**	**78,056**	**79,300**	**80,618**	**81,858**	**83,157**	**84,454**	**85,761**	**87,194**	**88,543**
	CAGR	*0%*	*1.63%*	*1.59%*	*1.66%*	*1.54%*	*1.59%*	*1.56%*	*1.55%*	*1.67%*	*1.55%*

Figure 27.1 Forecast of ESRD target patient population.

A key revenue driver for a financial model in the medical device space is market penetration rate. Market penetration rate is defined as number of devices purchased divided by total addressable market. Unlike markets such as consumer products and business services, there are no benchmarks that are readily available to determine medical device specific market penetration rates. With consumer products such as detergents and socks, for example, a market penetration range between 2% and 6% is reasonable depending on the number of market entrants. However, the wide range of medical device types, each with its own unique market dynamics, makes benchmarking penetration rates impractical. This is especially true of products that may change the standard of care or face increasing demand in emerging markets. Thus the analysis must be unique to each device and its targeted market segment.

Identifying factors that drive adoption is essential to determining market penetration rates. An example of factors that can influence market penetration rates including regulatory path, unit sale price, complexity of the device, willingness of physicians to adopt new treatment options, competing technology, reimbursement environment, to name a few. Primary market research specific to the company's technology is one of the more reliable tools to determine these important dynamics.

Reviewing competitors SEC filings is another means to determine the penetration rates of devices targeting the same addressable market. Caution should be used when incorporating data from large publicly traded companies as these firms have significant resources to drive adoption as compared to an early-stage company launching its first product. In addition, companies who are attempting to drive interest surrounding their technology have been known to exaggerate adoption values, so averaging more than one source for a data point will likely yield more defensible forecast.

Reasonable penetration rates should account for market limits, adoption rates of new products within the targeted market segment, company's internal constraints and access to capital as the company matures. The following guidelines may be useful in developing penetration rates:

1. Benchmark penetration rates based on firms that have developed a product within the same or similar market segment. The ideal company would be publicly traded and has products within that segment. An example would be Nevro, which only develops implantable neurostimulators.
2. Determine market share held by market leaders within a market segment. Market share held by leaders within a segment would be the upper limit of the penetration rate. It is highly unusual for a newcomer to take significant market share to match the market leaders within a 5–10 year forecast period. For example, if an established company has 25% of the market share, a newcomer may capture no more than 5%–10% of the market within the forecast period.
3. Determine internal constraints on the company to support the projected penetration rates as it launches its product and scales its operations such as manufacturing capability, funding to support sales and marketing efforts, etc.

Once determined, the market penetration rate can be used to determine the unit sales per year throughout the forecast period. The following figure illustrates how to apply penetration rates to the target population to determine unit sales (Fig. 27.2).

Pricing

The next driver of the financial forecast is unit price. Pricing medical devices can be quite complex as multiple factors have to be considered such as strength of patent portfolio, value proposition, cost-saving to the hospitals, reimbursement environment, promotional or volume discounts, budget cycles, product replacement cycles, channel strategy, and other factors. For the forecast model, the company can initially use average sales price ("ASP") of competing technologies as a basis to determine a price range for its device. Company executives can

	2021	2022	2023	2024	2025	2026	2027	2028	2029	2030
Target Population - VTE	164,869	168,562	172,239	176,158	180,019	183,934	188,064	192,057	196,099	199,870
Penetration Rate	*0.05%*	*0.85%*	*1.28%*	*2.32%*	*3.07%*	*3.97%*	*4.75%*	*5.27%*	*5.61%*	*5.72%*
Unit Sales	82	1,429	2,204	4,085	5,531	7,304	8,925	10,121	11,006	11,439

Figure 27.2 Sample market penetration rates used to determine unit sales within forecast period.

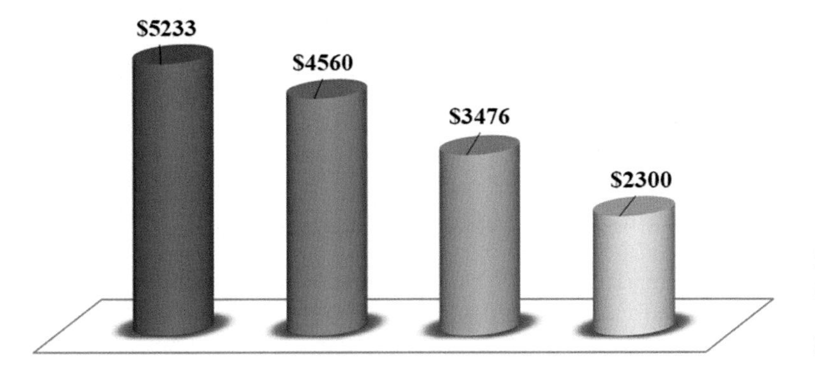

Figure 27.3 Price range for implantable device for top four competitors yielding an ASP of $3890.

then determine whether their innovative products should be priced at a premium or discount to competing products.

Below is an example of this approach where a review of 21 patient cases were analyzed to determine the ASP for an implantable device. Pricing for these devices ranged from $2300 to $5233 yielding an ASP of $3890. Note the data has been changed to comply with confidentiality requirements and are meant for illustrative purposes only (Fig. 27.3).

As the company conducts in-depth pricing analyses on its products, the pricing can be updated accordingly.

For novel technologies that do not have comparable products on the market, a methodology should be developed to build up a preliminary ASP. The methodology to estimate an ASP can incorporate current reimbursement rates for existing treatment options, gross margin threshold, cost savings to payers and providers, etc. A well-designed primary research project combined with internal pricing analysis will yield insight into what price the market will bear for the product.

Price of any device will likely erode over time as new competitors and technologies enter the market targeting the same patient population, reimbursement pressure, patent expiration, and similar factors. Thus, the ASP should be adjusted throughout the forecast period to account for price erosion, which may be as high as 30%–40% by the terminal year of the forecast period.

Unit sales and ASP can be combined to project gross revenues throughout the forecast period adjusting for changes in patient population and price erosion.

Expenses

Once the revenues have been determined, the focus will shift to determining the expenses incurred by the company. The first of these expenses are Cost of Goods Sold or COGS, which is the next line down from revenues in the income statement.

COGS is defined as direct costs incurred in the production of a product or service, such as raw material costs and direct labor. Common drivers of COGS include costs of raw materials and parts for the disposable and reusable components of the device, assembly, surgical kit, and packaging. At low volume, COGS tend to be higher as the device has not gone through design for manufacturing ("DFM") phase, which limits the opportunity to benefit from manufacturing at scale or bulk purchasing of raw materials. As sales increase, a reduction in COGS should be projected as the company realizes the advantages of large-scale manufacturing and capturing efficiencies of volume production. This cost reduction should be modeled into the financial forecast.

Below is an example of a COGS analysis for an implantable device that had both a disposable and a reusable component. The cost reduction was projected after a preliminary DFM study was completed and the results were then incorporated into the financial model (Fig. 27.4).

Operating expenses

The next section of the financial model will focus on projecting operating expenses. As defined earlier, operating expenses are ongoing expenses incurred in running daily operations of the company. For medical device companies, operating expenses can be categorized into sales, general and administrative or

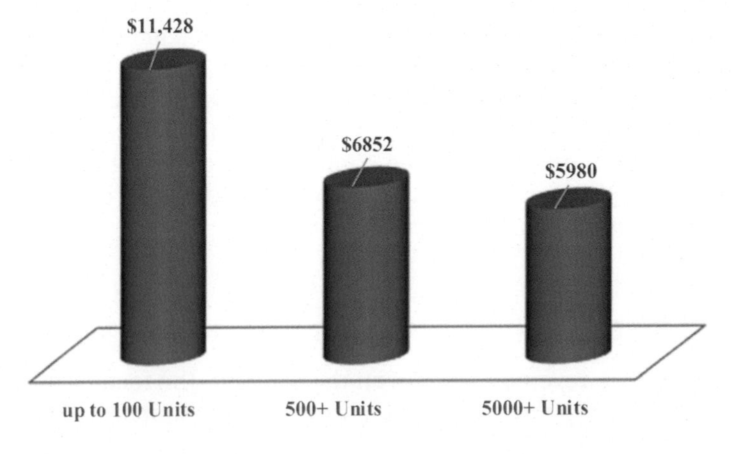

Figure 27.4 COGS erosion driven by increases in unit production.

SG&A, marketing and advertising, research and development or R&D, and regulatory and clinical.

For forecasting purposes, operating expenses are grouped into pre-product launch phase, where the majority of funds will be sourced from sale of equity to support R&D, regulatory and clinical, and prelaunch marketing efforts to bring the product to market. This is followed by postproduct launch phase, where the use of investment funds will be supplemented by sales revenues to support predominately product manufacturing, marketing and advertising, and SG&A.

The company's R&D budget will provide the basis for expenses during the prelaunch phase. Since R&D expenses are highly specific to each product, benchmarking these expenses offers limited value. As such, these figures need to be developed by the inventor and the executive team.

Regulatory and clinical costs are usually the largest single area of prelaunch operating expenses. Clinical trial costs are driven by several factors including therapeutic area, device classification, study size (number of patients), number of countries, number of clinical sites, and the specific tests and procedures needed per protocol, to name a few. Budgeting these costs poses a particular challenge to early-stage medical technology firms as many of these regulatory variables are highly device dependent and likely unknown early in the lifecycle of the company.

According to a 2010 Stanford study, the average regulatory and clinical cost to bring a 510(K) medical device to product launch through the 510(K) pathway is $31 million. The higher risk PMA pathway is estimated to average $94 million, triple the cost of a 510(K), and is often higher.

Researching the regulatory pathways and associated expenses for competing products on the market may yield insight into these costs. While not perfect, this is a good starting point until the company has defined its own regulatory strategy and can develop a more representative budget. Once completed, the model can be updated to better reflect the company's prelaunch expenses.

Postproduct launch, the company can benchmark its operating expenses against relevant publicly traded companies. A tiered expense structure based on annual reports of peer publicly traded companies can be used as a proxy for the company's expense structure as sales are projected through the forecast period.

The following example illustrates a postproduct launch operating expense analysis based on the tiered structure approach.

Table 27.2 Postproduct launch operating expense summary based on three tier revenue structure.

Expense category	Company size by gross revenues		
	$1–$70MM	$70MM–$300MM	$300MM +
COGS	33.27%	29.26%	31.06%
SG&A	81.43%	51.38%	35.88%
R&D	29.38%	18.57%	9.18%
Other expenses	0.00%	0.03%	7.27%
Operating expense	121.67%	88.25%	67.99%
Net margins	−46.32%	−8.34%	9.86%

The peer companies by gross revenue were identified and the expenses by category were calculated as a percentage of gross revenues. As the company's projected revenues increased throughout the forecast period, the basket of relevant companies were identified, and operating expenses were adjusted accordingly. Note that the peer companies used for this analysis report regulatory and clinical expenses as part of R&D line item (Table 27.2).

After COGS and operating expenses have been determined, the company's EBITDA can be calculated. EBITDA is equivalent to revenues less COGS and operating expenses. Lastly, net income can be calculated from EBITDA. For the purposes of the financial model, getting to Net Income from EBITDA requires assumptions, which are discussed in the following section. Below is an example of a forecast model based on the income statement for an early-stage medical device company using the methodology discussed herein.

(in $000)	Year 1	Year 2	Year 3	Year 4	Year 5	Year 6	Year 7	Year 8	Year 9	Year 10
Gross revenues										
US only										
Resueable	–	434	11,613	22,054	35,403	46,553	57,181	68,567	67,536	70,904
Disposable	–	1411	37,785	71,758	115,194	151,472	186,054	193,818	219,747	230,706
Gross Revenues	–	**1845**	**49,398**	**93,812**	**150,597**	**198,024**	**243,235**	**262,385**	**287,283**	**301,610**
Total										
COGS										

(Continued)

(Continued)

(in $000)	Year 1	Year 2	Year 3	Year 4	Year 5	Year 6	Year 7	Year 8	Year 9	Year 10
Resueable	–	256	5954	11,308	16,208	21,837	24,170	28,951	35,864	41,315
Disposable	–	683	10,314	19,586	31,442	39,324	49,492	55,283	58,916	60,088
Total COGS	–	**939**	**16,268**	**30,894**	**47,650**	**61,161**	**73,662**	**84,234**	**94,780**	**101,404**
Net revenues	–	**906**	**33,130**	**62,917**	**102,947**	**136,863**	**169,574**	**178,152**	**192,503**	**200,207**
Operating expenses										
SG&A	900	1502	40,226	48,203	77,381	81,750	98,981	107,974	129,992	131,727
R&D	7265	5572	10,511	17,419	27,962	36,768	45,163	40,559	30,407	33,698
Regulatory and clinical	333	5475	1517	1150	1070	995	925	860	800	744
Total operating expenses	**8498**	**12,550**	**52,254**	**66,772**	**106,413**	**119,513**	**145,069**	**149,393**	**161,199**	**166,169**
EBITDA	**$ (8498)**	**$ (11,645)**	**$ (19,124)**	**$ (3854)**	**$ (3466)**	**$ 17,349**	**$ 24,504**	**$ 28,758**	**$ 31,304**	**$ 34,037**
Tax expense	–	–	–	–	–	–	–	5045	6574	7148
Net income	**$ (8498)**	**$ (11,645)**	**$ (19,124)**	**$ (3854)**	**$ (3466)**	**$ 17,349**	**$ 24,504**	**$** 23,713	**$ 24,730**	**$ 26,889**

Valuation

After the forecast is completed, the next step is to perform the valuation analysis where the forecast created in the previous step is converted into estimates of the value of a firm.

While there are several methods to value established companies, there are limited number of methods that are applicable to valuing early-stage medical device companies. Because these companies lack historical financial performance and have not accumulated substantial assets. Thus this discussion will focus on three methods used to value development stage companies. These methods are as follows:

Valuation methods
I. Discounted cash flow
II. Enterprise value multiple
III. Return on invested capital—aka venture capital method

I. Discounted cash flow ("DCF")–The DCF analysis is based on the time value of money concept, which assumes the reduction in the value of money over time, and that the value of the firm's future earnings is worth less today due to the inherent risk of executing a business plan to capture future earnings. In other words, the value of a company is deduced by deriving the net present value of future cash flows within the forecast period. This method is suitable for early-stage medical device companies as it accounts for the early years when the firm will run at a loss due as well as the later years when the promise of higher growth rate with increasing sales is realized. The formula for the present value of an enterprise using the DCF method is as follows:

$$\text{Enterprise value} = \text{present value of future cash flows} + \text{present value of terminal value}$$

or

$$EV = CF_1/(1+r)^1 + CF_2/(1+r)^2 + CF_3/(1+r)^3 + \cdots + CF_n/(1+r)^n + \text{terminal value}/(1+r)^n$$

where

CF = cash flows within the period
n = time period, for example, years
r = discount rate

There are three main components of the DCF method:

A. Future cash flows,
B. Discount rate, and
C. Terminal value

A. The series of future cash flows or free cash flow (FCF) used in the DCF analysis can be derived from the company's forecasted income statement discussed earlier. FCF is defined as follows:

$$FCF = \text{net income} + \text{noncash expenses} - \text{increase in working capital} - \text{capital expenditures}$$

For early-stage medical device companies, several assumptions can be made so that the net income from the forecast model can be used as a proxy for FCF. These assumptions are as follows:

Category	Assumption
Net income	Net income is defined as EBITDA plus interest, tax expenses, depreciation, and amortization.

(*Continued*)

(Continued)

Category	Assumption
Interest expense	Assume the start-up will not have long-term debt during most of the forecast period as the company will likely sell equity to raise capital instead of taking on debt. As such, the interest expense can be ignored.
Taxes	While it is not necessary to do a complete tax analysis, a cursory estimate of the company's tax liability could be included in the forecast model. To determine if the tax liability will have a material impact on the valuation, the substantial loses in the early years of the business can be applied to the tax liability in the later years. If the tax expense is not significant, it can be ignored.
Noncash expenses—depreciation and amortization	Amortization and depreciation are noncash expenses that appear on the income statement. Since the purpose of the forecast is to determine the cash the business generates, these noncash expenses can be zeroed out.
Increase in working capital	The assumption can be made that all profits generated within a given year are reinvested back into the company's operations. As such working capital adjustment can be ignored.
Capital expenditures	Most early-stage device companies do not have in-depth capital budgets as decisions such as leasing versus purchasing, outsourcing cap ex extensive development and manufacturing functions, etc., may not have been made. Thus assuming all capital expenditures are expensed within the year it is incurred will eliminate the need for developing an extensive capital ex budget. In the event a capital budget is available, then it can be incorporated into the forecast model.

B. Discount rate—determining an appropriate discount rate is one of the most challenging elements of the DCF method. One approach to calculating discount rate is the Weighted Average Cost of Capital (WACC) method. WACC is a rate that is applied to discount the value of future cash flows to determine the present value of those same future cash flows. WACC is calculated by summing the cost of equity (COE) multiplied by the percentage of equity, and cost of debt ("COD") multiplied by the percentage of debt on the company's capital structure.

$$\text{WACC} = [\text{cost of equity} \times \% \text{ equity}] + [\text{after} - \text{tax cost of debt} \times \% \text{ debt}]$$

Cost of equity is defined as the expected return given the risk profile of an equity investment. Cost of equity can be calculated using the capital asset pricing model ("CAPM"), which is a methodology to quantify risk and translating that risk into estimates of expected return for equity holders. The CAPM formula is given as:

$$\text{Cost of equity} = \text{risk} - \text{free rate} + (\beta \times \text{market risk premium})$$

where

Risk free rate = rate of return if invested in a liquid and secure investment, for example, 10-year US Treasury Rate

β = Measure of systemic market risk of the asset relative to the market

Market risk premium = rate of return above the risk-free rate required to invest in riskier asset class

Below is an example of COE Model for a firm with the following risk profile (Fig. 27.5).

Based on the above risk profile, the COE would be as follows

$$\text{Cost of equity} = 1.20\% + (1.04 \times 15.87\%) = 17.70\%$$

While the CAPM formula captures the risk associated with expected return for securities traded on a public exchange, it does not adequately capture the additional risk inherent in investment in early-stage medical device companies. Consequently, additional risk premiums will need to be incorporated into this base formula to better estimate a more representative cost of equity. Examples of additional risk premiums include:

- Illiquidity premium—illiquidity is defined as the inability to convert the equity holdings to cash at their full value in a reasonable period of time. Shares of publicly traded stock can be sold at the market rate and cash received almost instantaneously. The same is not true of shares of private companies. As such, an illiquidity risk premium needs to be added.
- Company size premium—within the same industry sector, a smaller company tends to be riskier than its large competitors

Cost of Equity	US
Risk-free Rate	1.20%
Beta	1.04
Equity Risk Premium	15.87%
Cost of Equity	17.70%

Figure 27.5 Cost of equity for hypothetical medical device start-up.

as the latter have more resources to adapt to changing market conditions thereby mitigating exposure to this risk.

- Regulatory risk—difficult and complex regulatory pathways inherently have higher risk, for example, PMA versus 510(k).

Cost of debt are the expenses the company incurs to service its debts based on the rate a firm pays on each of the company's debt, for example, bonds, loans, etc. For the purposes of this analysis, an assumption is made that as the company matures, its capital structure will begin to mirror other companies within its market segment. Therefore a metric can be built to determine the rate to apply to the COD calculation. Since most publicly traded companies report their COD in their financial statements, a survey of appropriate group of medical device companies will provide a range of cost of debt. The formula for after-tax cost of debt is (Fig. 27.6).

$$\text{Cost of debt} = \text{cost of debt} \times (1 - \text{tax rate})$$

An example of cost of debt for a firm with the following profile is as follows:

$$\text{Cost of debt} = 2.75\% \times (1\% - 21\%) = 2.17\%$$

Taking the weighted average COE and COD and applying it to the capital structure of this hypothetical company, the WACC can be calculated.

$$\text{WACC} = [\text{Cost of equity} \times \%\ \text{equity}] + [\text{After} - \text{tax cost of debt} \times \%\ \text{debt}]$$

WACC

Cost of equity	17.70%
Equity to enterprise value	90.00%
Cost of debt	2.17%
Debt to enterprise value	10.00%
WACC	16.15%

$$\text{WACC} = [17.70\% \times 90\%\ \text{equity}] + [2.17\% \times 10\%\ \text{debt}] = 16.15\%$$

With the WACC calculated, the future cash flows from the forecast model can be discounted back at this rate to determine the present value of those cash flows.

Cost of Debt	US
Cost of Debt	2.75%
Corporate Tax Rate	21.00%
After-tax Cost of Debt	2.17%

Figure 27.6 After-tax cost of debt.

C. Terminal value ("TV") captures the value of all future cash flows beyond the last year of the forecast period as it is assumed the company will continue to operate and generate cash. There are several methods that can be used to determine TV. One such approach is the exit multiple method where the assumption is made that the firm will be valued at a multiple that is in line with similar company's EBITDA. As with most EV multiple methods, the collection of companies used to determine the multiple should be comparable to the firm as it is forecasted in the terminal year. The formula for EV/EBITDA multiple is as follows:

$$EV_{terminal\ year} = EBITDA \times EV/EBITDA$$

Example of TV assuming the forecast above and EV/EBITDA multiple is as follows:

Terminal value

EBITDA terminal year	$34,037,274
EV/EBITDA multiple	12.20
Terminal value	$415,254,739	

Once the components of the DCF model have been derived, the present value of the enterprise can be calculated. Recall that the enterprise value using the DCF method is:

$$\text{Enterprise value} = \text{present value of future cash flows} + \text{present value of terminal value}$$

or,

$$EV = CF_1/(1+r)^1 + CF_2/(1+r)^2 + CF_3/(1+r)^3 + \ldots + CF_n/(1+r)^n + \text{terminal value}/(1+r)^n$$

Using the cash flows from the forecast model, the calculated WACC and TV, the DCF model yields the following enterprise value:

DCF model

WACC	16.15%
EVterminal year	$415 million
Present EV	$83,379,717

This EV represents the present value of this medical device start-up based on the promise of the forecasted cash flows and terminal value. The NPV function in Excel is a useful tool to calculate net present value.

II. Enterprise value multiple—An enterprise value can be derived by determining the appropriate multiple of a number of financial metrics such as enterprise-value-to-sales (EV/sales), EV to EBIT, EV to EBITDA, to name a few. For early-stage medical device companies, EV/Sales maybe be preferred as the denominator is based on the top line of the income statement and therefore is less affected by differences in accounting practices amongst firms. The EV/Sales method assumes the company is sold for a multiple of its projected sales in the terminal year of the forecast period. The formula for EV/sales multiple is as follows:

$$EV_{terminal\ \ year} = sales \times EV/sales$$

An example of EV using this method is as follows:

Enterprise value multiple

Revenues terminal year	$301,610,342
EV/sales multiple	1.52
Enterprise value $458,447,720	

Discounting this EV by the discount rate of 16.15%, discussed earlier, yields a present value of $102.6 million.

The EVM approach relies on identifying a collection of companies that are comparable to the firm as it is forecasted in the terminal year. This is an important distinction as the terminal year of the financial model assumes the company has already enjoyed the abnormally high growth rate by successfully developing, obtaining regulatory approval for, and launching a novel medical device. Thus the company within the later years of the financial model should expect that its enterprise multiples will fall to levels that are in line with mature competitors within its market segment.

The comparable company analysis should include metrics from private and publicly traded companies. While financial metrics on publicly traded companies are readily available, they may be less comparable as most of these firms typically have multiple product offerings across several disciplines, established sales force, an international presence, and strong balance sheets with easier access to the capital markets. Therefore private company transaction data applicable to the targeted market segment are a better benchmark. Especially recent acquisition data of companies offering products within the same market segment will yield multiples that are more relevant. These companies tend to offer one or few products within a specific market segment, have a presence in specific countries or regions, have smaller sales force, may rely on distributors, etc.

Finding private company transaction data is challenging as this type of information is typically kept confidential. Researching press releases, SEC filings of acquirers, and industry specific publications and websites are useful resources for retrieving this information. There are also several commercially available databases that offer transaction data on private companies such as DealStats (formerly Pratt's Stats), Pitchbook, Perquin, to name a few.

III. Return on invested capital—Also known as the Venture Capital Method, the Return on Invested Capital was published in 1987 by William Sahlman at Harvard Business School and is driven by the investor's required return. The key elements of this method are the investment amount sought by the firm today, the investor's required rate of return (internal rate of return or "IRR") and the estimated value at an exit event, that is, IPO or acquisition, that occurs at a point in the future such as the end of the forecast period.

The first step entails determining a company's terminal value using a market multiple of revenues or net income. The company's terminal value is based on a success scenario where the company attains sales projections that facilitate an exit at a multiple of its sales or net income. A terminal value using the market multiple approach discussed earlier can be used for this analysis. Next, the required rate of return is used to derive at a postmoney valuation. The required returns for higher risk investments with longer term hold periods, typical of early-stage medical technology companies, range from 30% to as high as 80% and is dependent on the risk profile of the firm at the time of the investment. The terminal value is then discounted by the required rate of return to yield a present postmoney valuation of the firm, defined as value of the company after receiving the investment. Subtracting the amount of investment sought from the postmoney valuation yields the premoney valuation of the firm, which in turn is the value of the company before the new investment.

The formula for the VC method can then be written as:

$$\text{Postmoney value} = \text{terminal value}/(1 + \text{required rate of return})^{\text{Years}}$$

$$\text{Premoney value} = \text{postmoney value} - \text{investment}$$

To illustrate the VC Method, consider a scenario where an early-stage medical device company is seeking an investment of $5.0 million from a med tech fund to complete the product development and successfully launch its product. The company forecasts that in 7 years it will achieve sales revenues of $25 million. The company has determined that comparable companies are valued at four times annual sales and thus, at the end of the 7th year, it projects an exit value of $100 million. The med tech fund requires an IRR of 45% on its investment.

Venture capital method assumptions

Annual sales in terminal year	$25 million
Forecast period	7 years
Sales/EV multiple	4X
Initial investment	$5.0 million
Investor's IRR	45%

In this scenario, the postmoney value

$$= \text{terminal value}/(1 + \text{required rate of return})^{\text{Years}}$$

$$= \$100\text{ million}/(1 + 45\%)^7 = \$7.42\text{ million}$$

Then the premoney value = postmoney value—investment

$$= \$7.42\text{ million} - \$5.0\text{ million}$$

$$= \$2.42\text{ million}$$

The postinvestment equity ownership would be 67% ($5.0 million/$7.42 million) for the med tech fund and 33% ($2.42 million/$7.42 million) for the company. The basic VC method assumes that the firm will not require additional funds or issue additional shares to dilute the initial investment. As such, the payout at the exit for the med tech fund and the company would be $67 million and $33 million, respectively.

Conclusion

The financial modeling and the three valuation techniques discussed are designed to provide a window into the thinking involved when valuing an early-stage med tech company. The model should be regarded as a living document that is updated as the business matures, and the unknowns are determined. The enterprise value should increase as the business eliminates risks, such as, receivingregulatory clearance, successfully launching its products, etc. This bump in valuation will be a step function meaning that the value will jump, sometimes significantly, after reaching key milestones.

The valuation model enables the executive team to make data driven decisions using scenario analyses, refine funding strategies, and ultimately develop a better business plan. This analysis provides additional value by demonstrating to professional investors that the executive team has thought through the entire trajectory of the business.

Caution should be taken when undertaking a valuation exercise as value is in the eye of the beholder. The subjective component of the underlying assumptions leaves ample room for discussion and interpretation. Ultimately the value of the firm is the price agreed to between the buyer and seller.

28

Business operations

Neil Padharia Ray
Board Certified Pediatric Anesthesiologist, Raydiant Oximetry, Inc, Sacramento, CA, United States

Chapter outline

Abstract

A business plan serves as a tool for planning, management, and communication of key objectives for the company. Typically, a business plan consists of an operating plan, financial plan and staffing plan that outline how a company intends to achieve short-term and long-term milestones to mitigate risks and create value for the commercial opportunity. These objectives are communicated internally to the team to ensure alignment on progress necessary for success. In addition, these objectives are communicated to the board of directors, existing investors, and future investors/partners to ensure good governance, investment opportunity and long-term viability of the business proposition.

Keywords: Business plan; operating plan; top-down sales projections; bottom-up sales projections; P&L statements; net present value; cashflow statement; exit scenarios; EBIDTA; cost of goods

Innovation in Anesthesiology. DOI: https://doi.org/10.1016/B978-0-12-818381-6.00030-9

Key points

- The management team will need to determine how much capital, staff and time are required to take the concept to market. This is the essence of business operations.
- An operating plan and a profit and loss (P&L) statement conveys how resources will be used for concept development, risk-mitigation and shareholder value creation.
- A data-driven exit strategy and an understanding of cash flow projections will assist the management team convey to investors understand the potential return-on-investment (ROI) during a potential liquidity event.

Introduction

A business plan serves as a tool for planning, management, and communication of key objectives for the company. Typically, a business plan consists of an operating plan, financial plan and staffing plan that outline how a company intends to achieve short-term and long-term milestones to mitigate risks and create value for the commercial opportunity. These objectives are communicated internally to the team to ensure alignment on progress necessary for success. In addition, these objectives are communicated to the board of directors, existing investors, and future investors/partners to ensure good governance and long-term viability of the business proposition.

The founding management team should start the business planning with a solid understanding of the program costs expected to take the technology concept to market. A survey of 200 medical device companies in 2010 reported that the average program costs for a device seeking 510 (k) clearance was $20 million dollars and required 6 years of duration (see Fig. 28.1). In contrast, a device seeking PMA approval required $94 million dollars and 9 years of duration on average (see Fig. 28.2). Most founding teams will need access to capital to support the development program but unlikely that the total amount of capital required will be available up-front. Therefore the management team will need to determine with the current availability of capital/staff and the current stage of the technology, what milestones can be achieved to mitigate risks and facilitate capital raise the next tranche of funding to continue the development effort.

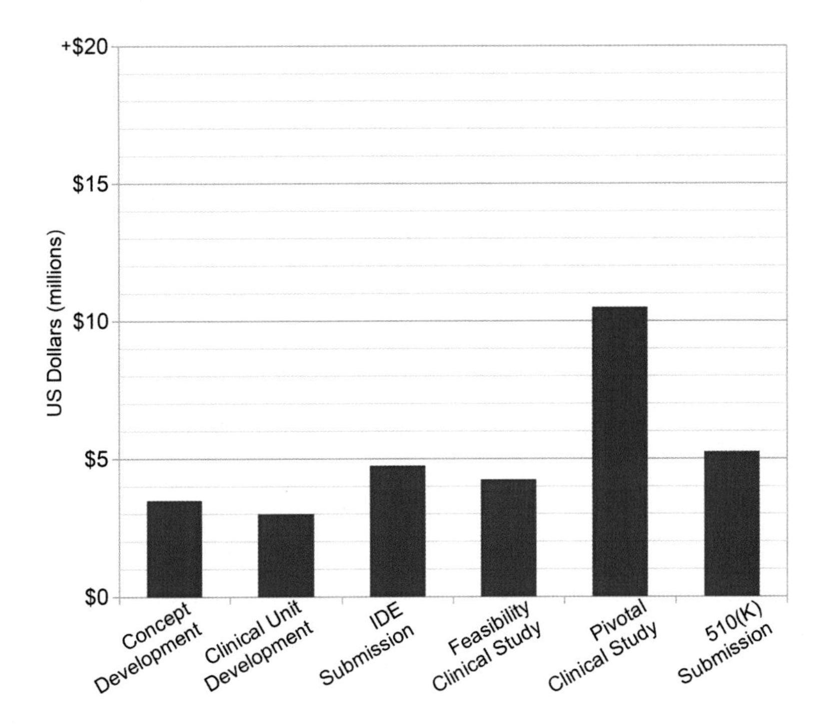

Figure 28.1 Average total expenditure by stage for 510(K) product.

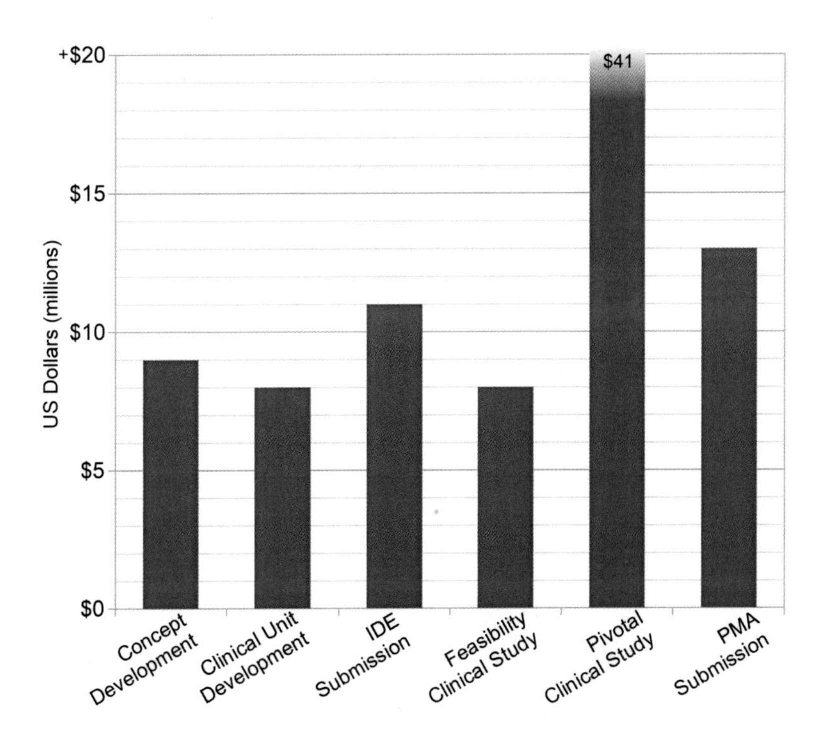

Figure 28.2 Average total expenditure by stage for PMA product.

Operating plan

The operating plan is a document that highlights the key activities necessary to develop the technology and mitigate risks. The operating plan also incorporates a timeline for expected deliverables and is typically separated into quarterly plans, annual plans and a 5-year plan. Each milestone in the operating plan can be further discussed in detail with a project plan for that specific milestone. The project plan is also associated with individuals within the organization who are accountable to ensure completion of each task in a timely fashion.

Staffing requirements

Staffing requirements outline who is responsible for completion of each task in the operating plan and what skill-sets are required to achieve that milestone. Companies need to determine whether to bring specific resources in-house or outsource that activity to a third-party. There are trade-offs between hiring in-house or outsourcing and the management team will need to assess their specific situation to determine which approach makes the most sense. Although outsourcing can appear costlier in the beginning, consulting firms often bring a tremendous amount of experience and collective thinking to a project. On the other hand, these third party firms will juggle many projects at once and smaller projects from early-stage startups may not get the attention of senior consultants. A staffing plan can be divided into three main categories:

1. R&D (research and development) staffing requirements
2. Manufacturing staffing requirements
3. SG&A (selling and general administrative)

Sales projections

The management team will need to assess the size of the market opportunity for their technology and how much money hospitals and third-party payers would be willing to pay for the technology. When estimating the size of the market opportunity, founders can approach the problem from a “top-down” perspective or a “bottom-up” perspective.

The “top-down” approach takes the total market size and segments the market opportunity into addressable markets that can be served by the technology. Statistical data from per reviewed journals are often used to make best guess estimates of the market segments; the "top-down" approach is useful to rapidly ascertain if the market opportunity is compelling.

Fig. 28.3 is an example of a top-down sales forecast from Raydiant Oximetry—a venture-backed medical device company in the women's health care space. Their technology addresses laboring pregnant women and specifically those that are experiencing Category II fetal heart rate tracings.

In contrast, the "bottom-up" approach is more useful to determine staffing requirements over time and requires a more nuanced understanding of market dynamics in the space. Some considerations when formulating a "bottom-up" market approach include:

1. What is the fundamental unit of the business opportunity?
2. How long is the sales cycle for that space?
3. What does a traditional adoption curve look like in that space?
4. How many sales reps does the company plan to start with?
5. What are the important market development factors such as the physician societies and third-party payers in the space?
6. What is the plan for future product evolution?
7. Are there unique circumstances in the space such as seasonal variation or geographic variabilities to consider?

Raydiant Oximetry, Inc.
P&L Summary 2021-2030
Top-Down Sales Forecast
Assumptions: FDA PMA Approval Mid-2025
Indication for Use: Category II Fetal Heart Rate Tracings

Revenue	2021	2022	2023	2024	2025	2026	2027	2028	2029	2030
L&D Beds (USA)					48,000	48,000	48,000	48,000	48,000	48,000
Labor attempts per month (USA)					285,000	285,000	285,000	285,000	285,000	285,000
Category II Births per month (USA)					240,000	240,000	240,000	240,000	240,000	240,000
Sensors per month					1,000	2,000	4,000	12,000	25,000	50,000
Stand-alone Modules					50	250	250	250	100	100
Stand-alone Modules Installed Base					50	300	550	800	900	1,000
OEM Modules							250	500	1,000	2,000
OEM Modules Installed Base							250	750	1,750	3,750
ASP Disposable Sensors					400	400	400	400	400	400
ASP Stand-Alone Module					10,000	10,000	10,000	8,000	8,000	8,000
ASP OEM Module							1,200	1,200	1,000	1,000
Gross revenue (Sensors)					2,400,000	9,600,000	19,200,000	57,600,000	120,000,000	240,000,000
Gross revenue (Stand-alone Modules)					500,000	2,500,000	2,500,000	2,000,000	800,000	800,000
Gross revenue (OEM Modules)							300,000	600,000	1,000,000	2,000,000
Total Gross revenue					2,900,000	12,100,000	22,000,000	60,200,000	121,800,000	242,800,000
Cost of Goods per Sensor					140	140	140	70	70	35
Cost of Goods per Stand-alone Module					5,000	4,000	3,000	3,000	3,000	3,000
Cost of Goods per OEM Module							400	300	250	250
Total Cost of Goods Sold					1,090,000	4,360,000	7,570,000	10,980,000	21,550,000	21,800,000
Market Penetration (Sensors)					0.21%	0.83%	1.67%	5.00%	10.42%	20.83%
Market Penetration (Modules)					0.10%	0.63%	1.67%	3.23%	5.52%	9.90%
Gross margin ($)					1,810,000	7,740,000	14,430,000	49,220,000	100,250,000	221,000,000
Gross margin %					62%	64%	66%	82%	82%	91%

Figure 28.3 Sample sales projections 2025-2030, Raydiant Oximetry, Inc.

Another important consideration for sales projections and staffing plan is whether to utilize a direct sale forces, an indirect sales force, or a combination of both. The indirect approach is capital efficient for the emerging medical device startup company but often the distributors will take a 30%–50% cut of the sales. Furthermore, sales reps from distributors have many products to sell and will often focus on products that generate the highest commissions. Sale reps from distributors may not be interested in customer feedback which is important to companies trying to learn and understand the market with new products. There are national distribution companies that have strong relationships with hospital systems, group purchasing organizations (GPOs), accountable care organizations (ACOs) and integrated delivery networks (IDNs). There are also specialized distributors that focus on single markets such as pediatrics and women's health. The sales reps from specialized distributors tend to better understand the clinical work flows in a particular area of care and have deep relationships with the clinician providers in that space. Finally, there are opportunities to partner with a potential strategic acquirer and utilize their existing sales force in the space. Strategic partnerships can lead to a cost-effective acquisition of the technology, but there have also been cases where the strategic partner violated the partnership by using inside information to copy the technology.

The alternative approach is to set up a direct sales team. Devices that shift the paradigm of care and/or are costly/complex are better suited for a direct sales force capable of educating users and building meaningful relationships with the customer. Emerging startups might launch a product with a small direct sales force (less than 10 reps) but ~300 sales reps are necessary to cover the United States, and this model ends up being prohibitively expensive. Companies interested in keeping a small direct sales force will often utilize a hybrid model with the in-house sales team educating and training the distributor reps. A hybrid model is often utilized for OUS (outside the United States) markets as the sales cycle and the process of closing accounts is vastly different in each country.

Budget requirements

The budget or operating statement/cost projections simply states how much capital will be required for operating expenses (OpEx) and manufacturing costs. The budget needs to tie into the staffing requirements and operating plan as outlined above.

Personnel salaries are often largest portion of the operating statement and the most valuable asset of a corporation is often the personnel. Hiring personnel who are subject-matter experts comes with a cost, but the cost to the company can be greater if the program is met with significant setbacks. There are significant geographic variations in salary requirements and start up companies may be competing with large public companies to attract a small pool of talented individuals. In these situations, offering equity and career advancing titles can help facilitate hiring.

Manufacturing costs include the cost-of-goods (COGS) for the raw materials. The COGS are estimated by a bill-of-materials (BOM) and the COGS will decrease significantly with scale of production. In addition, manufacturing labor costs and manufacturing facilities cost need to be accounted. Because the manufacturing of regulated medical devices is subject to quality control, audits, and inspections, a medical device start up may choose to outsource initial production to an FDA-approved contract manufacturer site.

Operating expenses (OpEx) account for the cost to run the business and develop the product. Initially when the company is precommercial, OpEx is the only expense. OpEx includes R&D costs + personnel, clinical trial costs, SG&A costs, and IP costs.

P&L statement

The P&L statement is simply an amalgamation of the staffing, operating and budgetary requirements into one document that founders use to support fundraising efforts. It is common for investors to request a 5-year or 10-year P&L to capture how future capital would be allocated to create a viable commercial business and create shareholder value.

Cash flow statement

Cash flow statements are relevant to companies that are in the manufacturing, sales and distribution cycle. A significant amount of capital is usually required to accumulate inventory of the product and depending on the sales cycle, there could be 1–2 years before capital expenditure is recovered. Understanding this cycle is paramount to ensuring that the company has reserved enough capital to support operations during these cash flow crunches.

When considering an acquisition of a technology, large corporations will calculate the Net Present Value (NPV) of a business to determine the potential value of the asset under consideration. Net present value is the present value of the cash flows at the required rate of return of the asset compared to the initial investment. NPV is also the method used by financial analysts to calculate a return on investment by acquiring a specific asset. The NPV allows a third-party to perform an "apples to apples" comparison of the investment upside of the business.

Other important metrics to consider are when the company will be "revenue positive" and "cash-flow positive."

Exit scenarios

In the medical device ecosystem, the most common liquidity events have been either a Merger/Acquisition (M&A) by a larger corporation or an Initial Public Offering (IPO). Investors and team members will want to know how and when their equity and/or stock options in the corporation could realize a return on investment.

First time founders often assume that their technology will get acquired upon FDA clearance but the market data as of 2020 has not supported that scenario. According to the 2020 Silicon Valley Bank (SVB) HealthCare Report on M&A activity in the MedTech space. Since 2015, of the 45 acquisitions that went through 510(k) clearance, 42 of the 45 acquired technologies were in commercial sales. For De Novo devices, 4 of the 4 acquired technologies were FDA cleared but precommercial and for PMA devices, 26 of the 30 acquired technologies were pre-FDA approval and precommercial. Using data to support the business plan gives confidence to investors that the business plan has been well thought out and often inspires confidence for investing in the management team.

Think like an investor

Investors have a fiduciary responsibility to deliver a significant return-on-investment to the LPs (Limited Partners) of the fund. By tying the operating plan with the business opportunity, founders can promote their opportunity in a manner that gets investors excited about the deal.

Seed investors typically expect a 15 × return on their capital investment while Series A investors tend to expect a 10 × return on investment. And Series B investors are attracted to deals that can deliver a 5 × return in less than 5 years.

Case example

Medical device start-up company "ANES" has developed a new device to improve the success with arterial-line placements. The technology was developed by an anesthesiologist and biomedical engineer working together at a university based medical center and they have negotiated an exclusive license from the university to commercialize the technology. Through SBIR grants and self-funding, the team has been able to build a prototype and demonstrate a proof-of-concept in a large animal model and cadaver studies. The team would like to raise a seed round of $1 million to build a prototype for a human feasibility study and meet with the FDA to develop a plan to achieve regulatory clearance.

The team is successful with the seed round raise and seed round milestones and now plans to raise a Series A round of $5 million to hire a team, complete the final product development and achieve FDA 510(k) clearance. The founding team has hired experienced MedTech executives who have a history of successfully taking early-stage medical devices to acquisition.

The team's exit plan is to get acquired by a large medical device corporation. Initial communication with several large corporations that distribute similar products reveals that they would like to see $20-$30 million dollars of annual sales before considering an acquisition of the assets. The recent comps from Silicon Valley Bank's Health Care report suggests that a device with 510(k) clearance and $20 million of annual sales could get acquired for $120 million.

The company achieves FDA 510(k) clearance from the Series A funds and develop an operating plan that targets $20 million of annual sales in 4 years. The company decided to raise a Series B round of $10 million to achieve "cash-flow positive" status by year 3.

Three years later the company is near cash-flow positive and a strategic acquirer decided to make an offer for the asset because they have a competing technology with an expiring patent. The company decides to use an investment bank to facilitate the transaction and the company is ultimately sold for $135 million. The transaction cost $3 million in legal and banking fees, and the university required a $2 million fee to transfer the license. Series B investors were paid $50 million for their $10 million investment (5× ROI). The founders, Seed and Series A investors are left with $80 million to divide up. The Series A investors get $50 million for their $5 million investment

(10 × ROI) and the Seed investors get $15 million for their early investment of $1 million (15 × ROI). The founders and company employees will then share the remaining $15 million remaining in this hypothetical case example.

This case has been simplified to illustrate specific concepts for educational purposes.

References

[1] Makower J, Meer A, Denend L. FDA impact on US medical technology innovation: a survey of over 200 medical technology companies. Arlington (Virginia): National Venture Capital Association; 2010.

[2] Silicon Valley Bank "Healthcare Investments and Exits", Annual Report 2023. Available from: https://www.svb.com/globalassets/trendsandinsights/reports/healthcare/2023/annual/healthcare-investments-and-exits-annual-report-2023.pdf.

29

Post launch physician relationships

Gary Haynes

Department of Anesthesiology, Tulane University School of Medicine, New Orleans, LA, United States

Chapter outline

Abstract

Worldwide the public embraces medical technology. It is credited for improving care and patients believe that newer tech is better and more likely to assure good health. As much as it is loved by the public, many authorities are concerned about the economic impact as medical technology and pharmaceuticals may be significant drivers of national healthcare costs. Medical products exist in a complex healthcare world that includes issues of privacy, patient-physician relationships, and highly competitive and regulated markets. In this environment, both the entrepreneur with a bright idea and the well-funded R&D department of a Fortune 500 company are competing for

Innovation in Anesthesiology. DOI: https://doi.org/10.1016/B978-0-12-818381-6.00028-0

the attention of physicians and acceptance of their ideas. Entrepreneurs face major challenges because innovation is not cheap: R&D costs for new technology and drugs can be staggering. There is no single estimate for the cost of developing a medical device, but with a wide range of products in development, entrepreneurs should expect considerable variability in costs. In contrast, bringing a new drug to market now costs nearly $1 billion. Building a relationship with the physicians who might use a new product is but one part of leading to a successful product. Planning how to build that relationship begins long before the launch of a new product.

Keywords: Medical technology; physicians; pharmaceuticals; R&D; healthcare; environment

Introduction

Worldwide the public embraces medical technology. It is credited for improving care and patients believe that newer tech is better and more likely to assure good health. As much as it is loved by the public, many authorities are concerned about the economic impact as medical technology and pharmaceuticals may be significant drivers of national healthcare costs. Medical products exist in a complex healthcare world that includes issues of privacy, patient-physician relationships, and highly competitive and regulated markets [1]. In this environment, both the entrepreneur with a bright idea and the well-funded R&D department of a Fortune 500 company are competing for the attention of physicians and acceptance of their ideas. Entrepreneurs face major challenges because innovation is not cheap: R&D costs for new technology and drugs can be staggering. There is no single estimate for the cost of developing a medical device, but with a wide range of products in development, entrepreneurs should expect considerable variability in costs. In contrast, to bring a new drug to market now costs nearly $1 billion [2]. Building a relationship with the physicians who might use a new product is but one part of leading to a successful product. Planning how to build that relationship begins long before the launch of a new product.

Elements of marketing to healthcare

For the experienced and the novice entrepreneur, marketing may be a vague concept somehow related to a business model.

In a classic article, marketing was defined as "the activity, set of institutions, and processes for creating, communicating, delivering, and exchanging offers that have value for customers, clients, partners, and society at large" [3]. Those activities are intertwined with the company's business model, and they should support the goals of demonstrating a product's value, distributing, and collecting on that value. More specifically, a *marketing strategy* identifies the market(s) of interest and how to convey the value proposition of a product or service. The goal is to demonstrate a competitive advantage for a new idea and secure sales. Once a product is launched, marketing continues until it is surpassed by more advanced products or is no longer competitive.

Our focus is on how to build relationships with physicians in healthcare markets. In most instances, physicians are the intended target since many products are designed for use by physicians, particularly devices intended for clinical practice applications. Not all new products reach the market by sales and distribution to end users, as some startup companies utilize other routes such as licensing or company acquisition. There is a series of phases in developing a marketing strategy to bring new products to market [4]. These phases can be organized as:

- Research phase—learning and identifying market(s) and set of customers and their concerns
- Focus phase—defining which market and customers to target
- Determination phase—stating the value proposition of the new product
- Formation phase—establishing a strategic plan to communicate benefits, competitive value, distribution, and compensation

Introducing new medical products can be very expensive. In 2019 a study published in the Journal of the American Medical Association documented marketing spending increased from $17.1 in 1997 to $29.9 billion by 2018, and most likely that was an underestimate [5]. Some part of that was used to market devices and technology, but there is little direct data that shows where the money is spent. Marketing to medical professionals may use several channels, and direct contacts with doctors and advertising account for much of that spending. This portion of the marketing effort increased from $15.6 to $20.3 billion over the 20 years and has undoubtedly increased even more in the years since then.

Transformation in healthcare is driven by gaining medical knowledge. As valuable as the progress in medicine has been, it

Table 29.1 United States healthcare expenditures and the US Gross Domestic Product, by decade.

	1960	1970	1980	1990	2000	2010	2019
NHE[a]	$27.1	$74.1	$253.2	$$718.8	$1365.6	$2589.7	$3795.4
US GDP[b]	$542.4	$1073.3	$2587.3	$5963.1	$10,252.3	$14,902.1	$21,433.2
NHE/GDP (%)	5.0	6.9	9.8	12.1	13.3	17.4	17.7

[a]National Health Expense.
[b]United States Gross Domestic Product, in billions [6].

has come with increased costs. The expansion of medical knowledge and the translational research that accompanied it partially explains why the United States spends much on medical care (Table 29.1).

The Rapid growth in US healthcare expenses began following the creation of Medicare and the combined federal-state Medicaid programs were created in 1965. They expanded healthcare coverage and fueled demand for more medical services. During the same period, federal investments were made in medical research that resulted in discoveries, spurring additional developments in medical care.

Market environment: a few basics about US healthcare

Some insight into the relationship between physicians and healthcare services is necessary to understand how relationships with physicians can be established. The public views physicians at the pinnacle of the healthcare system, and it would be natural to assume they should be the target of marketing strategy. In some instances that is true, but with the large healthcare systems that exist today they will likely be involved in buying decisions but lack final decision-making authority.

The organization of medical practice started changing in the 1980s. Medicare, Medicaid, and other third-party payer regulations resulted in doctors and practice administrators devoting more time and effort to managing professional charges and revenue collection. As the effort and expense required for business operations increased, doctors increasingly turned to experts with business experience. Larger and more specialized medical group practices formed with some growing into multispecialty groups.

Simultaneously, hospitals and academic medical centers expanded, leading to today's healthcare landscape dominated by large hospital systems covering several states. Implementation of the Affordable Care Act in 2008 and a changing economic environment over the past decade resulted in significant hospital consolidation, adding to the changing landscape.

During the past twenty years physician practices have experienced decreasing payments from Medicare and Medicaid programs. Decreasing payment for care has accelerated consolidation, leading to the formation of larger practice groups improving productivity, and gaining leverage when negotiating contract rates with health insurance companies. One result has been the growth of medical management companies, several of which now employ more than a thousand doctors. Private equity and venture capital firms have invested in some of these large practice management companies, and a few are now publicly traded corporations. Several hospital systems also directly employ thousands of doctors now. Hospital consolidation has also accelerated as provisions of the 2008 Patient Protection and Affordable Care Act have been enacted during the past decade. The economic fallout from the SARS-nCov-2 pandemic is now another catalyst promoting consolidation. Hospitals and management companies acquired 20,900 physician practices from 2019 to the end of 2020. In 2020 alone 48,400 physicians left independent practice for employment by hospitals, health systems, or corporate entities. In the United States, the greatest increase in hospital employment of doctors was in the Northeast US region, and the greatest increase in corporate-employed physicians was in the South. By the end of 2020, the effect of the social and economic disruption created by the pandemic resulted in almost 70% of physicians employed by healthcare systems or corporations [7]. The trend for physician employment by large systems continues to increase. With centralized management and physicians as employees, decisions about new technology and medical devices increasingly one requiring engaging many stakeholders.

Regardless of the healthcare system model, doctors almost always have input, express preferences, and make recommendations for new products. However, in large group practices, hospitals, and large integrated healthcare systems ultimate decisions for new products often rest with value analysis committees. Value analysis is typically assigned to an interdisciplinary committee that evaluates medical supplies and equipment. These committees most often involve a variety of stakeholders such as clinical leaders, service line leaders, operating room directors, financial directors, and others needed to implement new products. A value analysis committee considers clinical and financial considerations in addition to physician preferences when recommending the purchase of new items. Purchasing new products may include decisions about operationalizing and standardizing products throughout a hospital system. Successfully marketing depends on documenting cost-effective clinical use and building physician relationships is one aspect. Having physician champions and other healthcare providers for a product is a component of the process. System and hospital administrators must be convinced of the value that new technology delivers to the clinical ecosystem. The medical staff have influence even when they lack financial decision-making authority because administrators don't want to antagonize or alienate physicians, particularly when it involves equipment decisions that are key to doctors' clinical practice. The challenge for the entrepreneur is to engage physicians, allied healthcare providers, and administrators.

Marketing and the approval process

Medical device development and new drug discovery occur along very different paths and governmental approval for each occurs in distinctly different ways. Attaining FDA approval for new medical devices can be challenging, and physician involvement in the product development phase may benefit from gaining eventual approval.

New drug development occurs in highly structured research environments within pharmaceutical companies or in startup companies created to pursue drug development for a specific disease or investigate a related group of compounds. These firms are often financed by venture capital and other limited partner investors. New drugs are usually not developed by individuals alone since the process requires highly specialized technology. Pharmaceutical companies and biotech startup firms usually employ teams with advanced degrees in medicine, pharmacology, immunology, neuroscience, bioinformatics, and specialized branches of organic and biochemistry.

The US Food and Drug Administration (FDA) has a Congressional mandate to protect the public from products that may be harmful or lack proven efficacy. The FDA has at times been embroiled in controversies about its process and decisions, but it strives to high standards and their procedures have brought breakthrough drugs to the US market while ensuring safety [8]. The FDA uses a three-step process for new drugs. Once a promising compound is discovered, the first step is demonstrating its safety. A safe and effective dose is determined by administering a study drug to study subjects in escalating doses until a 33% incidence of toxicity is observed, then dropping back to the next lower drug dose, and accruing additional study subjects, all while observing for adverse events. This lower dose becomes the designated maximally tolerated dose. If safety and an effective dose are established, the investigative drug may progress to larger trials to demonstrate efficacy. For ethical and efficiency reasons the initial step involves the smallest number of study subjects possible. Both goals must be accomplished before moving to a later phase of clinical trials where a larger number of subjects are enrolled for proof of effectiveness for specific indications [9].

Regulatory approval: medical device approval in the United States and Europe

Ideas for medical devices often originate with individuals or small groups. Medical devices must also be approved, but many of them are not regulated as rigorously as pharmaceutical products because the approval process is based on an assessment of potential risk to the patient. Quite simply, a very low-risk device such as a tongue depressor does not require the same review as a pacemaker. In the US a device is classified to reflect the potential risk it carries to patients. In the EU a medical device is also evaluated based on a risk assessment scheme but using different criteria to determine the risk level [10]. In the United States, the FDA Center for Devices and Radiologic Health (CDRH) is responsible for regulatory approval of medical devices. The process differs from pharmaceuticals as approval of medical devices is based on the degree of risk posed by the device (Table 29.2). Class I devices pose a low risk and do not require pre-marketing approval. Class II devices have some risks

Table 29.2 US Physician Labor Force as of 2015.

Total number of US doctors and their activity level as of 2015	
Total Doctor of Medicine (MD and DO)	1,085,783
Total number of active doctors	870,900
Doctors actively involved in patient care	870,900

associated with their use and can be approved through the 510(k) process. This involves demonstrating a new device is substantially equivalent to another device already approved by the FDA. If equivalency to a predicate device is established, it is assumed the new device will be at least as safe and as effective. Many new medical devices are approved on this basis [11]. Unfortunately, there are notable examples where this process has not protected the public. However, the process remains in use [12]. Class III devices are considered high-risk devices and like drugs, their approval depends on demonstrating both safety and effectiveness. Demonstrating equivalence with a predicate device can facilitate gaining FDA approval for a Class II device. Nevertheless, the approval process is comprehensive and includes evaluation of safety data, and technical, non-clinical, and clinical data demonstrating effectiveness. Since many products seek 510(k) approvals, the FDA submission requirements are not the same. The FDA provides device-specific guidance for the information they expect in an approval submission. i

Clinical experts may serve as consultants at any stage of medical device development, and their familiarity and expert knowledge from early-stage development can be helpful when navigating the FDA approval process. Clinicians' understanding of a new product and its competitors may be invaluable when responding to concerns raised by FDA reviewers. Consultant physicians involved in the development and approval phases can bring credibility to the product and help make the case for new device approval.

Successful strategies for post launch physician relationships

Physician engagement strategies usually involve direct contact with doctors in clinics and hospitals, advertising in medical journals, and company presence at professional meetings.

These methods remain in common use in pharmaceutical marketing because frequent contact with physicians has been shown to result in greater use of a drug [13]. Whether this principle applies equally to medical devices is not certain.

Strategies to engage doctors require understanding their concerns and interests. Medical doctors are highly educated, with significant time devoted to post-graduate medical training. Entering the medical profession requires a minimum education in science and math, and many in clinical practice also have additional graduate degrees. Doctors have extensive clinical training, and they tend to solve problems using analytical approaches. Physicians will judge new products based on the supporting clinical evidence. When considering new products medical doctors will consider several factors in addition to the cost of purchase including how the use of a new product, or technology may enhance their value in the medical community, and the time and effort required to learn how to use new technology.

Healthcare systems are competing for patients, particularly in surgery and other procedural specialties. There is also competition among surgeons and other specialists, such as cardiologists and gastroenterologists to develop outstanding reputations in their fields. A business with a new product can exploit its ambitions by identifying early adopters for new products and technologies. Finding early adopters to be product champions can establish credibility and gain attention in the physician community.

Understanding customers: physicians, hospitals, and hospital systems

The United States has more than 1 million individuals with medical degrees (Table 29.3). Of those approximately 870,900 are actively working in healthcare, with 827,261 involved in patient care.

Statistical data reporting physician specialties from 2021 shows the distribution of physicians among medical specialties.

Table 29.3 Distribution of physicians in major medical specialties.

Specialty	Number	Percent
Psychiatry	55,878	10.1
Surgery	53,872	9.8
Anesthesiologists	50,921	9.2
Emergency medicine	58,836	10.7
Radiology	48,767	8.8
Cardiology	33,368	6.0
Oncology (cancer)	21,409	3.9
Endocrinology, diabetes, and metabolism	8377	1.5
All other specialties	220,882	40.0
Total specialty	552,310	100.0

Altogether these groups represent approximately 60% of the medical doctors in active practice (Table 29.3).

This number is segmented into many small groups, and the estimated number of anesthesiologists is 50,921.

Developing a market strategy to build relationships with physicians requires an understanding of their work environment. Throughout the United States and Europe, physicians are experiencing increasing workloads and economic and administrative requirements. Worldwide physicians are frustrated with electronic health records, societal expectations, and government regulations. A 2018 survey conducted on behalf of The Physicians Foundation, an organization of state medical societies, reported that only 30% were working independently as individuals or in groups, 78% experienced physician burnout, and 80% were working at full capacity or overextended. Almost half of the physicians surveyed planned a career change. A marketing strategy should consider how their products can improve the lives of overworked, stressed professionals who are anxious to make a career change (Figs. 29.1 and 29.2).

In bringing a new product to market the entrepreneur needs to identify who is ultimately responsible for final decision-making. While initial marketing efforts may be directed to physicians, it is hospitals or healthcare systems that often pay for it. Hospital administrators are presented continuously with requests from doctors for new equipment and supplies, so understanding how spending is prioritized matters. About 70% of hospital revenue comes from the operating room and procedural areas. These are the areas and services with positive margins. Hospital administrators realize older technology must be updated to retain surgeons and other proceduralists, and they want community recognition that their hospital offers the most advanced technology. For economic reasons, some doctors have more influence over spending decisions because they largely determine if the institutions are profitable (Fig. 29.3). Clinical service areas are also difficult and intense areas of medical practice. Administrators recognize state-of-the-art technology is important to maintaining a high-performing and satisfied medical staff.

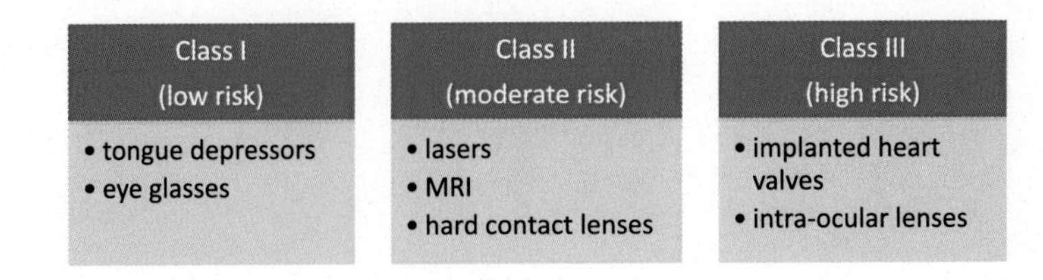

Class I
(low risk)
• tongue depressors
• eye glasses
Class II
(moderate risk)
• lasers
• MRI
• hard contact lenses
Class III
(high risk)
• implanted heart valves
• intra-ocular lenses

Figure 29.1 US FDA risk categories for medical devices with examples.

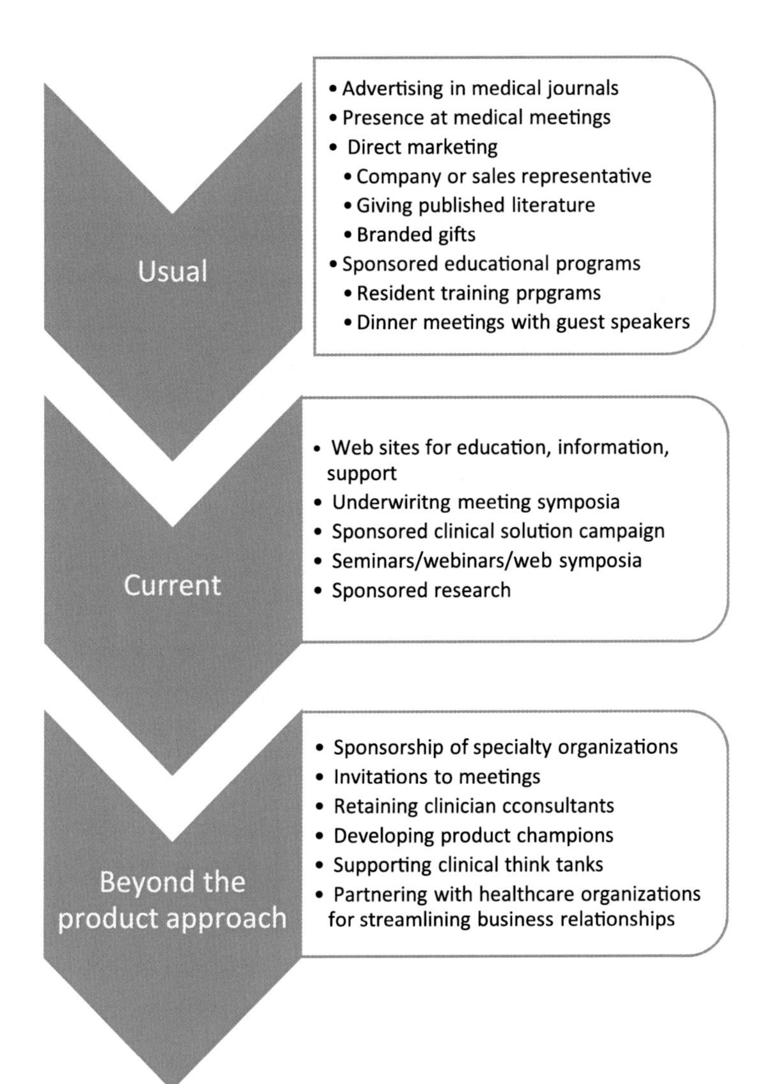

Figure 29.2 Marketing approaches for new medical products.

Understanding how a new product impacts clinical practice is important. New medical devices usually are not disruptive to other physicians, but their implementation can alter the environment. As an example, the introduction of laparoscopic surgical procedures provided advantages for patients but also resulted in longer surgeries and required more OR time. New products may generate pre-launch discussions and even controversy among many other physician groups. Consideration in advance about how a new device may affect other physicians, nursing staff, and hospital resources can avoid controversies following the product launch.

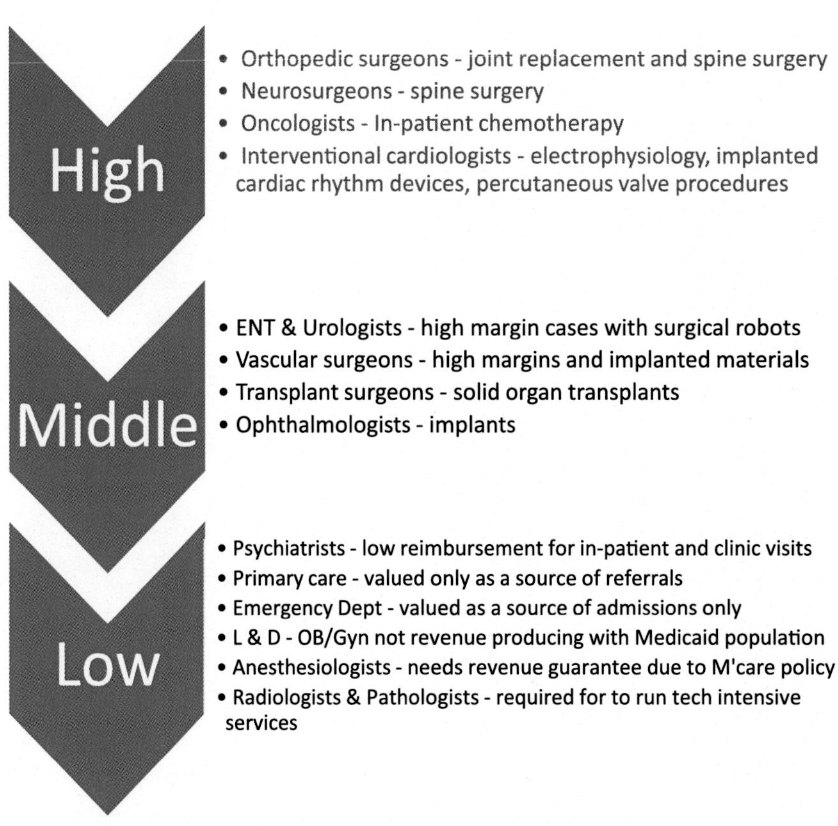

Figure 29.3 Relative value ranking of physician groups essential to hospital function.

Physician attitudes toward industry are important and they are aware the public may be skeptical about the relationships between physicians and industry. The public will hold a positive view of new medical devices provided they believe they truly improve medical care. When products are misrepresented by physicians, or if there is the appearance of a hidden financial agenda benefiting a doctor, public sentiment can sour very quickly. On the other hand, most physicians have a positive view of medical device companies, and among doctors, there is less controversy with the medical device industry than with the pharmaceutical industry.

Ethics and physician attitudes toward industry

When marketing to physicians it is important to know the legal and ethical boundaries that must be respected. Doctors are exposed to medical product marketing from the beginning of their careers. Pharmaceutical reps contacting medical students and residents in training was at one time a universal practice. Relationships between drug reps and young physicians started with free drug samples, branded gifts, and providing breakfasts and lunches at the hospital. The medical device industry tended not to use this approach with physicians in training as they had virtually no influence on purchasing decisions. Direct contact marketing to physicians came under scrutiny when healthcare costs began increasing, and the spotlight has focused primarily on the pharmaceutical industry. Concerns emerged that conflict-of-interest situations were influencing clinical decisions and drug prescribing practices. Professional organizations subsequently developed guidelines restricting access to trainees.

More marketing opportunities are possible once doctors complete training. Companies have built relationships with physicians and surgeons by involving them as consultants, researchers, and educators. The relationships range from basic agreements to complex relationships closely integrated with clinical practice or research programs. Examples of strategies include paying doctors for speaking on behalf of products, reimbursing them for educational meetings, participation in panel discussions, and consulting fees. Such relationships become problematic when the activities begin to blur the boundaries between responsibilities to patients and maintaining integrity in research and education.

Entrepreneur physicians who develop ideas for new devices can be justifiably compensated as founders and business partners. The compensation to physicians must be for real contributions or work done on behalf of a product. Ethical and legal issues can emerge regarding payments to physicians when the standard is not met and failure to meet standards of legitimacy raises the question if payments are nothing more than sham consulting fees, or a kick-back arrangement for using their products. This has been a subject with marketing drugs, but more recently there have been many investigations into practices with implanted surgical devices, particularly involving orthopedic surgeons and implanted surgical devices in spine surgery. In the past decade, there have been more than 100 federal fraud and whistleblower actions in the United States related to spine implants alone. Additional concerns have been raised in these cases whether patients need surgery if incentives were paid to a surgeon for implanting a device. When issues are raised about questionable marketing of prescribed drugs an immediate solution is the patient can stop taking the drug. When issues arise about implanted surgical hardware the resolution is not a simple matter [14,15].

In the US pharmaceutical and medical device manufacturers must report payments they make to physicians, teaching hospitals, nurses, physician assistants, and other specialized clinical personnel to the Open Payments Program operated by CMS (openpaymentsdata.cms.gov). Reporting payments made to covered recipients is available online for public review. The types of payments reported are general payments, research payments, ownership, and investment interests. Reported payments must include honoraria, grants and research contributions, medical education programs, facility fees, travel and meals, and entertainment expenses. Any ownership interest by a physician or a physician's family member is also reported. The data is easily searchable and is used frequently by the media in reporting healthcare issues. Open Payments is designed to be easily used by the public to provide greater transparency in US healthcare.

Physician–industry relationships

Graduate medical programs have always had limited financial support, and funding for educational programs has come from clinical revenues unless outside support could be found [16]. Pharmaceutical companies stepped in to use this funding gap to build relationships with doctors. Medical device companies have done so to a lesser extent, but medical device and instrumentation companies typically occupy a smaller market niche in procedural areas and surgery. In a survey among a group of New York hospitals, physician attitudes toward both the medical device and pharmaceutical industries were studied. Physicians surveyed held a positive view of both pharmaceutical and device manufacturers. They had no issues with small gifts related to medical practice, textbooks, and modest meals but larger gifts and vacations were inappropriate. Advertising professionals will not be surprised that the respondents believed that it was only other physicians, and not the surveyed doctors, who were influenced by industry marketing. Surgical specialties had a more favorable attitude toward industry support of residency education programs than primary care specialists. Among physician groups, pediatricians had the least favorable view of industry support of their educational programs.

As the use of implanted devices has become more common, surgical specialists and hospitals allow company reps to be present during surgery to provide materials and to serve as a technical resource during surgery. To maintain professional boundaries, US medical schools and the American Association of Medical Colleges developed policies that place restrictions on industry practices. Despite this, physicians are often not aware of their institutional policies regarding interactions with industry [16] and many high-profile violations of recommended practices have occurred [17].

Developing a market strategy—product development and promotion meeting customer needs

Product development

Ideas for medical devices range from a simple solution to an everyday task to devices for treating a specific disease or injury. With such a wide-ranging possibility there is no uniform approach to engaging physicians. Basic marketing approaches may be enough for simple devices and disposables. Relationship marketing directed at doctors, advertising, and presence at medical meetings may be adequate to gain their attention. Products originating from intense research and design efforts might involve physicians in the early stages of development. In these cases, the doctors involved will likely emerge as thought leaders in the field and they may continue as advocates, educators, and mentors to physicians following product launch.

The entrepreneur's problem is how to compete for the attention of intelligent, discerning buyers. If any prospective customer requires a compelling value proposition, it is doctors. Their decisions require information from clinical studies that confirm what they are told about a product. It is a wonderful advantage for the entrepreneur when their new product has no competitors. Unlikely as that may be, a new device must stand out from the rest of the field. A value proposition that goes right to the point is necessary to gain the attention of doctors. They want to hear how it will solve a problem and why your device is better than a competitor's. Delivering the value proposition clearly and directly distinguishes your device from all others in the marketplace. For game-changing devices, the message needs to signal to doctors they need to invest time and energy into learning about a new product. Learning new techniques requires effort and unless there is a compelling reason to do so, doctors do not want to change how they learned to perform procedures during residency training. Surgeons are competitive, and they want to be recognized for using the most advanced methods in their practice. How the value proposition for a new product is communicated can inspire them to change.

Physicians who are part of the product development team may participate in clinical trials for regulatory approval. These doctors are entrepreneurs and early adopters who can best communicate value to other physicians. They can best explain the supporting data and the ones who have the most experience with a new product. Communicating their experience builds a bond with potential physician customers. Their experience, especially when described with specific stories of how a new product made a difference, carries considerable weight with other doctors.

Devices with the potential to change medical practice may involve new technology that is expensive and complicated to use. Implanted devices, laparoscopic, and robotic surgical equipment are representative examples. Products like these must show demonstrated value for improved patient outcomes. Potential hospital system and physician customers are obligated to thoroughly evaluate their value since their patients who will receive these procedures or devices can never evaluate the options themselves. This is an example where doctors involved in pre- and post-market clinical trials can be effective peer-to-peer advocates. All physicians value hearing first-hand the experience of other physicians when it comes to medical practices.

Promotion—online interaction to gain the attention of physicians and hospital systems

The medical device market historically relied on face-to-face meetings at medical conferences and sales reps to make direct contact with physicians. The Internet expanded direct physician marketing to provide wider dissemination of medical information (Fig. 29.4). Online presence is essential to business operations. Websites are a source of advertising, education, feedback, and customer support. Direct-to-patient promotion of drugs began in the mid-1990s when the Federal Government relaxed rules regulating pharmaceutical advertising. The medical device industry once lagged the pharmaceutical industry in direct-to-patient advertising but now ads on television, in social media, and on other internet platforms promoting various medical devices are common. Numerous medical devices are now sold over the counter, such as blood pressure monitors and glucometers, or by prescription, for example, contact lenses. The US Food and Drug Administration shares responsibility for oversight of the direct-to-consumer advertising of medical devices with the Federal Trade Commission [18].

Analytic tools generate data by measuring interest in a product, how markets respond to messaging, and guiding changes

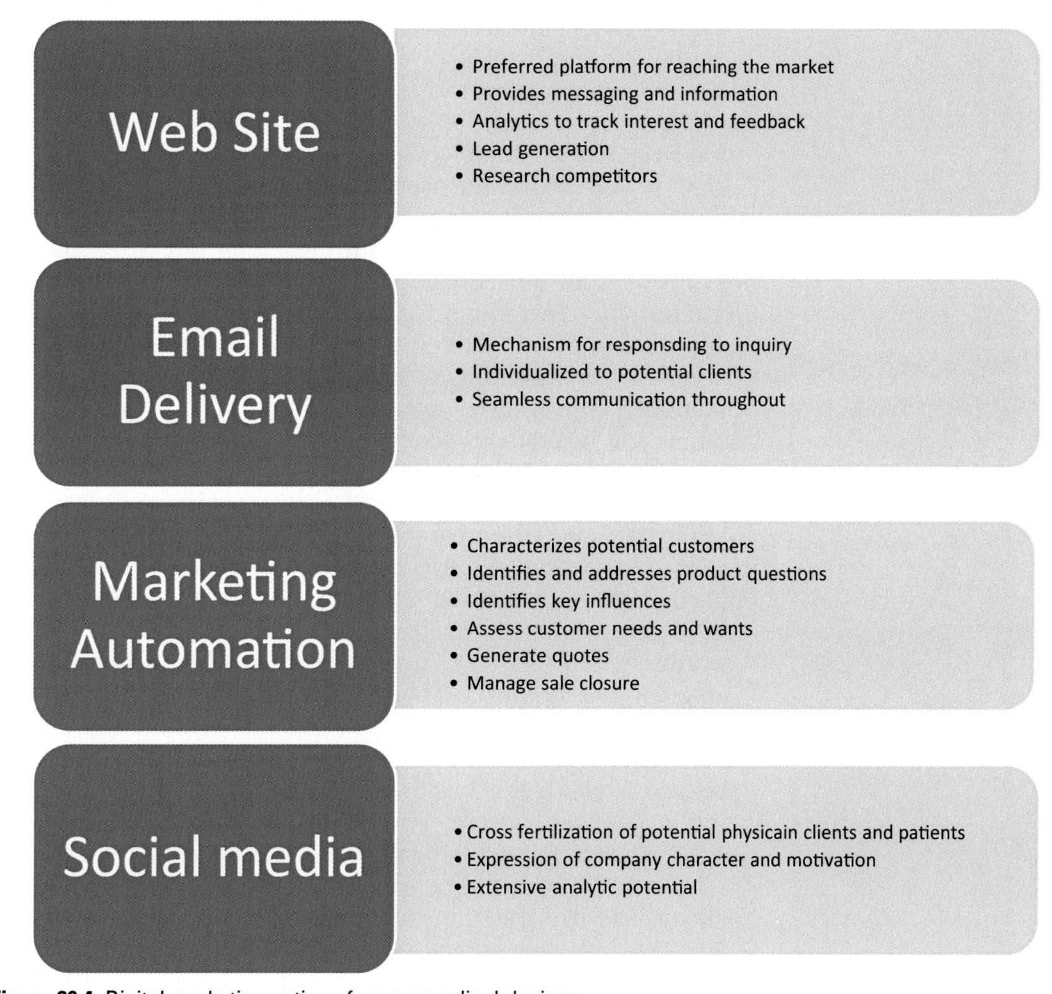

Web Site
• Preferred platform for reaching the market
• Provides messaging and information
• Analytics to track interest and feedback
• Lead generation
• Research competitors
Email Delivery
• Mechanism for responsding to inquiry
• Individualized to potential clients
• Seamless communication throughout
Marketing Automation
• Characterizes potential customers
• Identifies and addresses product questions
• Identifies key influences
• Assess customer needs and wants
• Generate quotes
• Manage sale closure
Social media
• Cross fertilization of potential physicain clients and patients
• Expression of company character and motivation
• Extensive analytic potential

Figure 29.4 Digital marketing options for new medical devices.

that make messages more effective. Data from clinical trials and expert opinions can be disseminated to inform clinicians in the way most convenient for customers.

Medical specialty organizations rely on revenue generated from technical exhibits at annual meetings to support their work. The travel and meeting restrictions caused by the COVID-19 pandemic would have had a greater impact on these organizations if they had not pivoted to virtual meetings that included corporate technical exhibits. Advanced meeting software and the incorporation of virtual meeting platforms allowed the business of healthcare to continue. Virtual meeting platforms are now yet another way to inform doctors and healthcare providers about new products.

Social media is the newest digital route for marketing new products. Rather than rely on physician-centered marketing alone, social media and websites go beyond to reach patients directly. Social media strategies can build brand recognition by reaching patients, who will in turn ask their doctors about new technologies for their medical care. Social media provides a platform for physicians and hospitals to inform the public when they adopt new medical devices, providing a way to stand out from the competition. Estimates are that more than 80% of the global population with Internet access have used it to search for information and healthcare recommendations. Not only are searches for information about medical products, but the public searches for the experience others have with doctors and treatment options. In one example a study on Internet use in North America and Asian countries revealed that 50% of Americans and 80% of Koreans and Hong Kong citizens exchanged peer-to-peer information about medical experiences on Facebook and Twitter. Two-thirds of Americans, and more than 85% of Koreans and Hong Kong residents used blogs for health information [19].

Medical problems are very personal and emotionally charged issues. While all products employ digital marketing, when the message is about our health it must be clear and accurate and express care and concern. A social media platform can be used to engage physicians and patients, but it can be used by the entrepreneur to describe the personal character of their company, so patients feel like they are interacting with an empathetic company. When using social media platforms, the interactions between followers and a product, or company are quantified. This information can show trends and develop sales leads.

Customer input is usually spontaneous and often anonymous. Monitoring responses can provide insight into a product's strengths and weaknesses. These platforms provide a measure of positive or negative feedback that can be used to improve the product and advertising messaging.

The next step in med-tech-physician relationships

Major medical device companies are creating partnerships with hospital systems and academic medical centers in the United States and Europe. Healthcare executive-initiated partnerships with medical tech companies are a strategy to engage industry leaders to decrease medical expenses and improve quality. These developments are not led by physicians but require medical technology companies to build expanded relationships with physicians. This may be challenging since physicians may have a long-standing relationship with competitors. This movement began before the COVID-19 pandemic and is now expected to continue rapidly. The strategy involves aligning and integrating a company's suite of products to improve efficiency. Consequently, it is not a matter of a single product but connecting their range of products with software solutions and implementing the entire package. In part this is driven by health system executives frustrated dealing with multiple representatives from a single company. They want a streamlined presence from med tech companies that provide fully integrated solutions. Some examples include the partnership between Stryker and a large Texas hospital system designed to improve safety. The hospital will implement the full suite of Stryker products in what is called a "Journey to Zero" in medical complications and errors. Another example is Medtronic promoting its "Integrated Health Solutions" partnerships to improve cost and efficiency.

This hospital system strategy may create tension between major healthcare systems and medical technology companies reflecting some of the deep problems in changing health care. They also need to be aware of this dynamic and how it may influence efforts to establish relationships with physicians. The fledging entrepreneur attempting to enter the medical product space needs to be aware the desire for integrated solutions could present an opportunity for them. Whatever their service or device, entrepreneurs may need to convince customers their products can align with other devices and that their product possesses the expected interoperability. As this business model evolves it likely favors larger, established companies. A new entrepreneur may need to convey to physicians that they can respond faster and integrate their products better than a large corporation [20].

Striking contrasts: failure and success with physician relationships

Entering as a new entrepreneur often means working alone, or as part of a small group to create a new product that solves a problem. It involves establishing a company to develop and distribute a new idea. This is no small task, even if the new product is a simple one. Entrepreneurs face challenges with financial backing and business expertise and unfortunately, most fail. The odds are stacked against a new start-up.

However, a large corporation is not guaranteed success. Established companies may have more resources and experience with bringing new products to market, but they can make mistakes marketing their products and large size and bureaucratic corporate structures may not be an advantage.

New products can fail for many reasons and marketing issues may not fully explain why a new device doesn't make it to the market. However, two examples with different outcomes illustrate a contrast of how physician relationships were shaped during product development and after product launch.

Sedasys—computer-controlled sedation

In the early 1990s, a new drug to induce and maintain general anesthesia was introduced to the US market. Propofol had been in use in Europe and its approval was anxiously awaited in the United States. Propofol was superior to thiopental (Pentothal) which was the principal drug for inducing general anesthesia. When used for general anesthesia Propofol allowed patients to emerge faster from anesthesia and when titrated slowly it could also be used to sedate patients to keep them comfortable and free of anxiety during procedures. Propofol proved to be an ideal drug for sedating patients as well as a good drug for general anesthesia. Propofol quickly displaced thiopental and with increasing experience, it became widely used for sedation during diagnostic procedures.

During the 1990s there was also a national initiative to prevent colon cancer. Colon cancer has been a major health problem and screening colonoscopy was a way to identify and remove small benign polyps from the lining of the colon before they became malignant. The national effort aimed to prevent colon cancer using screening colonoscopies which quickly became standard procedures. Because the procedures are uncomfortable and, in some instances painful, the demand for moderate to deep sedation during colonoscopies also increased. Anesthesiologists routinely manage such situations since the FDA largely limited Propofol use to anesthesia personnel since they are trained to manage issues of respiratory depression and airway obstruction. Anesthesiologists found it difficult to cover this new expanding need because of staff limitations, problems coordinating scheduling with gastroenterologists, and low payments for services. Nurses could provide moderate sedation, but they lacked the qualifications to use Propofol as it is restricted to use by anesthesiologists. Consequently, sedation results were often unsatisfactory.

However, a device using complex pharmacokinetic models to administer propofol and fentanyl, a very potent opioid analgesic, could solve administrative problems and decrease the expense by automating sedation to colonoscopy patients. Ethicon Endo Surgery, a unit of Johnson and Johnson acquired a computer-driven system developed by Scott Laboratories (Lubbock, Texas) that could control the administration of propofol and fentanyl to patients while using patient vital signs and patients' feedback to a mild stimulus to sedate without overdosing. The system was called Sedasys and J&J's value proposition directed to hospitals and gastroenterologists was that Sedasys could eliminate the need for anesthesiologists and safely sedate patients for endoscopy procedures at a cost of about $150 per case. This price was much less than the hundreds, and sometimes even thousands of dollars in anesthesia fees when anesthesiologists and nurse anesthetist were involved [21].

In May 2009, an FDA advisory committee gave preliminary approval to the device provided safety considerations were met. Ethicon expected final approval would follow and sales would begin later in the year. The committee vote, however, was not unanimous with two nationally recognized academic anesthesiologists on the committee raising questions about safety. In the weeks and months following preliminary approval, the device was debated within the ASA and in public media. Articles appeared in the Wall Street Journal and NBC's The Today Show focused on safety issues with Sedasys. Discussions within the ASA raised questions about device safety and the qualifications of those who sought to use it. In response, the FDA issued a ruling in April 2010 soundly rejecting the recommendation of their advisory committee and informed Ethicon that their device was "not approvable." Ethicon did not give up but made refinements, and ultimately gained premarket approval for Sedasys. In May 2013. However, the approval came with restrictions limiting it to use only in healthy endoscopy patients. The ASA provided recommendations in early 2014 that anesthesia department directors should monitor quality issues associated with it, and recommended anesthesia personnel must be immediately available when it is in use. These actions nullified some aspects of the proposed advantages of using the system to replace anesthesiologists.

Da Vinci Surgical Robot—better surgical outcomes The da Vinci surgical robot was developed by Intuitive Surgical, Inc. (Sunnyvale, California) to provide surgeons with greater control with flexibility than is possible with traditional surgery. A da Vinci surgical robot has multiple arms that manipulate custom surgical instruments through ports surgeons insert in the patient. This requires only small incisions and avoiding a large surgical incision is one of the principal advantages of robotic surgery. The robot is directed by a surgeon sitting at a council. The control of the surgical instruments is very precise and can even compensate for a surgeon's shaky hand. The robot provides the surgeon with an excellent visual field and control of the cutting instruments. Precise control allows surgeons to carefully remove disease tissue, minimizing some of the undesired side effects of surgery. The value proposition for surgical robots is twofold: clinical and economic. Clinically patients have shorter hospital stays, fewer complications, less bleeding and transfusion requirements, lower rates of infection, fewer post-op readmissions, and more importantly, faster return to normal activity. The da Vinci is intended for cardiac, colorectal surgery, gynecologic, head and neck, urologic, and general surgery applications.

Sedasys was introduced in a controlled product launch at the ASA annual meeting in October 2014. Ethicon had planned to introduce Sedasys slowly in selected medical centers while involving anesthesiologists in its implementation. The plan did not capture interest and sales failed to take off as the company had expected. With skepticism from the anesthesiology community, Ethicon announced that it was halting sales of its automated system in March 2016 [22].

Da Vinci Surgical Robot—better surgical outcomes

The da Vinci surgical robot was developed by Intuitive Surgical, Inc. (Sunnyvale, California) to provide surgeons with greater control with flexibility than is possible with traditional surgery. A da Vinci surgical robot has multiple arms that manipulate custom surgical instruments through ports surgeons insert in the patient. This requires only small incisions and avoiding a large surgical incision is one of the principal advantages of robotic surgery. The robot is directed by a surgeon sitting at a council. The control of the surgical instruments is very precise and can even compensate for a surgeon's shaky hand. The robot provides the surgeon with an excellent visual field and control of the cutting instruments. Precise control allows surgeons to carefully remove disease tissue, minimizing some of the undesired side effects of surgery. The value proposition for surgical robots is twofold: clinical and economic. Clinically patients have shorter hospital stays, fewer complications, less bleeding and transfusion requirements, lower rates of infection, fewer post-op readmissions, and more importantly, faster return to normal activity. The da Vinci is intended for cardiac, colorectal surgery, gynecologic, head and neck, urologic, and general surgery applications.

The surgical robot was introduced in 2000 and the market for these devices has grown to $20 billion. Intuitive is now the market leader with 60% market share and has a market capitalization of $117 billion and more than 4,400 installations. The starting price for a da Vinci robot begins at more than $1 million, and the supplies and surgical instruments required to use it in surgery are expensive. Intuitive Surgical receives more revenue from the sale of disposable supplies, instruments, and service arrangements than from sales of surgical robots (Fig. 29.5) [23].

Surgical robotic technology is not without controversy, and its expense and the effect on increasing national healthcare expense is concerning. A report from the *New England Journal of Medicine* describes how surgical robots increased the cost of surgical procedures by 13%, adding more than $2.5 billion each year to the national healthcare system [24]. Despite the controversy associated with it, robotic surgery for prostatectomy procedures has grown rapidly (Fig. 29.6). Its use has expanded to several additional procedures and indications. Robotic hysterectomy surgery has also grown significantly and although it also adds to the hospital cost, the increase is modest (Fig. 29.7) [25].

LinkedIn

- Direct to business professionals and companies

YouTube

- Allows visual showcasing of medical devices and companies

Twitter

- A route for fast dissemination of information

Facebook

- Targets most relevant individuals

Sermo, Doximity, etc

- Targets medical professionals

Figure 29.5 Social media platforms of interest in medical device marketing.

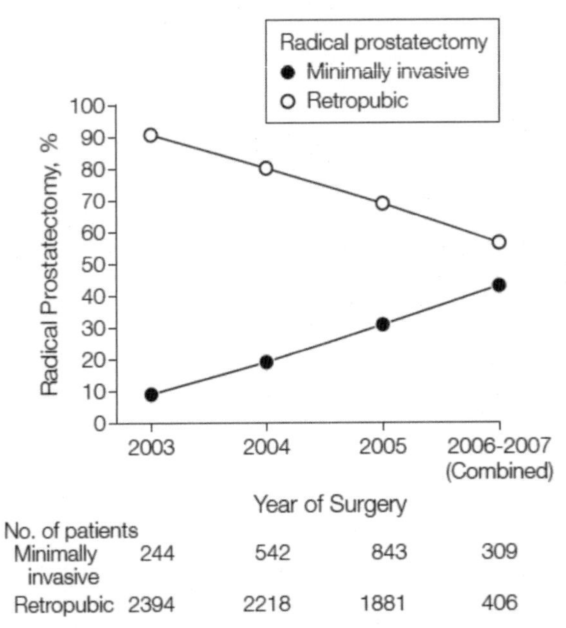

Figure 29.6 Increase in minimally invasive prostate surgery using a surgical robot compared to open surgical prostate surgery (Hus, JAMA, 2009).

Sedasys™		da Vinci™
Johnson & Johnson		Intuitive Surgical
Computer controlled drug delivery	Operating Principle	Minimally invasive surgery and enhances surgical dexterity
Automated, computer-controlled sedation of patients for GI endoscopy procedures • Safely sedates patients while • Eliminates need for anesthesiologist or CRNA • Reduces cost for screening colonoscopies	Value Proposition	Better surgical outcomes ▪ Less blood loss ▪ Less post-op pain ▪ Faster recovery time and return to normal activity ▪ Shorter hospital stays
"You won't need Anesthesia"	Physician Appeal	"You can be among the very best"
Gastroenterologists Hospitals	Marketing Target(s)	Surgeons Hospitals
None	Patient Marketing	Public campaign featuring surgeons
Limited	Product Launch	Methodical, selected surgeons and medical centers
Gastroenterologists • Widely aimed at Gastroenterologists Anesthesiologists • Negative messaging	Physician Relationships	Surgeons • Initially focused on selected Urologists • Used to build/expand practices • Hospitals – used to capture larger market share

Figure 29.7 Comparison of marketing success and failure: computer-controlled patient sedation with Sedasys versus surgeon-enhanced surgery with the da Vinci surgical robot.

Intuitive Surgical launched the da Vinci surgical robot in 1999 and received FDA approval as the first all-encompassing surgical robot for use in general laparoscopic surgery in 2000. The value proposition for using da Vinci is that it would result in better surgery measured by fewer complications, less blood loss, shorter hospital stays, and most of all, faster return to normal life for patients. More than simply citing the clinical benefits, their marketing campaign aimed a positive message right at surgeons: our device will make you better, or at least as good, as the very best surgeon. The message for hospitals was that it will set you apart from the competition. Many believe this was the primary message while messaging the clinical benefits was a secondary consideration.

Intuitive Surgical initially released its product to only a small number of hospitals and surgeons. One of the first was the Henry Ford Health System in Detroit, where the first robotic prostatectomy was performed in 2000. Once the surgeons and institutions in the initial launch sites mastered prostatectomy with the robot, they established a competitive advantage over other surgeons and hospitals in their community. As the others began losing patients, they recognized the value of this new technology and those groups started pursuing Intuitive Surgical for the da Vinci device. This created growing demand with surgeons clamoring to get the device and training so they too, would be considered among the best surgeons in their communities. Hospitals sought the device, despite its expense, to retain surgical patients and grow surgical volumes. The developers met growing demand with elaborate booth displays at national surgical meetings that served as education centers. Surgeons were invited to watch procedures broadcast live from those select hospitals that had the da Vinci robot. The surgeons who were the early adopters were enlisted by Intuitive Surgical to become educational proctors for other surgeons. The company utilized the teaching ability of surgeons with robotic experience but avoided ethical and conflict of interest issues by limiting financial compensation for proctoring sessions to $2,500. The compensation was not so much that it alone justified time away from their practice, the prestige, and gains in professional reputation for those doctors was compensation itself.

Marketing Contrast

The success or failure of a new product should not be attributed to marketing strategy alone because the design, function, and real value of a new device of course matters. However, when comparing devices that were successful with those that had potential, there are lessons in how relationships with physicians affect product outcomes. These cases demonstrate two lessons. Both devices had technical limitations but the value proposition of the da Vinci surgical robot was stated clearly with benefits for patients, surgeons, and hospitals. No one was excluded. The messaging for Sedasys did not convey benefits for patients, only there would be greater convenience for gastroenterologists and lower costs associated with sedation. A value proposition for patients was never included in the strategy while the message to anesthesiologists was, that they weren't needed. The company looked for a way to involve them only after a protracted, uphill regulatory battle that ended with a requirement for anesthesia involvement. The marketing plan for the Sedasys was still taking shape when the product was released, whereas the da Vinci robot was well-planned in advance and methodical. In both cases, the companies avoided the trap of overly incentivizing doctors to use their device.

References

[1] Ortiz SE, Rosenthal MB. Medical marketing, trust, and the patient-physician relationship. J Am Med Assoc 2019;321(1):40–1.

[2] Wouters OJ, McKee M, Luyten J. Estimated research and development investment needed to bring a new medicine to market, 2009–2018. J Am Med Assoc 2020;323(9):844–53.

[3] Grönroos C. Defining marketing: a market-oriented approach. Eur J Mark 1989;23(1):52–60.

[4] Denault, J.-F.. (2018). The handbook of marketing strategy for life science companies. Taylor & Francis Group, New York. https://learning.oreilly.com/library/view/the-handbook-of/9781351235280/xhtml/B10_fm1.xhtml.

[5] Schwartz LM, Woloshin S. Medical marketing in the United States, 1997–2016. J Am Med Assoc 2019;321(1):80–96.

[6] Centers for Medicare and Medicaid Services. (2021). National health expense, in billions of US$. National Health Expenditure Data. https://www.cms.gov/Research-Statistics-Data-and-Systems/Statistics-Trends-and-Reports/NationalHealthExpendData/NationalHealthAccountsHistorical.

[7] Physician Advocacy Institute (2022). Covid-19's impact on acquisitions of physician practices and physician employment 2019–2021. http://www.physicianadvocacyinstitute.org.

[8] Sharfstein JM, Charo A. The promotion of medical products in the 21st century: off-label marketing and first amendment concerns. J Am Med Assoc 2015;314(17):1795–6.

[9] Ivy SP, Siu LL, Garrett-Mayer E, Rubinstein L. Approaches to phase 1 clinical trial design focused on safety, efficiency and selected patient populations: a report from the clinical trial design task force of the National Cancer Institute Investigational Drug Steering Committee.

Clin Cancer Res 2010;16(6):1726–36. Available from: https://doi.org/10.1158/1078-0432.CCR-09-1961 March 15.
[10] Aronson JK, Heneghan C, Ferber R. Medical devices: definition, classification, and regulatory implications. Drug Saf 2020;43:83–93.
[11] Challoner DR, Vodra WW. Medical devices and health – creating a new regulatory framework for moderate-risk devices. N Engl J Med 2011;365(11):977–9.
[12] Curfman GD, Redberg RF. Medical devices – balancing regulation and innovation. N Engl J Med 2011;365(11):975–9.
[13] Hadland SE, Cerdá M, Li Y, Krieger MS, Brandon DL, Marshall BDL. Association of pharmaceutical industry marketing of opioid products to physicians with subsequent opioid prescribing. JAMA Intern Med 2018;178(6):861–3.
[14] Schulte F, Lucas E. Device makers have funneled billions to orthopedic surgeons who use their products. Kais Health N 2021. Available from: https://khn.org/news/article/spine-surgery-implants-device-makers-orthopedic-surgeons-kickbacks/.
[15] Schulte F. Surgeo*ns cash in on stakes in private medical device companies.* Kais Health N 2021. Available from: https://khn.org/news/article/spinal-tap-orthopedic-surgeons-cash-in-on-stakes-in-private-medical-device-companies/.
[16] Korenstein D, Keyhani S, Ross JS. Physician attitudes toward industry. Arch Surg 2010;145(8):570–7.
[17] Johnson J, Brumbaugh B. Conflict of interest in biomedical research and clinical practice. The Hastings Center; 2022. Available from: https://www.thehastingscenter.org/briefingbook/conflict-of-interest-in-biomedical-research.
[18] GAO. (2023). Medical Advertising: Federal Oversight of Devices. In gao.gov. Retrieved October 14, 2023. Available from https://www.gao.gov/assets/gao-23-106197.pdf.
[19] Song H, Omori K, Kim J, Tenzek KE, Hawkins JM, Lin W, et al. Trusting social media as a source of health information: online surveys comparing the United States, Korea, and Hong Kong. J Med Internet Res 2016;18(3):e25. Available from: https://doi.org/10.2196/jmir.4193 Published 2016 Mar 14.
[20] Gulati M, Henry J, Llewellyn C, Peters N, Simon C, Tolub G. Creating "beyond the product" partnerships between providers and medtech players. McKinsey Co 2019. Available from: https://www.mckinsey.com/business-functions/operations/our-insights/creating-beyond-the-product-partnerships-between-providers-and-medtech-players.
[21] Rockoff JD. September 26. J&J's Sedasys Puts Challenge to Anesthesiologists. Wall Str J 2013.
[22] Rockhoff JD. March 14. J&J to Stop Selling Automated Sedation System Sedasys: poor sales from a product that was opposed by anesthesiologists. Wall Str J 2016.
[23] Trefis Team. A look at in*tu*itive s*urgic*al' s *reven*ue s*o*urce*s* a*nd outlook.* Forbes 2018. Available from: https://www.forbes.com/sites/greatspeculations/2018/09/28/a-look-at-intuitive-surgicals-revenue-sources-and-outlook/?sh = 662c3b2c4635.
[24] Barbash GI, Glied SA. New technology and healthcare cost-the case for robotic assisted surgery. N Engl J Med 2010;363(8):701–4.
[25] Wright JD, Ananth CV, Lewin SN, Burke WM, Lu Y-S, Neugut AI, et al. Robotically assisted vs laparoscopic hysterectomy among women with benign gynecologic disease. J Am Med Assoc 20 2013;309(7):689–98.

30

Lessons from first inventions

Ioana Pasca[1,2] and Ashish Sinha[1,2]
[1]Loma Linda University at Riverside University Health System, Moreno Valley, CA, United States [2]University of California Riverside, Riverside, CA, United States

Chapter outline

Abstract

New inventions begin with thorough research of existing patents, procedures, and literature related to your patented product. Using Dr. Sinha's nasogastric tube introducer patented in April 2011 (USPTO patent number 7918841) we exemplify the design, production, patenting, and marketing of an invention from conceptualization to patent expiration. We then describe the limitations faced in commercializing the product before the end of patent life.

Keywords: Nasoenteral tube; perforated introducer; marketing; 3D printer; utility patent; preliminary patent; intellectual property; commercialize; patent attorney

Innovation in Anesthesiology. DOI: https://doi.org/10.1016/B978-0-12-818381-6.00018-8

Key points

- Before starting your patent, search the United States Patent and Trade Office website at http://www.USPTO.gov for patents of a similar theme.
- Do not discuss your product in a public forum which could limit your ability to patent the product.
- Understand your employer's rules regarding personal patents and royalties.
- Your patent has a limited life so once you have the patent, put your marketing plan into play.

Glossary

Intellectual property —an individual product of original thought
Preliminary patent —protects a potential patent for up to 12 months while the inventor is testing a product and applying for an official patent
Utility patent —a patent that protects a new way in which a process is used
Commercialize —to emphasize the profitable aspects or to offer for sale
3D printer —a device used to print three-dimensional objects
Nondisclosure (NDA) agreement —prevents parties from discussing information included in the agreement
Patent attorney —a lawyer with expertise in intellectual property law and patent application for new inventions

Why it matters

New inventions begin with thorough research of existing patents, procedures, and literature related to your patented product. Using Dr. Sinha's nasogastric tube introducer patented in April 2011 (USPTO patent number 7918841) we exemplify the design, production, patenting, and marketing of an invention from conceptualization to patent expiration. We then describe the limitations faced in commercializing the product before the end of patent life.

Background

In 2006 Dr. Sinha came across a paper by Dobson describing his success with using perforated endotracheal tubes (ETTs) to accurately insert nasogastric tubes (NGT), or nasoenteral tubes, into the most difficult of patients [1]. A nasoenteral tube once placed may be used to decompress the stomach or place liquid medications into the stomach in a patient not tolerating oral intake. The body of the tube can be modified to measure temperature, assess cardiac function, and respiratory sounds, or deliver an electric shock close to the heart. Dobson inserted the NGT through the nares and out the mouth with the assistance of McGill forceps [1]. Next, an ETT with regular perforations was blindly passed into the esophagus. The nasogastric tube was inserted into the ETT and into the stomach with the ETT then split away using the precut perforations (*see* Fig. 30.1). Dr. Sinha began experimenting with this novel technique but found the passage of the nasogastric tube into the oropharynx sometimes more challenging then advancement into the esophagus itself.

Dr. Sinha then began a literature search of similar techniques and published his findings in the *Journal of Parenteral and Enteral Nutrition* in January 2011. The publication found that malposition rates of NGT placement varied widely from 0.3% to

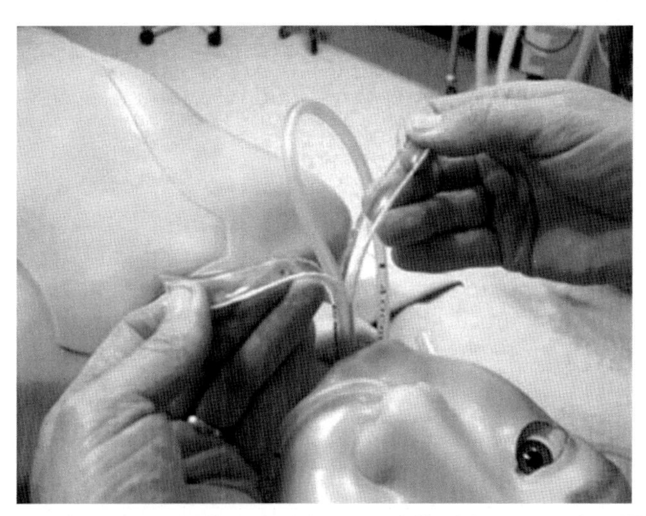

Figure 30.1 The description of nasogastric tubes placed successfully with the use of an ETT blindly inserted into the esophagus [1]. Dobson A.P. Nasogastric tube insertion-another technique. Anaesthesia 2006;61(11):1127. https://doi.org/10.1111/j.1365-2044.2006.04853.x. Permission granted by Wiley Publishing 07/11/22.

Table 30.1 Incidence of complications resulting from small-bore nasoenteral tubes and associated factors [2].

Study	Feeding tube placements, *n*	Malpositions	Complications	Associated factors
Sorokin and Gottlieb [3]	2000	l.3%–2.4% overall	In 26% of malpositions, including two deaths	Altered mental status, preexisting endotracheal tube
McWey et al. [4]	1100	1.3% overall	Pneumothorax, hydrothorax, empyema, pneumonia, mediastinitis, esophageal perforation	Preexisting endotracheal tube
Ghahremani and Could [5]	340	2.1% in airway	Pneumothorax in one patient (0.3%)	Critical illness
Valentine and Turner [6]	28-month observation period	0.3% intrapleural	Pneumothorax, pleural effusion, pneumonia	Altered mental status, preexisting endotracheal tube
Harris and Huseby [7]	71	Not reported	4% with airway complications	Preexisting endotracheal tube
Rassias et al. [8]	740	2% in airway	0.7% major complications, two deaths	Altered mental status, preexisting endotracheal tube
Ellett et al. [9]	201 (pediatric)	15.9% on first radiograph	Not reported	Age, altered mental status, abdominal distention, vomiting, orogastric tube

15.9% and complications from malposition included bleeding, pneumothorax, pleural effusions, pneumonia, and even death (see Table 30.1). Patients with the highest risk had altered mental status, preexisting endotracheal tubes, and critical illness. The awake patient coughs, gags, tries to remove the tube, or complains of pain or difficulty breathing with improper NGT placement. Furthermore, in patients with altered mental status due to critical illness, sedation, or anesthesia, the feedback regarding a malpositioned nasogastric tube is inadequate.

Journal of Parenteral and Enteral Nutrition/ Vol. 35, No. 1, January 2011

Patients with preexisting endotracheal tubes under sedation or general anesthesia easily tolerate an orogastric tube. However, awake patients with a gag reflex, who require gastric tubes, need nasal placement, which is better tolerated. Nasal bleeding and the pain and pressure from the sphenopalatine ganglion, or nasal ganglion, may make this procedure intolerable. Therefore, lubricant with local anesthetic is used to decrease the nasal discomfort and the patient is asked to swallow to facilitate passage into the esophagus. Air is then inserted into the nasogastric tube and auscultation over the stomach is used to signify successful placement. Eventually, most institutions adopted the use of X-ray to verify the correct placement of the NGT under the diaphragm and away from the trachea in an attempt to decrease complication rates.

Dr. Sinha began to consider other methods for placement of the nasogastric tube, especially using the model of a nasal airway. He began modifying the nasal airway to point posteriorly once placed and lengthening it to guide the NGT into the esophagus rather than the trachea. He created a perforated introducer so that once the NGT was placed, it was easy to tear off the nasal airway introducer and remove it from the patient. Alternately, the proximal one inch can be saved to protect the nares from the long-term pressure of the NGT in sedated ICU patients. Dr. Sinha then applied for a grant that allowed further research time for testing the nasogastric introducer. The grant funding allowed 3D printing which drove modifications of lengths and diameters with the assistance of an otolaryngologist. Dr. Sinha eventually decided that by making the proximal end in a D-shape rather than an O-shape and providing a little width at the rim and an adhesive, it was now possible to easily affix this to the upper lip (Fig. 30.2).

Once the design was optimized, Dr. Sinha, who was on faculty at the University of Pennsylvania's Perelman School of Medicine at the time of the invention, required collaboration with the medical school. The process at the University of Pennsylvania requires that any patent invention be presented to the Center for Technology Transfer, CTT, giving the medical school the right of first refusal to patent rights. CTT at Penn chooses which inventions to keep as their intellectual property. This invention was considered appropriate for the University of Pennsylvania to keep within its fold. The advantage of this approach was that Penn fronted all the costs of the patent using a legal firm in New York. After multiple negotiations between the patent attorney and the inventors, the idea was submitted to the USPTO and immediately granted a preliminary patent. Finally, a formal final patent was granted on April 5, 2011.

At Penn, 30% of the royalty would be assigned to the inventors, 15% to the inventor's research efforts in the future, and 15% to the Penn department the inventor represents. The rest would be distributed between the school of medicine and the university for further educational and research endeavors.

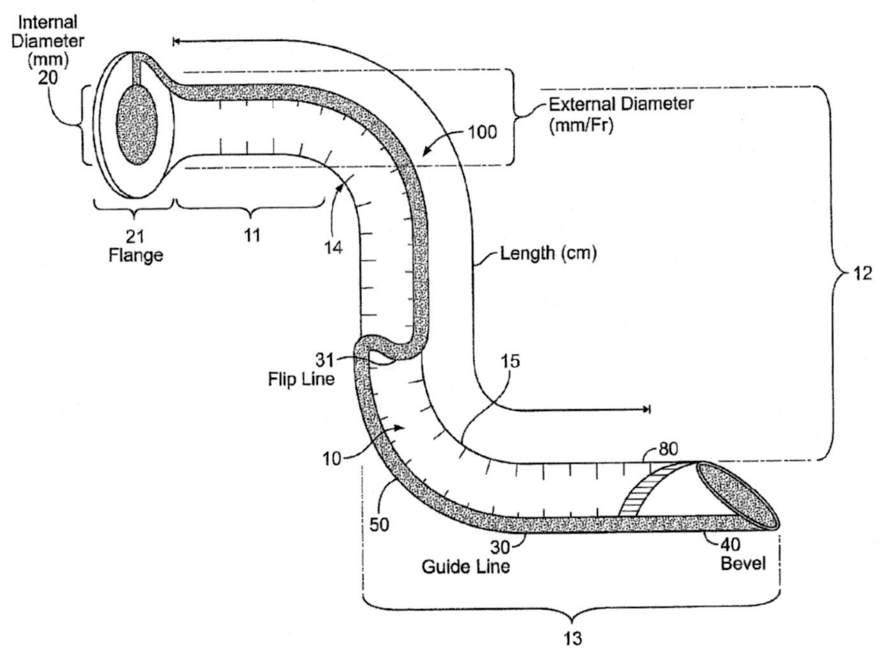

Figure 30.2 The nasogastric introducer.[2,10] Images reproduced from the original patent. *United States Patent and Trade Office*, accessed 30 May, 2022. <https://patft.uspto.gov/netacgi/nph-Parser?Sect1 = PTO1&Sect2 = HITOFF&d = PALL&p = 1&u = %2Fnetahtml%2FPTO%2Fsrchnum.htm&r = 1&f = G&l = 50&s1 = 7918841.PN.&OS = PN/7918841&RS = PN/7918841>; Halloran O., Grecu B., Sinha A. Methods and complications of nasoenteral intubation. J Parenter Enter Nutr 2011;35(1):61–66.

Shortly after the patent was granted, marketing began to ensure the sale of the product before the 7-year patent expired. In this instance with this device, the obvious market was a company that makes NG tubes. Mallinckrodt was approached and signed the nondisclosure agreement (NDA), and an agreement was signed allowing Mallinckrodt to sell their NG tubes with or without an introducer. Unfortunately, due to the vagaries of the market (the Great Recession of the 2000s), companies were loath to invest in any new products but were consolidating their product lines. Therefore, Mallinckrodt never sold NG tubes with the introducer, and eventually in 2020 through a complex series of events unrelated to the nasogastric introducer, the Mallinckrodt company filed for bankruptcy.

Getting started with practical advice

If you get an idea to improve a process, be brave enough to use your time and energy to research the topic. You must face the risk of not only rejection but also of lost time and money. By applying for a grant early in the process, Dr. Sinha was able to secure extra funding to cover the costs of research and design. He also used the preliminary background research to publish a review paper on the complications of nasogastric tube placement.

During his preliminary research, Dr. Sinha was reminded not to discuss his idea in a public forum, but especially in a publication, because then his invention may not be considered for patent. This is the challenge of having an idea which one person considers the next "*Velcro*." Some will easily dismiss the idea having little knowledge of the subject, while others will discourage the idea because it appears too complicated. Neither group offers useful criticism and support. Dr. Sinha discussed his idea with only one highly respectable colleague who gave it the best possible accolade, "brilliantly simple." This encouraged him tremendously to proceed with the process.

The USPTO website was an easy place to begin Dr. Sinha's search regarding similar products already in existence. If doing this personally, unless you are a patent attorney, it is possible to get a reasonable quote for the appropriate search done professionally. This step is essential before committing energy, resources, and emotion to a product that has already been described and patented.

Even if the product already exists, a utility patent may still be applicable. A utility patent protects a novel way in which a product is used. For example, if prior knowledge exists about electricity and you have a way to treat cancer using electricity, the new process of using electricity to treat cancer could be patentable, but electricity would not.

Once designed, it is important to have a physical 3D model to establish proof of concept before marketing the product for which there are now 3D printers in most academic institutions. It is important to commercialize the product and to have a supplier in mind as your patent is granted to increase its impact and possibly provide recognition and financial reward to the inventors in the short patent life. Make plans for commercialization in advance to begin commercialization as soon as the patent is granted.

However, keep in mind limitations like the expiration of patents as short as 7 years when your product will need to be adopted into standard medical use for profitability. Also, you may need to patent and market your invention in multiple countries for success. It is important to target the countries where you believe the bulk of the market will exist. These are usually North America and Europe. Products can easily be replicated in countries not using US patents and undercut the price outside the US. However, markets in countries outside North America and Europe are usually limited.

The satisfaction of creating a patent is immense but should be driven by more than financial considerations. Once the product is patented, the sense of accomplishment can result in a state of complacency that does not pursue the commercialization of the product. If the product is truly useful and unique, it should be pursued for the benefit of the targeted users, not just its commercial potential. The process of patenting, though long and sometimes unpredictable, is very rewarding in and of itself because you would have created something novel that you can be proud of and that can improve the lives of your patients and fellow caregivers. Furthermore, the process of technology development opens new avenues for collaboration, research, and education with interdisciplinary industry teams that can enrich your personal development as an anesthesiologist, researcher, educator, and administrator.

We wish you the best of luck on your first invention!

Getting started summary

1. Search http://www.USPTO.gov for similar existing inventions or utility patents for novel product use.
2. Avoid discussing your invention in public forums or with unreliable colleagues.
3. Apply for a preliminary patent as you test your concept.
4. Search for grants to support your research endeavors.
5. Present the patent to the appropriate local institutional committees.
6. Apply for a US patent with a patent attorney.
7. Market your invention before the patent ends with well-made plans for early commercialization.

Patent history from USPTO.gov:

Patent number: 7918841
Type: Grant
Filed: Apr 3, 2009
Date of Patent: Apr 5, 2011
Patent Publication Number: 20090306626
Assignee: The Trustees of the University of Pennsylvania (Philadelphia, PA)

Inventors: Ashish C. Sinha (Philadelphia, PA), Owen J Halloran (Philadelphia, PA)
Primary Examiner: Kevin C Sirmons
Assistant Examiner: Michael J Anderson
Attorney: Pearl Cohen Zedek Latzer, LLP
Application Number: 12/418,464
Current U.S. Class: To Or From The Intestines Through Nasal Or Esophageal Conduit (604/516); Nozzle Insertable Into Body Orifice (604/275); With Body Soluble, Antibactericidal Or Lubricating Materials On Conduit (604/265)
International Classification: A61M 39/00 (20060101)

References

[1] Dobson AP. Nasogastric tube insertion--another technique. Anaesthesia 2006;61((11):1127. Available from: https://doi.org/10.1111/j.1365-2044.2006.04853.x.
[2] United States Patent and Trade Office, accessed 30 May 2022, <https://patft.uspto.gov/netacgi/nph-Parser?Sect1 = PTO1&Sect2 = HITOFF&d = PALL&p = 1&u = %2Fnetahtml%2FPTO%2Fsrchnum.htm&r = 1&f = G&l = 50&s1 = 7918841.PN.&OS = PN/7918841&RS = PN/7918841>.
[3] Sorokin R, Gottlieb J. Enhancing patient safety during feeding tube insertion: a review of more than 2000 insertions. J Parenter Enteral Nutr 2006;30(5):440–5.
[4] McWey R.E., Curry N.S., Schabel S.I., Reines H.D. Complications of nasoenteric feeding tubes. Am J Surg 1988;155(2):253-257. Available from: https://doi.org/10.1016/s0002-9610(88)80708-6. PMID: 3124652.
[5] Ghahremani G.G., Gould R.J. Nasoenteric feeding tubes. Radiographic detection of complications. Dig Dis Sci 1986;31(6):574-585. Available from: https://doi.org/10.1007/BF01318688. PMID: 3086062.
[6] Valentine RJ, Turner Jr. WW, Borman KR, Weigelt JA. Does nasoenteral feeding afford adequate gastroduodenal stress prophylaxis? Crit Care Med 1986;14(7):599–601. Available from: https://doi.org/10.1097/00003246-198607000-00001.
[7] Harris M.R., Huseby J.S. Pulmonary complications from nasoenteral feeding tube insertion in an intensive care unit: incidence and prevention. Crit Care Med 1989;17(9):917-919. Available from: https://doi.org/10.1097/00003246-198909000-00016. PMID: 2504541.
[8] Rassias A.J., Ball P.A., Corwin H.L. A prospective study of tracheopulmonary complications associated with the placement of narrow-bore enteral feeding tubes. Crit Care 1998;2(1):25-28. Available from: https://doi.org/10.1186/cc120. PMID: 11056706; PMCID: PMC28998.
[9] Ellett ML, Maahs J, Forsee S. Prevalence of feeding tube placement errors & associated risk factors in children. Am J Matern Child Nurs 1998;23:234–9.
[10] Halloran O, Grecu B, Sinha A. Methods and complications of nasoenteral intubation. J Parenter Enter Nutr 2011;35(1):61–6.

31

From idea to commercialization: development of a breathing-controlled gaming device for pediatric anesthesia induction

Abby V. Winterberg[1,2], Aniruddha P. Puntambekar[3] and Anna M. Varughese[4,5]

[1]Department of Anesthesiology, Cincinnati Children's Hospital Medical Center, Cincinnati, OH, United States [2]College of Nursing, University of Cincinnati, Cincinnati, OH, United States [3]Cincinnati Children's Innovation Ventures, Cincinnati Children's Hospital Medical Center, Cincinnati, OH, United States

[4]Department of Anesthesiology, Johns Hopkins All Children's Hospital, St. Petersburg, FL, United States [5]School of Medicine, Johns Hopkins University, Baltimore, MD, United States

Chapter outline

Abstract

Children undergoing medical procedures requiring general anesthesia commonly experience anxiety during anesthesia inhalation induction. High anxiety during induction is associated with poorer outcomes after surgery. This case study presents the product development process for a breathing-controlled app/device designed to easily engage children in the process of anesthesia inhalation induction. The tablet-based

Innovation in Anesthesiology. DOI: https://doi.org/10.1016/B978-0-12-818381-6.00015-2

gaming app includes a breath sensor (embedded in the tablet case) that familiarizes children with an anesthesia mask preoperatively and promotes relaxation during induction through breathing-controlled gameplay. The product is currently in routine clinical practice at our hospital, beta-testing at an external hospital is complete, and a commercialization partner has been found. In this case study, we share the product development journey from idea to commercialization. We also present the unique considerations for the development of a product designated by the FDA as a "general wellness device."

Keywords: Pediatric anesthesia induction; anxiety; breathing-controlled app; general wellness device

Problem/opportunity

In the United States, there are approximately 3.9 million pediatric surgeries annually [1]. Researchers reported that over 35% of children exhibit anxious behaviors during anesthesia induction [2–4]. Induction anxiety has been associated with poorer postoperative outcomes [5]. Mask play/acclimation, digital distractions, and video games have been shown to help reduce anxiety [6–9], but there was no commercially available device to directly incentivize mask acclimation.

While working in the preanesthesia consult clinic at Cincinnati Children's Hospital, Abby Winterberg Hess, APRN (advanced practice registered nurse), envisioned a breathing-controlled gaming app that would decrease induction anxiety by engaging children in the process of breathing into the anesthesia mask. She developed a simple prototype and submitted an invention disclosure to the hospital's commercialization department, which provided funding for initial product development. With the help of her mentor (Anesthesiology Division Chief) she began the initial phases of product design and testing.

Our commercialization office (Cincinnati Children's Innovation Ventures, CCIV) conducted a competitive analysis via web search and patent databases. While multiple existing products were identified, no commercial products specifically addressed pediatric mask-based anesthesia induction. To estimate market size, multiple stakeholders identified settings where mask-based anesthesia is administered (same-day surgery, procedure center, dental clinic, imaging center, and proton therapy center as illustrative examples) to define the number of devices needed per hospital based on hospital size. With approximately 250 US pediatric hospitals [10];

we estimated a fairly modest market potential. However, the product could also be used in adult hospitals caring for pediatric patients, increasing the potential market size.

Concept

For initial product ideation and design, we partnered with the Live Well Collaborative (LWC), a nonprofit academic-industry research incubator that combines design-thinking methodologies with multidisciplinary talent to translate research into consumer-based innovations [11]. A detailed description of the design phase and product requirements is described by Winterberg et al. (2022) [12]. During a semester-long partnership, families and medical staff engaged in product development activities. Initial product requirements included easy implementation, high patient comfort, engagement, and sense of control, adaptability to different environments, and infection control compliance. The team developed a prototype design for a zoo-themed tablet-based gaming app. The app directly connects a video game to the anesthesia mask. This provides the opportunity to engage patients in the process of anesthesia induction and transforms a scary piece of medical equipment into a fun game controller.

We tested the prototype in the operating room, and refinements were defined through iterative feedback from patients, parents, and clinical staff. The final concept (Fig. 31.1) included a standard Samsung tablet with a custom tablet case that contained the breath sensor that connects to standard anesthesia

Front of tablet
Back of tablet
Product includes a standard Samsung tablet with a breath sensor built into the back of a custom tablet case
The product is designed to facilitate patient engagement & mask acceptance resulting in smoother induction experiences
Standard anesthesia equipment
Standard anesthesia equipment connects to a breath sensor built into the back of a custom tablet case

Figure 31.1 Breathing-controlled gaming device. Credit: Cincinnati Children's Hospital Medical Center.

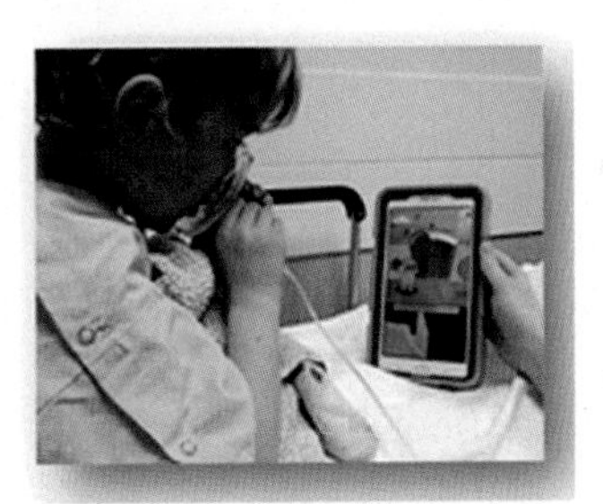

The zoo-themed game requires breathing into the mask to win. Preoperatively, the child becomes comfortable breathing through the mask for 3 to 5 minutes

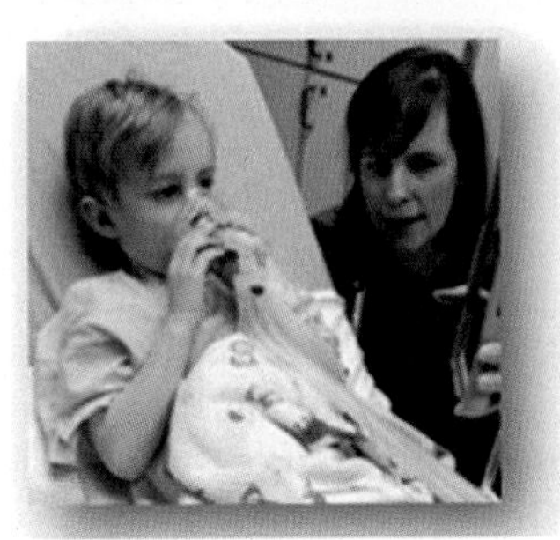

The child looks forward to playing the next game during Anesthesia Induction. During induction, the breathing sensor is connected to the anesthesia circuit. Calm breathing provides children with a sense of control and promotes relaxation as they drift off to sleep

Figure 31.2 Patient interaction with breathing-controlled gaming device. Credit: Cincinnati Children's Hospital Medical Center.

equipment. Patients complete gaming challenges (e.g., blowing up balloons) by breathing into the anesthesia mask (Fig. 31.2).

Product development

We partnered with a local product development company, Kinetic Vision to advance from design concept to functional prototype (app and custom breath sensor) for clinical use. To extend funding and continue iterating on the product design, Kinetic Vision worked with our hospital's clinical engineering team to improve the case design and develop new prototypes.

To obtain unbiased product feedback and develop our marketing strategy, we partnered with the marketing firm Lead From Insights. They collected internal usability data from internal/external stakeholder interviews to further improve the product. External beta-testing supported interest in external adoption. We subsequently sought a partner for final commercial development.

Commercialization pathway

Following prototype development, CCIV worked with a patent attorney to draft and submit a provisional patent application and later filed a nonprovisional US patent application.

As an academic medical center, our institution does not directly commercialize products. The initial addressable market for this product is fairly modest. Thus, we determined that licensing the technology to an established partner was preferable to forming a startup. The potential licensee has an array of digital-health products that each have modest market potential but collectively can support a small business's operation.

Funding

Funding pediatric product development can be challenging due to relatively small markets, low returns on investment, unique design considerations for ages/developmental levels, regulatory complexities, and lack of research and development infrastructure [13]. The commercialization department within an academic medical center can be a key partner in product development and offer crucial funding for early-stage development. Our team received $200,000 from an internal Innovation Fund grant [14] for prototype design, development, initial product testing, and partial salary support for the project lead and a research assistant. A $10,000 internal grant provided partial support for product design and LWC partnership [15].

These internal opportunities funded the development of a clinically tested functional prototype needed to apply for external funding. We were subsequently awarded $50,000 from the Johnson & Johnson Nurses Innovate QuickFire Challenge [16]. The final development was funded by the State of Ohio Technology Validation and Startup Fund ($75,000, with $75,000 institutional match) [17]. The CCIV team remained key partners in meeting commercialization grant milestones and identifying funding opportunities. Fig. 31.3 outlines the product development process, timeline, and funding.

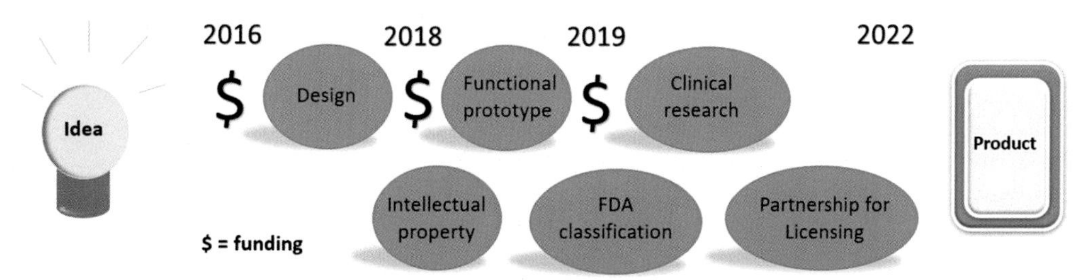

Figure 31.3 Product development process and timeline for commercialization. Credit: Cincinnati Children's Hospital Medical Center.

Regulatory

We partnered with a regulatory consultant to submit a formal request to the US FDA (per Federal Food, Drug, and Cosmetic Act Section 513(g)) for the classification of the product. Feedback from the FDA indicated the product would be considered a general wellness device (GWD), not a medical device.

As shown in Table 31.1, GWDs are not subject to the FDA regulation required for products classified as medical devices and do not require FDA approval (or the large efficacy studies that support such approval) before marketing. Manufacturers are not required to comply with FDA pre- or postmarket requirements, and there are no specific regulatory requirements for product manufacturing. Therefore, commercialization costs are lower for GWD than for medical devices.

Product claims for GWD cannot include medical outcomes. For our product, referencing reduction of stress/anxiety and improvement of family experience are acceptable outcomes to study/market. Of note, using the same product to support the improvement of a medical outcome, such as improvement in anesthesia delivery, would change the product classification to medical device. For more detailed information, refer to the FDA's General Wellness: Policy for low-risk devices [19].

Table 31.1 Commercialization considerations for FDA regulated medical device versus general wellness device.

Type of device	Research outcomes	Product claims	Regulatory requirements	Commercialization costs
FDA regulated medical device	May study medical outcomes	Product claims may reference medical outcomes (e.g., treat obesity, treat an anxiety disorder).	Many regulatory requirements [18]	Significantly higher development/ manufacturing costs due to quality management system (QMS) requirements
General wellness device (not FDA regulated)	Studies to support product claims should focus on outcomes that are in alignment with the FDA determination of "general wellness device"	Claims may not reference medical outcomes. Product claims refer to general wellness (e.g., promote or maintain a healthy weight, promote relaxation, or manage stress) [19]	N/A	Relatively lower costs

Commercialization-focused protocols are not yet common within academic medical centers. Throughout product development, we had discussions with internal regulatory leadership to navigate how our research and development fit into the academic research framework. The product and all clinical testing protocols were reviewed by our hospital's equipment evaluation committee and Institutional Review Board (ethics review).

Challenges, solutions

One of our greatest product development challenges was a lack of familiarity with the relatively new "general wellness" designation and how to develop a GWD within an academic medical center. Medical systems have rigid processes and policies designed to support high-quality research outcomes and minimize risk to patients. These policies are neither designed for the development of low-risk digital products nor are they conducive to testing these kinds of products.

Because GWDs do not diagnose or treat disease, the required research is less structured than research typically conducted in an academic setting. We initially found it challenging to align our iterative commercialization-focused human subject protocols (e.g., design research, market research, usability testing, external beta-testing) with rigid clinical research procedures. We navigated these challenges through discussions with clinical and regulatory leadership to determine the approach at each stage of development. Even with regulatory leadership support, we continued to encounter system barriers that significantly delayed progress.

CCIV was instrumental in working through FDA classification, intellectual property, development partnerships, and commercial pathway identification. Finding project mentorship outside of CCIV to support product development/testing in the clinical setting was challenging, as this skill set is not common in academic medical centers. A centralized clinical innovation hub that provides mentorship and guidance for navigation of the ethics review and clinically oriented aspects of similar projects would likely help support future innovation.

Advice

New product development is a long journey that requires tremendous and varied support for success. Academic medical centers can provide exceptional resources if they can be successfully adapted to support and accelerate product development. Clinicians have many key qualities/skill sets that translate well to leading innovation projects. Clinicians regularly employ creativity and adaptability to meet the needs of their patients. Their deep understanding of hospital systems and workflows allows them to intuitively understand product strengths and weaknesses. Clinicians are experts in collaborating across many different disciplines, and their empathy for patients/families and passion for improving outcomes fuels the collaboration and persistence needed to navigate product development. Given the right education, resources and institutional support, clinicians, and healthcare providers can play a key role in designing and developing products that will revolutionize healthcare.

References

[1] Rabbitts JA, Groenewald CB. Epidemiology of pediatric surgery in the United States. Pediatr Anesth 2020;30(10):1083–90.

[2] Chorney JM, Kain ZN. Behavioral analysis of children's response to induction of anesthesia. Anesth Analg 2009;109(5):1434–40.

[3] Varughese AM, Nick TG, Gunter J, Wang Y, Kurth CD. Factors predictive of poor behavioral compliance during inhaled induction in children. Anesth Analg 2008;107(2):413–21.

[4] Winterberg AV, Ding L, Hill LM, Stubbeman BL, Varughese AM. Validation of a simple tool for electronic documentation of behavioral responses to anesthesia induction. Anesth Analg 2020;130(2):472–9.

[5] Yuki K, Daaboul DG. Postoperative maladaptive behavioral changes in children. Middle East J Anaesthesiol 2011;21(2):183–9.

[6] Aydin T, Sahin L, Algin C, et al. Do not mask the mask: use it as a premedicant. Pediatr Anesth 2008;18(2):107–12.

[7] Dwairej D, Obeidat HM, Aloweidi AS. Video game distraction and anesthesia mask practice reduces children's preoperative anxiety: a randomized clinical trial. J Spec Pediatr Nurs 2020;25(1):e12272.

[8] Walker KL, Wright KD, Raazi M. Randomized-controlled trial of parent-led exposure to anesthetic mask to prevent child preoperative anxiety. Can J Anesth 2019;66(3):293–301.

[9] Chaurasia B, Jain D, Mehta S, Gandhi K, Mathew PJ. Incentive-based game for allaying preoperative anxiety in children: a prospective, randomized trial. Anesth Analg 2019;129(6):1629–34.

[10] Casimir G. Why children's hospitals are unique and so essential. Front Pediatrics 2019;7:305.

[11] Live Well Collaborative. "Who We Are." Live Well Collaborative, <https://www.livewellcollaborative.org/who-we-are> [accessed 18.10.23].

[12] Winterberg AV, Lane B, Hill LM, Varughese AM. Optimizing pediatric induction experiences using human-centered design. J Perianesth Nurs 2022;37(1):48–52.

[13] Espinoza J, Cooper K, Afari N, Shah P, Batchu S, Bar-Cohen Y. Innovation in pediatric medical devices: proceedings from The West Coast Consortium

for Technology & Innovation in Pediatrics 2019 Annual Stakeholder Summit. JMIR Biomed Eng 2020;5(1):e17467.

[14] Ventures CCsI. Cincinnati Children's employees are innovators. 2022. <https://innovation.cincinnatichildrens.org/innovators>. [accessed 15.02.2024].

[15] Training CfCTS. Design Thinking Research Award. <https://www.cctst.org/funding/design-thinking-research-awards>. [accessed 29/3/2022].

[16] Innovation J.J. Johnson & Johnson Nurses Innovate QuickFire Challenge. 2022. <https://nursing.jnj.com/nursing-news-events/nurses-leading-innovation/meet-the-johnson-johnson-nurses-innovate-quickfire-challenge-awardees>. [accessed 15.02.2024].

[17] Development ODo. Technology Validation and Start-up Fund. <https://development.ohio.gov/business/third-frontier-and-technology/technology-validation-start-up-fund>. [accessed 15.02.2024].

[18] FDA. <https://www.fda.gov/medical-devices/device-advice-comprehensive-regulatory-assistance/overview-device-regulation>. 2020. [accessed 18.10.2023]

[19] Administration USFD. General Wellness: Policy for Low Risk Devices. 2019. <https://www.fda.gov/regulatory-information/search-fda-guidance-documents/general-wellness-policy-low-risk-devices>.

32

Case study: improve the safety and compliance of drug preparation and administration with barcode technology

Cristy Berg, Gary Keefe and Ross Goodman
Codonics, Middleburg Heights, OH, United States

Chapter outline

Abstract

Codonics is a privately-held U.S. company focused on the development, manufacturing and sales of medical devices. Years of successful brand building coupled with a proven record of releasing breakthrough medical products opened a unique opportunity in 2009. Massachusetts General Hospital (MGH) SIMS Lab was in search of a technology partner to productize a game-changing drug labeling safety solution for the operating room (OR). Realizing the product would save thousands of lives, Codonics partnered with Massachusetts General Hospital by licensing the technology and investing in the development of what became known as the Codonics Safe Label System. Through rigorous development, testing and regulatory processes, the Safe Label System received FDA Class II Medical Device approval as the first-of-its-kind predicate device in this class. Codonics released this transformative safety system in 2011 and used a combination of direct sales and bundling with anesthesia drug cart OEMs to open an entirely new

Innovation in Anesthesiology. DOI: https://doi.org/10.1016/B978-0-12-818381-6.00021-8

marketspace for fast, safe and compliant electronic labeling of medications in the OR.

Keywords: Medication safety; patient safety; compliant labeling; label printer; drug labeling system; integrated labeling system; Codonics; Safe label system

Company introduction

Founded in 1982, Codonics is a privately-held U.S. company focused on the development, manufacturing and sales of medical devices. The initial products were centered on hardcopy printing of medical images on radiographic film and paper and expanded to later include label printing solutions for drug preparation of syringes and IV bags used by anesthesia providers.

Company background

Codonics introduced the world's first DICOM radiology printer in 1996. Since then, Codonics has worked closely with the radiology community and diagnostic imaging system manufacturers of CT, MR, Ultrasound and X-Ray systems to produce leading-edge imaging storage and hardcopy printing devices. Today, the company is represented in 110 countries with offices in the U.S., France, and China, and has sold over 60,000 medical devices worldwide. Codonics expertise includes selling healthcare solutions direct to hospitals, through OEM partners, and local reseller channels. This multi-faceted approach to delivery of medical solutions has enabled the company to have a global reach while meeting the needs of individual countries and markets (see Fig. 32.1).

As both a Food and Drug Administration (FDA)-registered manufacturing facility and developer of medical devices, and an ISO (International Organization for Standardization) 13485 QSR (Quality Systems Registrars) certified company, Codonics has vast experience with FDA 510K and MDR submissions and compliance. Codonics also distributes products internationally and complies with international regulatory filings including EU MDR (Medical Device Regulations). The company's core competencies in product invention, innovation, global sales and distribution coupled with an entrepreneurial culture has enabled it to deliver advanced medical solutions to thousands of customers around the world. Codonics has successfully

Codonics Product Solutions

Figure 32.1 Codonics patient safety and medical imaging product line.

developed, manufactured, commercialized and marketed medical devices on a global scale for over a quarter-century.

Years of successful brand building coupled with a proven record of releasing breakthrough medical products opened a unique opportunity in 2009. Massachusetts General Hospital (MGH) Learning Laboratory (SIMS Lab) was in search of a technology partner to productize a game-changing drug labeling safety solution for the operating room (OR). Multiple companies were evaluated, and Codonics was ultimately selected. Realizing the product would save thousands of lives and align with our corporate mission to build revolutionary products, Peter O. Botten, Codonics owner and CEO, partnered with MGH by licensing the technology and investing in the development of what became known as the Codonics Safe Label System (SLS) [1]. Through rigorous development, testing and regulatory processes, the Safe Label System received FDA Class II Medical Device approval as the first-of-its-kind predicate device in this class. Codonics released this transformative safety system in 2011 and used a combination of direct sales and bundling with anesthesia drug cart original equipment manufacturers (OEMs) to open an entirely new marketspace for fast, safe and compliant electronic labeling of medications in the perioperative environment. Today, SLS is highly awarded for innovation and is the standard of care in over 950 of the world's leading hospitals. It has been used in more than 57 million procedures and over 230 million drug preparations helping to prevent over 1.7 million medication errors [2] across 19 countries. These statistics grow every day as the SLS continues its main function of saving lives by making OR drug labeling fast, safe and compliant.

Product definition

In the operating room, studies show there is one drug administration medication error for every 133 anesthetics delivered, and one in 250 of those errors is fatal [2]. Massachusetts General Hospital Learning Laboratory's initial concept was prompted by their understanding of this problem and the desire to provide safer drug preparation and administration in the operating room using barcode technology for drug identification. Their prototype was demonstrated in 2009 at the American Society of Anesthesiology (ASA) Annual Meeting.

After licensing the concept from MGH, Codonics spent the next two years developing and refining the design into a medical device now known as the Safe Label System (SLS). It needed to quickly scan the National Drug Code (NDC) barcode on the drug container in the provider's hand, verify it using a hospital-specific formulary database, provide visual and audio confirmation to the clinician of the drug name and concentration, and print a full-color Joint Commission compliant label. All of this needed to be done in the few seconds between scanning the medication container barcode and filling the syringe, so the label could be applied immediately and be ready for safe patient administration. It was also important to consider the fast-paced nature of the OR and ER environments where drugs are administered first and documented in the patient record later, sometimes referred to as the "wild west." While SLS was a boon to address the problems of safety and compliancy that existed for many years, it needed to do much more than just print a label. SLS had to simplify and improve the clinicians' workflow.

There were several critical and challenging design requirements identified early in the product's life. Realizing the operating theater is a tight and equipment-heavy space, a small device footprint needed to be developed. Some of the technical challenges included integrating a specialized color label printer along with a barcode scanner, touchscreen user interface and a computer to support the drug formulary database, plus SLS application software in a compact, "easy to site" appliance that also could provide audible and visual feedback anywhere drugs are being prepared. Ultimately a device about the size of a bread-box was needed to fit in the clinicians' rapid workflow. Multiple design concepts were explored before settling on a configuration that balanced user efficiency, maintainability and cost.

With the system design complete, Alpha and Beta testing were conducted in 2010–11 and final FDA 510K clearance as a Class II medical device was received. Codonics released the Safe Label System to market in May 2011. Customers immediately realized the product's fundamental benefits—to reduce the three most common medication errors in the operating room, including (1) vial and ampoule swaps (2) mislabeling and illegible labeling and (3) syringe swaps. The SLS also provides 100% compliance with Joint Commission drug labeling requirements [3]. The system eliminates the need for handwritten labels while also complying with American Society of Anesthesiologists (ASA) guidelines. Full-color labels include the drug name, concentration, dilution and diluent (if required), date and time of preparation, date and time of expiration, the preparer's initials, and any hospital-specific messages or warnings needed. A barcode is also included on SLS's printed label that can be scanned during drug administration to transfer information to the electronic medical record (EMR) or anesthesia information management system (AIMS) for patient documentation, billing and additional safety checks (see Fig. 32.2).

A key factor in the success of SLS is the way it brings together anesthesia and pharmacy, *often for the for the first time,* to improve the safety of drug preparation and administration in the hospital. SLS has two distinct elements that separately support the needs of these departments but combine to produce a comprehensive solution with greater safety and control than either could have achieved on their own. The first is the SLS Point-of-Care Station (SLS PCS), the hardware component used by anesthesia providers to perform the actual printing of labels for syringes and IV bags (see Fig. 32.3). The other is the SLS Administration Tool (SLS AT), the software application that enables Pharmacy to create formularies of approved drugs and establish the safety and configuration settings used by the SLS PCS that correspond to the hospital's best practices and preferred use conditions (see Fig. 32.4) [4]. The SLS AT enables compliancy and accreditation with the

Prints full-color labels that comply with The Joint Commission NPSG.03.04.01*, and meet the intent of ISO 26825, ASTM 4774 Standards & ASA Guidelines

*Prints diluent and dilution if required

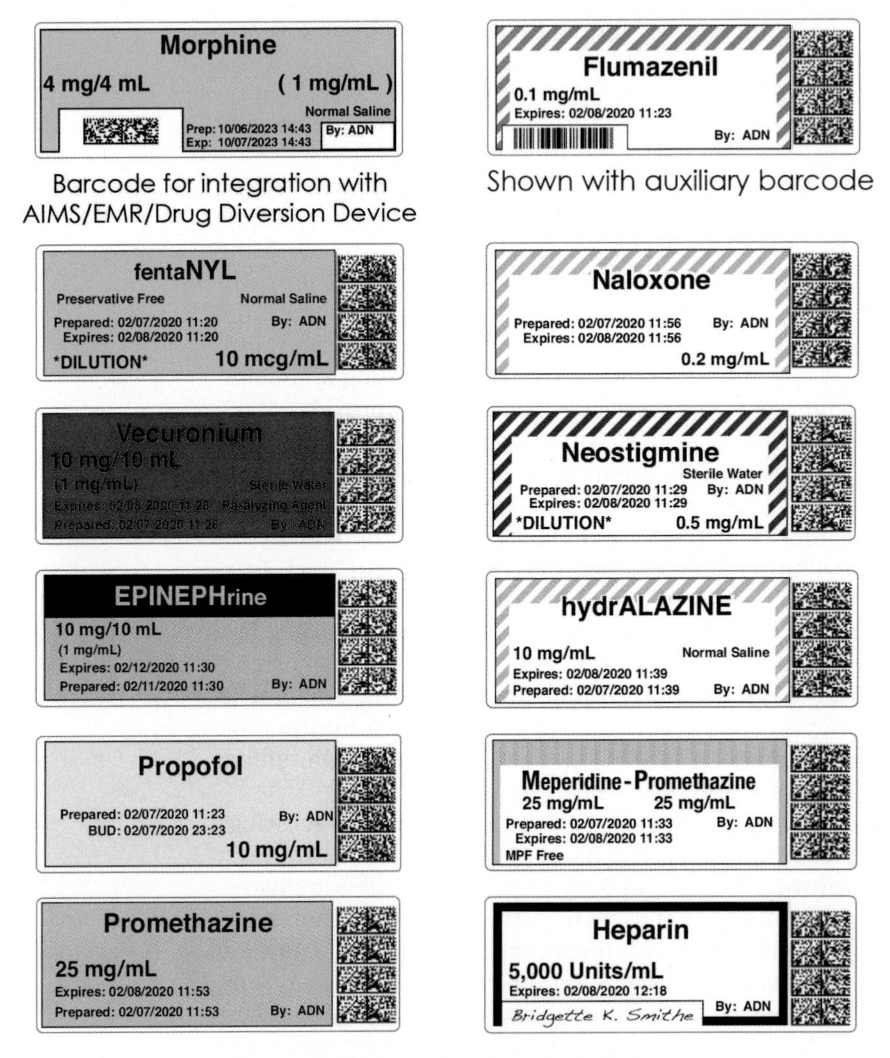

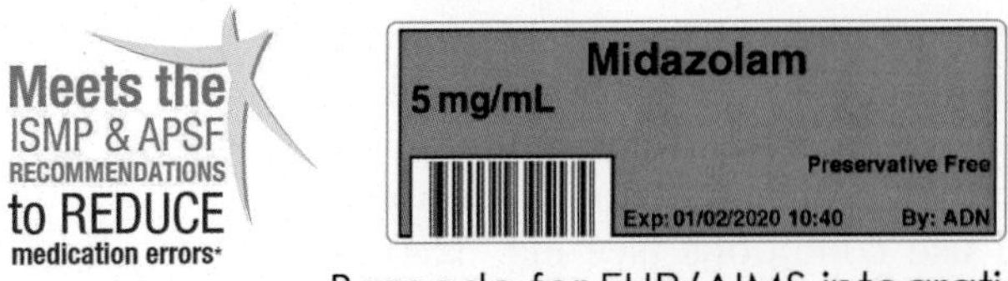

Barcode for EHR/AIMS integration

Figure 32.2 Safe Label System label output is compliant, easy to read and includes a barcode for integration with AIMS/EHR, interoperable syringe/LVP pumps, and drug diversion device.

Safe Label System Point-of-Care Station (SLS PCS)

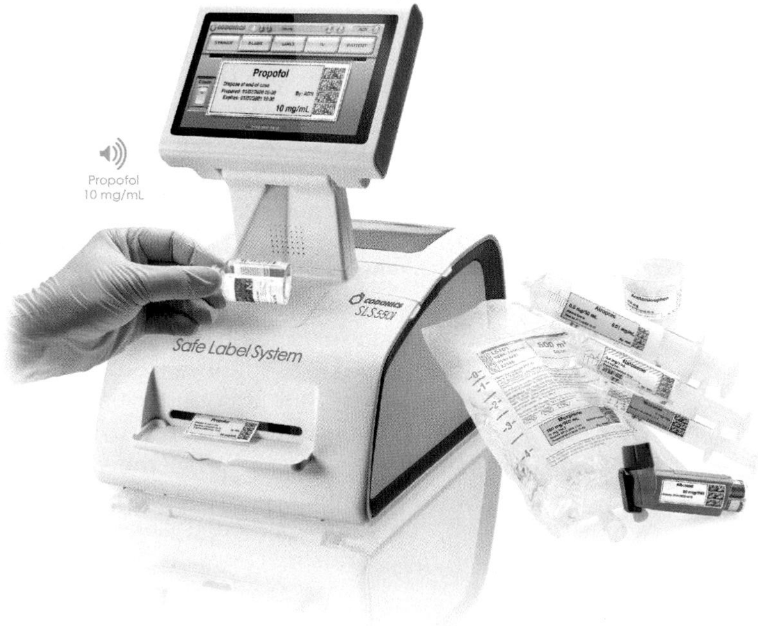

Figure 32.3 Safe Label System provides visual and audible confirmation of the drug-in-hand.

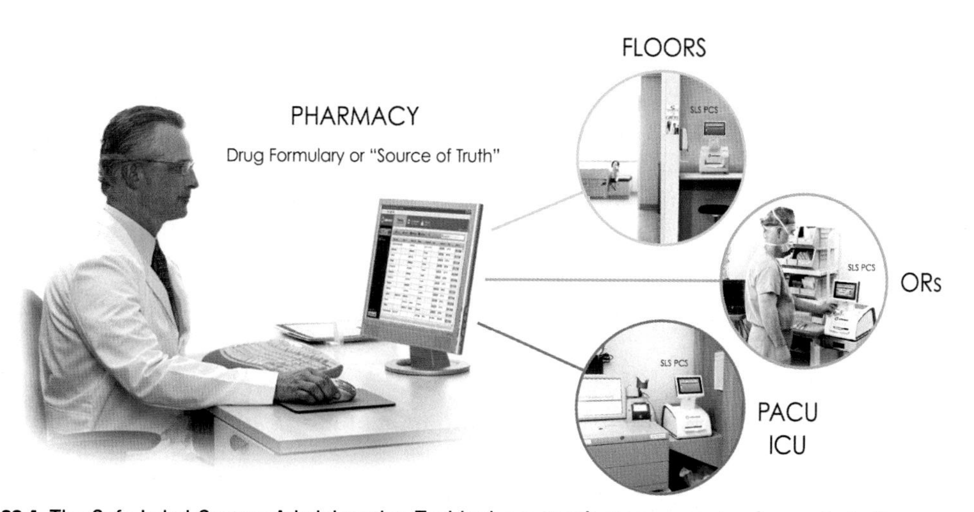

Figure 32.4 The Safe Label System Administration Tool is the system's management software that allows pharmacy to easily create and control a drug formulary, including site-approved medications, concentrations and rules for use, expanding pharmacy's scope of influence to the point of medication preparation.

complicated set of TJC, USP, and ASA guidelines [5,6], complete with pharmacy-controlled formularies that can be network-deployed to SLS PCSs anywhere medications are prepared in the hospital. The SLS AT can manage multiple formularies specific to the OR, ER, PACU, ICU, Labor and Delivery and other patient care areas as required to meet hospital needs.

The SLS PCS provides real-time verification of drug containers and applies rules for medication preparations consistent with hospital policies and best practice guidelines, eliminating the need for handwritten syringe labels. Visual and audible confirmation of the actual drug vial and final prepared concentration ensures that the correct drug is being prepared and the correct information is on the label. For example, when needing to differentiate look-alike, sound-alike drugs such as Ephedrine and Epinephrine, the SLS announces the drug and concentration and simultaneously displays the correct drug information and associated drug-class color so the clinician knows exactly what container is in their hand. As a final step, SLS PCS delivers a full-color, compliant barcoded label with all relevant drug information in seconds that is ready for application to a syringe or IV bag.

Beyond using the barcode on drug containers to facilitate the safe, compliant labels for syringes, SLS labels include a barcode which is the lynchpin to improved interoperability and AIMS/EHR documentation accuracy in the perioperative and procedural areas that enables pharmacy to realize ROI (return on investment) goals and helps close the loop on medication use.

The Safe Label System has been recognized by The Joint Commission as a "best practice" during audits at several leading hospitals [7]. Various case studies performed by clinicians using the system show significant labeling compliance, accuracy and charge capture improvements, as shown in the following graphics excerpted from customer case studies (see Figs. 32.5 and 32.6).

Continued development

Shortly after introducing SLS to the market, customer feedback started coming in that the product needed to "integrate" with other hospital systems. However, the requests for integration were not straightforward. Each hospital seemed to have their own concept of what integration meant. Codonics spent significant time researching and understanding the market

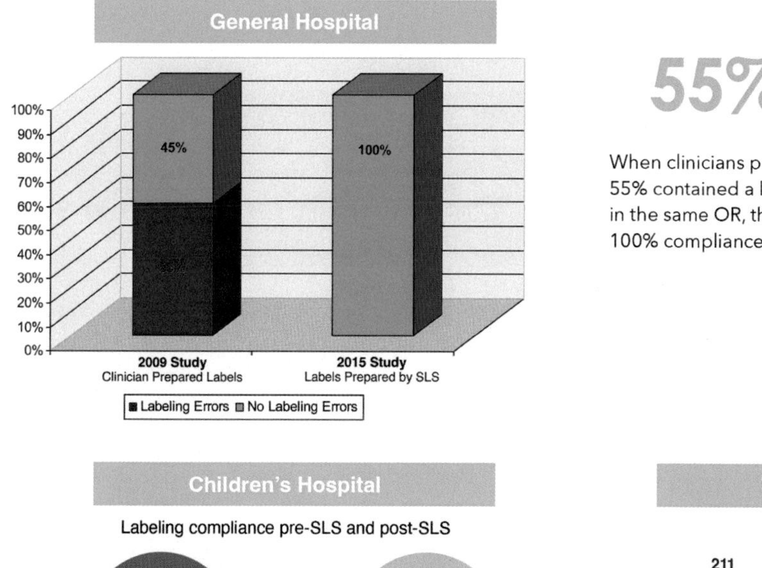

55% to ZERO

When clinicians prepared syringes manually in the OR, 55% contained a labeling error. When SLS was used in the same OR, there were zero labeling errors and 100% compliance.[5]

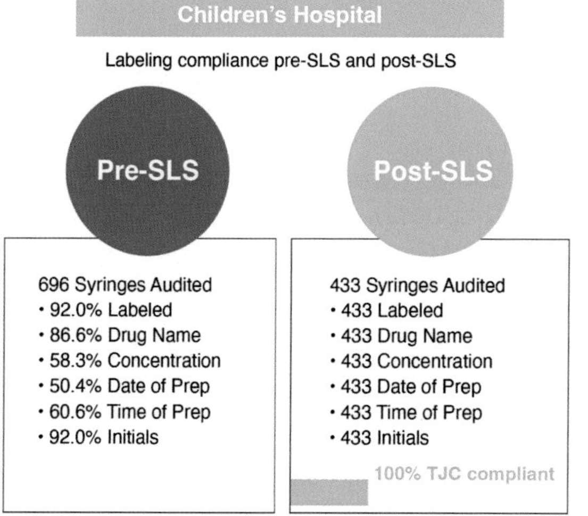

Children's Hospital
Labeling compliance pre-SLS and post-SLS
Pre-SLS
Post-SLS
696 Syringes Audited
• 92.0% Labeled
• 86.6% Drug Name
• 58.3% Concentration
• 50.4% Date of Prep
• 60.6% Time of Prep
• 92.0% Initials
433 Syringes Audited
• 433 Labeled
• 433 Drug Name
• 433 Concentration
• 433 Date of Prep
• 433 Time of Prep
• 433 Initials
100% TJC compliant

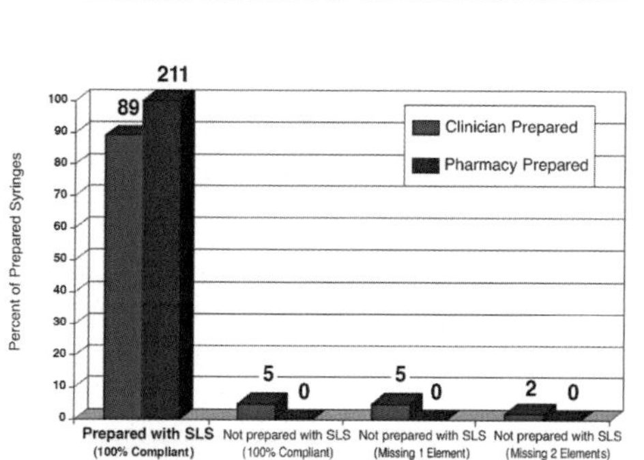

Post SLS installation compliance.
Percent of prepared syringes in each group displayed by compliance and the number of label elements missing.[6]

[5]Nanji, Karen C., M.D. M.P.H., etal. Evaluation of Perioperative Medication Errors and Adverse Drug Events, Perioperative Medicine. January 2006. [6]Jelacic, Srdjan MD; Bowdle, Andrew MD, et al. A System for Anesthesia Drug Administration Using Barcode Technology: The Codonics Safe Label System and Smart Anesthesia Manager™ Anesthesia & Analgesia 2015.

Figure 32.5 Customer success in a glance.

requirements for integration in order to meet customer needs. Broadly speaking, clinicians wanted to simplify their workflow by reducing manual tasks and capture more precise and timely drug administration information to improve recordkeeping in the AIMS/EMR. It became very apparent hospitals did not want another stand-alone piece of equipment that would not communicate with existing systems. They wanted interoperability and standardization in the OR.

As a first step, Codonics product management worked with industry-leading automated drug cart manufacturers such as

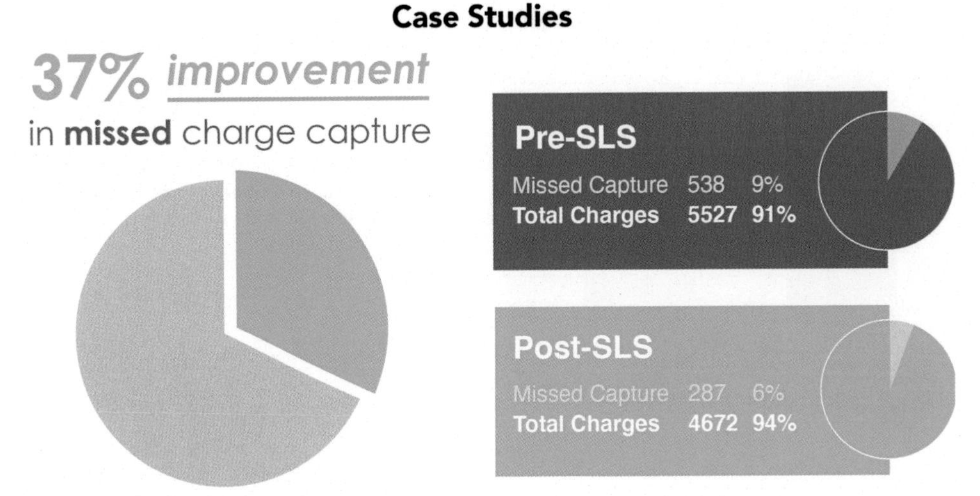

Figure 32.6 Case study results show improved charge capture in the operating room.

BD (formerly Carefusion) and Omnicell to integrate SLS with their anesthesia medication carts. For the perioperative area, this meant implementing improvements related to drug preparation such as "single login" enabling a user login on the cart to automatically login the same user on the SLS. It also enabled a "single scan" of the drug vial on SLS that would send the medication information to the cart to improve inventory management and significantly reduce out-of-stock events.

As time went on, the need for integration expanded. New requirements related to drug administration and billing were requested. One unique feature of SLS is the way it identifies every drug by scanning the NDC barcode on the parenteral vial/ampoule label. This provides very detailed information about the specific drug vial used on a patient and solved a big pharmacy billing problem in the OR related to 340B drug accountability. Pharmacy receives substantial discounts on specific drugs used on specific patients under a federal program called 340B, if supported by documentation. SLS captures the NDC drug information required for 340B but needed an easy way to include that information in the patient record. In 2017, Codonics introduced SLS-WAVE [8], an accessory to SLS PCS, that integrates with AIMS such as Epic Anesthesia and Cerner SurgiNet. The SLS-WAVE is a robust, high speed barcode scanner that reads the barcode from SLS-prepared syringe labels and most prefilled syringes to facilitate drug information capture into the patient record. Anesthesia providers "wave" the syringe label barcode under the SLS-WAVE scanner and information about that drug is automatically sent to the EHR/AIMS. In addition to capturing the required information for 340B and general billing, the same workflow allows additional safety checks on many AIMS during drug administration such as drug allergy checks and confirmations that the right drug is being administered at the right time. The latest version of Epic not only provides a visual safety check, it also provides an audible safety check, helping to eliminate syringe swaps during administration. This also reduces several of the manual clicks of a clinician in the AIMS, automatically presenting a drug dose entry screen so the provider can enter the specific dose administered to the patient (see Fig. 32.7).

Integration also takes anesthesia infusion safety to the next level. Codonics Safe Label System together with interoperable syringe pumps and Epic's EHR system are transforming infusion pump programming and safety in the OR. This dynamic, event-driven setting where barcoded syringes do not always come from the Pharmacy with a patient-specific label, but are prepared by the anesthesiologist in the OR based on patient conditions at the time, is no longer an "out of scope" activity. The combination of a barcoded syringe with EMR to pump interoperability, and the ability for the anesthesiologist to enter an order in real time allows the OR to be part of a hospital-wide implementation that raises patient safety. Until now, EMR to

Exact NDC matching vial to syringe

Captures NDC of parenteral vial, providing 100% accurate documentation for charge capture and the exact NDC for 340B accountability

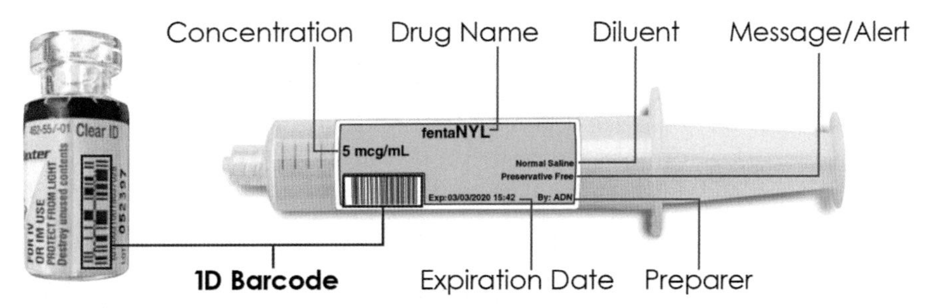

Figure 32.7 Every SLS label includes a barcode that can be scanned at administration for documentation into the local AIMS/EHR in the operating room to record the the U.S. National Drug Code (NDC) number of the drug being administered to the patient.

pump interoperability in the OR has been impossible, but this integrated triad enables a *first-of-its-kind standardization*, delivering the same safety protocols for bi-directional interoperability found on the ward floors, and enabling BCMA throughout the entire healthcare system. Once time-consuming tasks are reduced to just four "clicks," enabling the pump to be programmed and infusion administration documentation to be captured in the EHR. Each titration or any interaction with the pump throughout the procedure is also captured in the patient's anesthesia record without separate clinician interaction. The NDC of the drug being administered is automatically documented for charge capture and 340B accountability.

As part of our expanded integration capabilities, Codonics and BD (Becton, Dickinson and Company) have signed an agreement for joint design and development of technology aimed to help improve medication safety in the perioperative area. Through this collaboration, the companies will develop a new solutions that aim to help healthcare workers to intercept medication errors and automate medication administration documentation for IV bolus injections by leveraging BD Intelliport technology and Codonics Safe Label System technology. Both products have received clearance from the U.S. Food and Drug Administration (FDA) and have been honored with Medical Design Excellence Awards and Best Practice recognitions from Frost & Sullivan.

To help detect and deter drug diversion, Safe Label System prepared syringes also integrate with drug assay systems to prevent drug diversion in healthcare. The integration enables a wasted or returned syringe label barcode to be scanned, automatically identifying and documenting the drug name, concentration and preparer information to improve the drug assay workflow.

Integration has increased the value proposition that SLS provides to hospitals. What started as a medication safety and compliance device is now an integral part of hospital medication management and documentation accuracy. More integrations are planned for SLS to further expand its value and improve the ROI for the hospital. Some of these integrations will be with new medical devices and forward-looking applications that ensure a bright future for SLS moving forward (see Fig. 32.8).

Through the unique integration of technology, evolving best practices and enterprise-wide implementation, SLS is having a significant impact on hospital cultural change by improving quality and safety, closing the medication loop from preparation

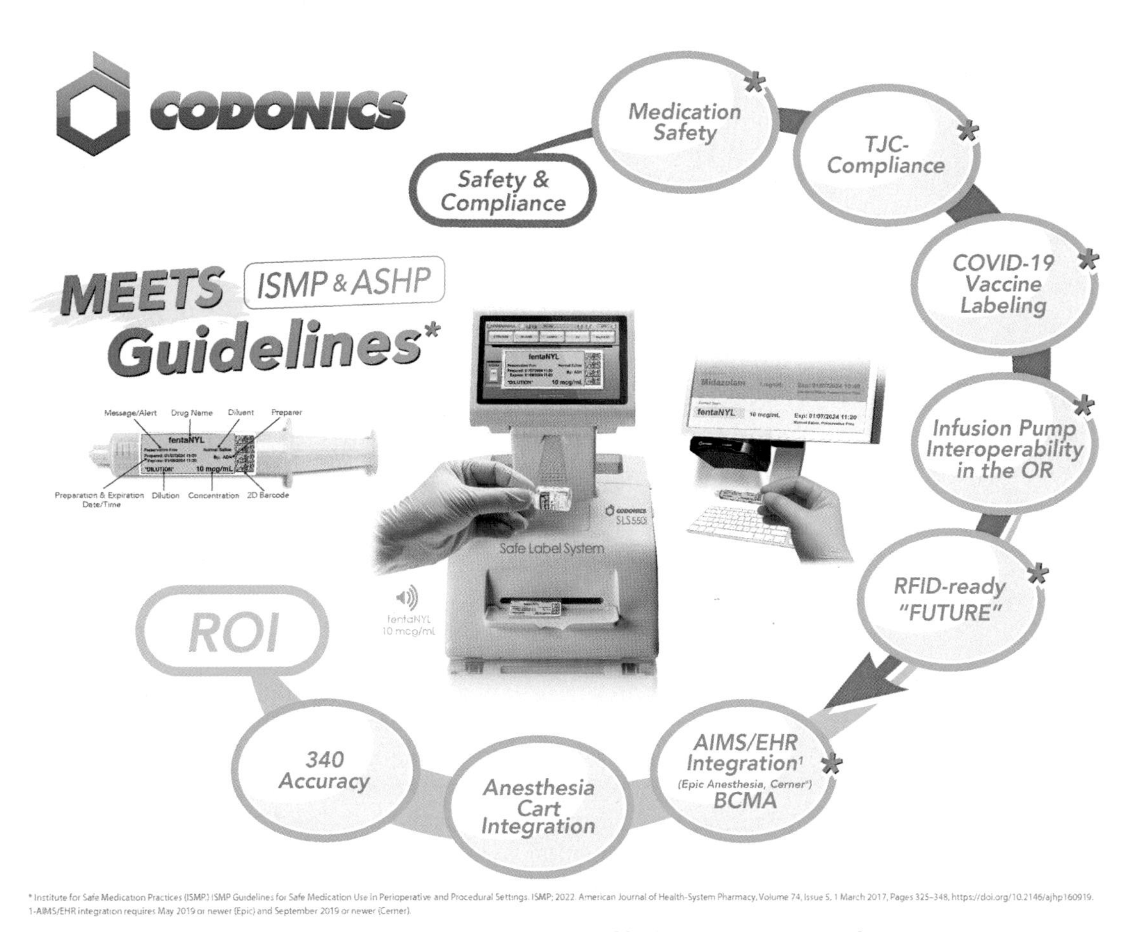

Figure 32.8 Safe Label System: The center of an in Integrated Medication Management System.

through administration to documentation. Done correctly, proper drug preparation and "smart" medication labeling ensures the right dose is prepared, documented and administered in compliance with drug expiration timelines, with the type of accountability demanded by modern day healthcare professionals. SLS brings all of the best practices for medication safety together to ensure guidance-driven compliancy and improved patient care in a game-changing product that has helped prevent hundreds of thousands of medication errors.

Strategic insights

SLS was invented by a group of anesthesiologists focused on solving their drug labeling problem in the OR. The role of pharmacy, as the overseer responsible for all drugs used in the hospital, was not recognized as a critical factor of SLS success until well into the development cycle. The need to understand and meet pharmacy requirements resulted in the development of the SLS AT that accounted for about half of the SLS project efforts. Making the SLS formulary management as easy and flexible as possible has been an ongoing effort. Understanding the relationship between pharmacy and anesthesia providers took time, but was necessary for a successful product launch and planning ongoing product improvements.

Part of the design philosophy behind the initial release of SLS was identification of requirements for the minimum viable product. As safety and efficacy were primary considerations, the design allowed for autonomous use of SLS PCS stations without dependency on a network or communication with a server. Software updates to SLS were later released that added network capabilities to both the SLS PCS and SLS AT while preserving autonomous operation of the SLS PCS for maximum uptime to ensure performance of its primary function of preparing safe and compliant labels for drug administration.

In the time since networking support was added to SLS, the incidents of hospital network attacks and data breaches have grown and become a major concern for their IT security staff. Hospitals increasingly require that medical devices must be hardened from being hacked or used as a vector of attack inside the hospital. Codonics has responded by adopting policies and best practices that provide regular and ongoing monitoring of threats to SLS and established guidelines for software releases that include security improvements, penetration testing, vulnerability testing, security patches, centralized user management, data encryption and enhanced network authentication for connectivity. This applies to both wired and wireless networks.

Another related area that has impacted SLS is the trend of military and hospital-specific security questionnaires that must be completed before new equipment or software releases are allowed on hospital networks. These questionnaires take a significant investment of time and technical resources to fill out, requiring answers to hundreds of questions that differ from hospital to hospital with no standardized format.

Ongoing challenges such as these require medical device companies like Codonics to implement product improvements and adapt internal processes to address changing market needs and regulatory requirements. While the specific nature of these challenges cannot be predicted, the general need for product design philosophies and an organizational structure that is flexible to accommodate change is a key to long-term success in this day and age.

Financial and risk

Codonics is a privately-held company in business for more than 40 years with over 25 years of experience developing, manufacturing and selling medical devices. All funding of new product development continues to be handled internally without financing from outside investors. Being self-funded, Codonics has the autonomy to pursue products with the right mix of innovation, revenue potential, sales channels, alignment with core business competencies, longevity in the market and intellectual property protection. These are the key factors that all Codonics products share in the healthcare market. But overarching is the CEO's focus on finding ideas that are revolutionary. New medical devices typically require years of development and tens of million dollars to reach initial launch. Given the costs and risks of bringing new medical devices to market, the ability to identify game-changing products along with the business and technical acumen to develop them so they reach their true potential is one of the greatest strengths of the company. Codonics prides itself on its financial independence, solid management team, world-class engineering capabilities, U.S. manufacturing facilities and global sales and service reach. These qualities have enabled the company to successfully weather many market and economic challenges by finding new avenues of growth and business opportunities. Codonics continues to seek out innovations that will drive the next generations of products.

References

[1] https://www.codonics.com/wp-content/uploads/2023/03/900-500-018.01.pdf: Safe Label System.

[2] Based on Merry AF, Peck DJ. Anesthetists, Errors in Drug Administration and the Law. N Z Med J 1995;24:185–7.

[3] https://www.jointcommission.org/-/media/tjc/documents/standards/national-patient-safety-goals/2021/npsg_chapter_ahc_jan2021.pdf: NPSG 03.04.01; 2021

[4] https://www.codonics.com/wp-content/uploads/2019/05/900-581-005_Rev04.pdf: Codonics SLS AT.

[5] The Joint Commission NPSG and Medication Management, Joint Commission International, ASA, APSF, ISO26825/ASTM4774, ISMP and APIC.
[6] https://www.asahq.org/standards-and-guidelines/statement-on-labeling-of-pharmaceuticals-for-use-in-anesthesiology: ASA color labeling guidelines.
[7] Recognized by The Joint Commission as a 'best practice': Texas Children's Hospital, Shriner's Hospital for Children, Madison Health Hospital.
[8] Codonics SLS-WAVE: https://www.codonics.com/wp-content/uploads/2019/05/900-734-004.03.pdf.

33

Revolutionizing blood transfusion: HemaLogiX-optimizing clinical appropriateness and resource utilization

James Lee Hill and Jennifer Dawson
University Hospitals Health System, Cleveland, OH, United States

Chapter outline

Abstract

Blood transfusion is a common procedure in the United States representing substantial opportunity to improve patient safety and cost effectiveness. Hemalogix was created to improve the utilization of transfusion by applying a clinical appropriateness algorithm to enhance performance awareness. Following successful proof of concept, and in collaboration with key stakeholders, a company was created to develop the commercial product.

Keywords: Transfusion; patient blood management; ventures

Innovation in Anesthesiology. DOI: https://doi.org/10.1016/B978-0-12-818381-6.00025-5

Key points

- Blood transfusion is a common procedure in the United States representing substantial opportunity to improve patient safety and cost effectiveness.
- Hemalogix was created to improve the utilization of transfusion by applying a clinical appropriateness algorithm to enhance performance awareness.
- Following successful proof of concept, and in collaboration with key stakeholders, a company was created to develop the commercial product.

Problem/opportunity

Over 16 million blood transfusions are administered in the United States annually with the supply increasingly being constrained [1,2]. Healthcare professionals suggest approximately 50% of blood transfusions are not clinically indicated [3]. The cost of blood is variable with packed red blood cells (pRBC) ranging between $522 and $1183 per unit, including storage, labor, and waste [4,5]. Blood transfusions have the potential to create patient risk in the short term such as acute transfusion reactions, and also long term hazards including transfusion related immune modulation, myocardial infarction, and renal failure [6]. While the immediate cost of transfusion implies large scale opportunity, when viewed through the lens of population health the problem is further magnified.

Concerns of liberal transfusion have led to the development of patient blood management programs to reduce unnecessary blood transfusions [7–9]. Prior to automated analytics, manual chart review was used to gather data, which was labor intensive and prone to error. Analytic solutions existed in the market utilizing large data warehouses, or applying one measure to all transfusions. Analytics centered heavily on pRBC transfusion. Errors in reporting and failing to consider patient-specific factors lead to concerns about validity of the reports and questioned credibility [10]. Many complex patient populations were excluded from analysis. There was a desire for high quality analytics to support the change management process for reducing inappropriate blood transfusion in all patient scenarios.

Concept

The team at University Hospitals (UH) was tasked with building an analytic platform that would assist in change management efforts to reduce unnecessary transfusions and patient harm. An extensive amount of near-real time data is entered into the electronic record on every treated patient. With access to clinical specialist in nearly every area of adult medicine, our team chose to utilize an algorithmic approach to analyze blood transfusions. We assembled a group of 30 clinical specialist to review literature and practice standards in their respective clinical area and report back patient variables or unique transfusion requirements for all blood components. This approach would harness the power of evidence-based medicine to examine every transfusion event to determine clinical appropriateness. This led to the development of a Transfusion Appropriateness Algorithm (TAA) that assesses the appropriateness of every transfusion based on over 80 patient-specific variables. After development of the algorithm, extensive commitment was extended to ensure data was being pulled from the electronic medical record accurately and populated into reports to serve various clinical and administrative users. Every transfusion to an adult patient was made available with a series of filters to help the end user focus each report. Dashboards were emailed biweekly to clinicians who ordered a transfusion as seen in Fig. 33.1. The data was also sent for peer reference to division chiefs, chairpersons, and hospital administration to provide context for change management efforts as seen in Fig. 33.2. The software system was branded HemaLogiX 1.0. A successful pilot was conducted at University Hospitals' (UH) large academic medical center showing a reduction in the use of pRBC and platelets [11].

Pathway

Following the successful pilot of HemaLogiX 1.0 in reducing blood utilization and increasing clinical appropriateness, our team collaborated with UH Ventures to develop a minimum viable product and plan for commercialization. HemaLogiX 1.0 was specifically developed for UH use with the Allscripts (Chicago, IL) electronic medical record (EMR). Realizing the limitation that this posed from a commercialization perspective,

Figure 33.1 Dashboard sample.

Provider Name	Department	Visits	Units Transfused	Units Appropriate	% Appropriate	Units per Visit	Average HGB
Provider A	General Surgery	3	5	3	60 %	1.67	7.00
Provider B	General Surgery	10	32	24	75 %	3.20	7.78
Provider C	General Surgery	7	18	15	83 %	2.57	6.98
Provider D	General Surgery	4	17	15	88 %	4.25	8.77
Provider E	General Surgery	11	25	23	92 %	2.27	6.52
Provider F	General Surgery	6	16	15	94 %	2.67	8.73
Provider G	General Surgery	17	43	41	95 %	2.53	7.28
Provider H	General Surgery	2	2	2	100 %	1.00	6.80
Provider I	General Surgery	5	9	9	100 %	1.80	6.17
Provider J	General Surgery	1	2	2	100 %	2.00	6.45

Figure 33.2 Provider comparison.

HemaLogiX 2.0 was developed to be EMR agnostic. UH Ventures assisted the team with market analysis, the protection of intellectual property, and the examination of different business models. Ultimately, stakeholders established a company known as Hemaptics, LLC, which was granted a license to commercialize HemaLogiX.

Building a company

Hemaptics is owned by UH. The company is managed by UH employed inventors and UH Ventures staff in partnership with an Executive-in Residence serving as a part-time CEO. The company's expenses have been funded by UH Ventures and multiple grant sources who were impressed by the initial pilot results and support prospective new company development. Aside from the CEO, one of the inventors also works as a contractor for the company to assist prospective clients and to manage ongoing product development. Use of part-time contractors has allowed the company to minimize cash burn.

Regulatory

This product does not require FDA approval. It provides retrospective analysis to enhance awareness about evidence-based practices and improved future performance. It does not direct practice at this time.

Launch

When the COVID-19 pandemic began and the blood supply was strained, UH expanded the use of HemaLogiX from one hospital to being used across the larger UH system. HemaLogiX helped to identify opportunity and manage the use of transfusions. After the initial study quantified a 10% reduction during the pilot year at one large hospital [11], a similar study was conducted at a community location demonstrating similar results 12 months following the expansion as seen in Fig. 33.3. Hemaptics began leveraging relationships to share the success of HemaLogiX to identify pilot sites beyond the UH system. During this time, legal contracts and other documentation were developed to support pilot arrangements and to attract additional investors. Following the pilot periods of the first 2–3 hospitals, Hemaptics will license HemaLogiX to healthcare systems and continue to develop additional features for additional clinical decision support. The go-to-market strategy involves a direct sales force and possible distribution agreements through other healthcare analytics companies. Hemaptics plans to raise additional funds through a convertible note and by launching a Series A round.

Challenges/solutions

The biggest challenge in developing HemaLogiX was time and resources. The time necessary to invent, develop, pilot and market the software solution extended well beyond routine responsibilities. The time required to create a company

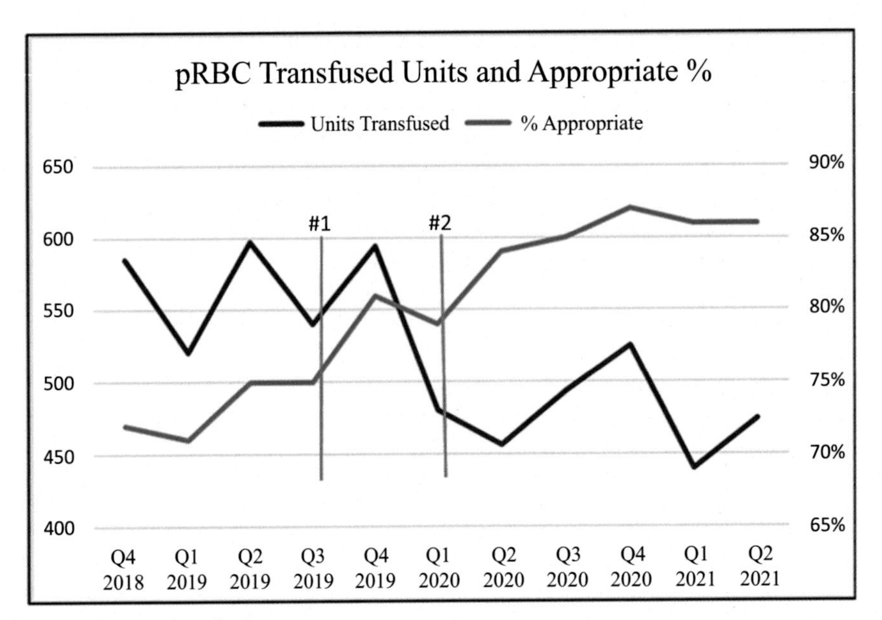

Figure 33.3 Interventions and performance.

and connect with investors and customers was intensified by the strains of the current pandemic, but would likely have been a challenge at any time. The team has consistently devoted time to fundraising from both grants and investors. One specific challenge has been the availability of IT resources both internally and for customers. Despite interest in the solution, prospective clients face limits in their ability to move forward due to IT resources. With so many competing priorities, the team developed compelling presentations detailing the value of HemaLogiX and clarifying the technical, operational, and clinical resources required to implement the solution. Communications were tailored to specific stakeholders.

Advice

There is great value in patience, persistence, and collaboration. Start small, demonstrate success, and move forward to expand and develop. If there is a potential problem or concern, stop to analyze and address it before moving forward. Set realistic expectations regarding timing and adoption rates within the healthcare market. Partner with a ventures team or other experts for legal, funding, and market analysis support. Stay positive.

References

[1] Jones JM, Sapiano MRP, Mowla S, Bota D, Berger JJ, Basavaraju SV. Has the trend of declining blood transfusions in the United States ended? Findings of the 2019 National Blood Collection and utilization survey. Transfusion 2021;61(Suppl 2):S1–10. Available from: https://doi.org/10.1111/trf.16449.

[2] Doughty H, Green L, Callum J, Murphy MFNational Blood Transfusion Committee. Triage tool for the rationing of blood for massively bleeding patients during a severe national blood shortage: guidance from the National Blood Transfusion Committee. Br J Haematol 2020;191(3):340–6. Available from: https://doi.org/10.1111/bjh.16736.

[3] Shander A, Fink A, Javidroozi M, et al. Appropriateness of allogeneic red blood cell transfusion: the international consensus conference on transfusion outcomes. Transfus Med Rev 2011;25:232–46. Available from: https://doi.org/10.1016/j.tmrv.2011.02.001.

[4] McQuilten ZK, Higgins AM, Burns K, et al. The cost of blood: a study of the total cost of red blood cell transfusion in patients with β-thalassemia using time-driven activity-based costing. Transfusion. 2019;59(11):3386–95.

[5] Shander A, Hofmann A, Ozawa S, Theusinger OM, Gombotz H, Spahn DR. Activity-based costs of blood transfusions in surgical patients at four hospitals. Transfusion. 2010;50(4):753–65.

[6] Redding N, Plews D, Dodds A. Risks of perioperative blood transfusions. Anaesth Intensive Care Med 2022;23(2):80–4. Available from: https://doi.org/10.1016/j.mpaic.2021.12.001.

[7] Hasler S, Kleeman A, Abrams R, et al. Patient safety intervention to reduce unnecessary red blood cell utilization. Am J Manag Care 2016;22(4):295–300.

[8] Frank SM, Thakkar RN, Podlasek SJ, et al. Implementing a health system-wide patient blood management program with a clinical community approach. Anesthesiology 2017;127(5):754–64. Available from: https://doi.org/10.1097/ALN.0000000000001851.

[9] Staples S, Salisbury RA, King AJ, et al. How do we use electronic clinical decision support and feedback to promote good transfusion practice. Transfusion 2020;60(8):1658–65. Available from: https://doi.org/10.1111/trf.15864.

[10] Soril LJJ, Noseworthy TW, Stelfox HT, Zygun DA, Clement FM. Facilitators of and barriers to adopting a restrictive red blood cell transfusion practice: a population-based cross-sectional survey. CMAJ Open 2019;7(2):E252–7. Available from: https://doi.org/10.9778/cmajo.20180209.

[11] Dawson JL, Schwabl C, Pronovost PJ, et al. Transfusion utilization and appropriateness: thinking differently at a tertiary academic medical center. Phys Leadersh J 2022;9(4):18–23. Available from: https://doi.org/10.55834/plj.114687726.

34

FirstLook LTA from Guidance Airway Solutions

Joshua Herskovic[1,2]

[1]Baptist Health Bethesda Hospital, Boynton Beach, FL, United States

[2]Veterans Affairs Medical Center, West Palm Beach, FL, United States

Chapter outline

Abstract

The focus of this chapter is how can someone turn a back of the napkin idea into a reality. I will guide you through the process I used to overcome this challenge. In order to remain brief I will focus on the early stages for bringing the FirstLook Device to reality in lieu of the later stages after the initial outside funding was procured. A current update can be found at: https://guidanceairway.com/FirstLook/.

Keywords: FirstLook LTA; nozzle; pathway; USPTO; CAD; 3D printing; intubation

Joshua Herskovic is a practicing anesthesiologist with a background in mechanical engineering. He is the inventor and founder of Guidance Airway Solutions. The focus of this chapter is how can someone turn a back of the napkin idea into a reality. He will guide you through the process he used to overcome this challenge. The focus will be on the early stages for bringing the FirstLook Device to reality in lieu of the later stages after

Innovation in Anesthesiology. DOI: https://doi.org/10.1016/B978-0-12-818381-6.00008-5

the initial outside funding was procured. A current update can be found at: https://guidanceairway.com/FirstLook/.

FirstLook LTA combines a low-profile nozzle with an LTA device capable of acting as an Endotracheal tube (ETT) stylet. The ability to intubate and provide LTA in one step decreases the possibility of the user losing sight of the airway, and may reduce many of the risks associated with this beneficial procedural step. FirstLook LTA does this without introducing new practitioner procedural steps or efforts and is faster and easier to set up than standard LTA devices (Fig. 34.1).

Early in my career out of residency I noticed most anesthesiologist in the practice were using a device called the LTA, laryngo tracheal anesthetic applicator, prior to intubation. There are several benefits to using this device and I started to use it as well. However, I encountered issues during placement and I found that using this device can produce more trouble than benefit. A golden rule in Anesthesia is your first look is your best look. This device was essentially requiring 2 intubations. Direct laryngoscopy, LTA placement, ETT placement. Sometimes, immediately after LTA placement you lose site of the airway, the vocal cords move, you're fumbling for the ETT, etc. Hence, this was the motivation for a device that can provide the benefits of topical anesthesia from LTA on your first pass, your "FirstLook" while simultaneously placing the ETT and securing the airway.

Concept

During early brainstorming stage an entrepreneur should be considering as many ideas as possible without restraint. No matter how crazy. Perhaps, someone on your team can add value to the idea to make it possible. At this stage we should think of anything; possible or impossible. At the next stage we will eliminate the nonpossibilities.

For FirstLook LTA, my first step was to define the problem:

Problem: Conventional LTA requires prolonged direct laryngoscopy secondary to its 2-step intubation process. Prolonged direct laryngoscopy places considerable stress on the patient. Extended and repeated direct laryngoscopy increases the chances of losing sight of the airway.

Solution: A device that consolidates the number of steps required for LTA and ETT placement.

It is important at an early stage to define your product requirements. I used to work as a mechanical engineer for

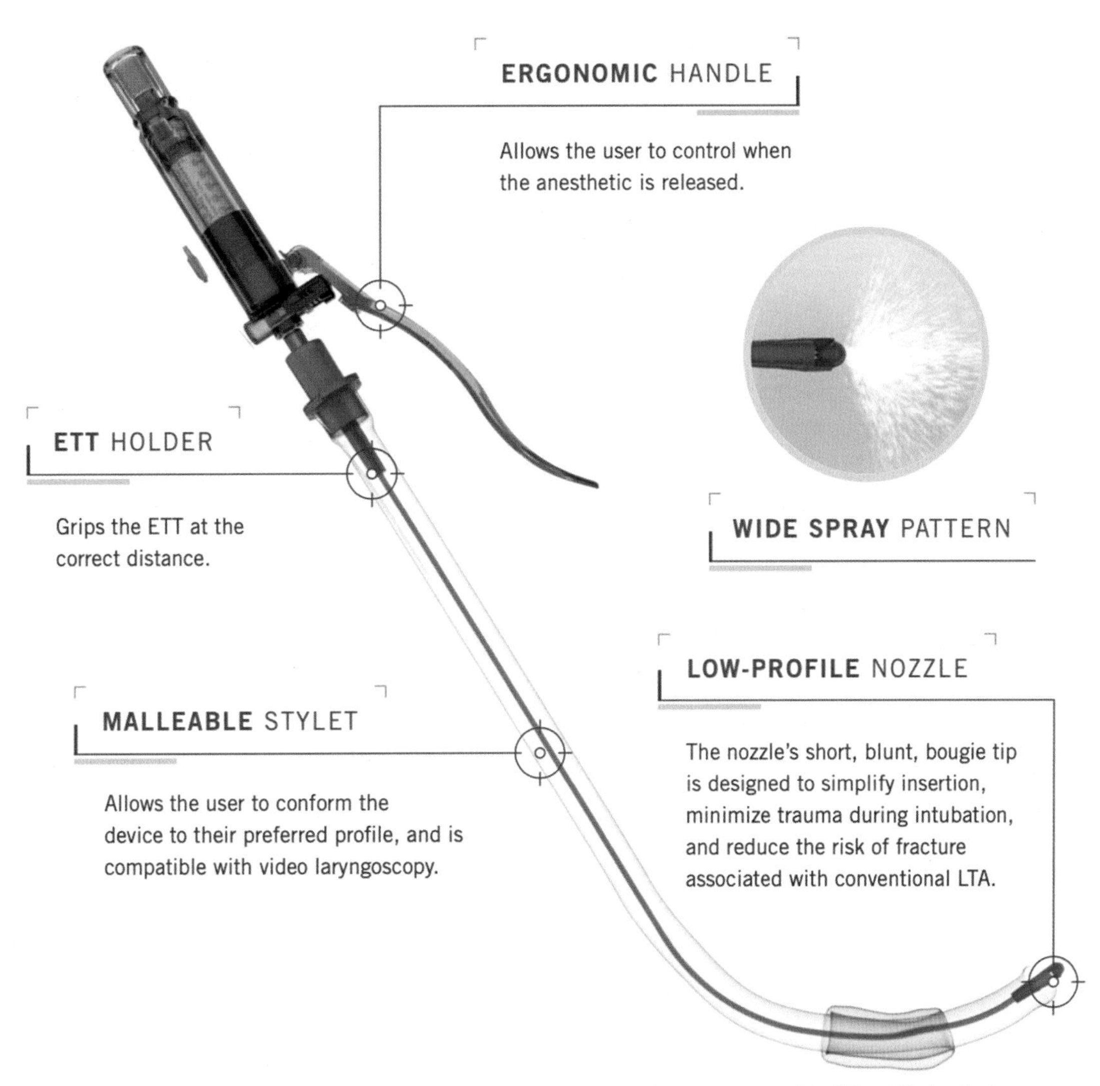
ERGONOMIC HANDLE
Allows the user to control when the anesthetic is released.
ETT HOLDER
Grips the ETT at the correct distance.
WIDE SPRAY PATTERN
MALLEABLE STYLET
Allows the user to conform the device to their preferred profile, and is compatible with video laryngoscopy.
LOW-PROFILE NOZZLE
The nozzle's short, blunt, bougie tip is designed to simplify insertion, minimize trauma during intubation, and reduce the risk of fracture associated with conventional LTA.

Figure 34.1 Illustrative Rendering: A disposable device that allows the user to provide LTA while intubating. FIRSTLOOK LTA combines a low-profile nozzle with an ETT stylet.

Terumo, a medical device company, prior to becoming a physician. I can speak first hand for the lack of understanding especially when it comes to how is the device used from the engineering perspective. I am also aware of the limitations engineers face, that a physician may not understand when introducing an idea. Thus, an important criterion is I did not want the anesthetist to require additional procedural efforts or to require

VARIANT	PROCEDURAL EFFORTS BY PROVIDER	ERGONOMICS/ SIZE& WEIGHT	DEVELOPMENT COMPLEXITY	UPFRONT COST TO MANUFACTURE	UNIT COST IN VOLUME	ADAPTS SYRINGE OR VIAL	DISPOSABLE	TOTALS
SLIDER	—	—	++	+	++	Both (++)	Partly (+)	6
MOTORIZED	+	—	—	+	++	Both (++)	Partly (+)	5
CO2 CHARGED	+	+	—	—	+	—	Full (++)	2
SPRING LOADED	+	+	+	+	+	Vial (+)	Full (++)	8
PNEUMATIC	+	+	++	++	+	Both (++)	Full (++)	11

Figure 34.2 Analysis table: Simple weights are assigned to predetermined categories to determine the concepts with the most potential.

a steep learning curve. No matter how cool the product is; no-one will use it if requires more effort than the conventional process. Additional important criteria included complexity, size, cost to manufacture, and product cost when in production. I then assigned grades for each concept based on their performance in each category to weed out the various concepts to the top 3 (Fig. 34.2).

Pathway

Now that I have a concept, what are my next steps? This is the pathway I chose for FirstLook; this was also a learning process. Would I have altered the sequence of steps if I were to do this again? Yes!

My strategy: I needed to talk about this idea to gather feedback and gather interest. Thus, I decided to file a provisional patent since an NDA does not provide much protection. I had several variations of how this device will work, but I had not actually made a prototype yet. A provisional patent will give me 1 year to file a utility patent and will hold my place in line. Essentially, it's a bookmark of when I came up with the idea. Most important, it gave me time to work out the idea before making the larger financial commitment to proceed with an utility patent and device development. The USPTO is a first inventor to file system and no longer a first to invent system. There are some downsides to filing a provisional patent early which may not be obvious that I will discuss later.

To reduce cost, you can write the provisional yourself and find a patent attorney or an online service to file for you. A patent agent can assist you as well. You can also file it yourself, which is the method I chose. However, this takes some effort to be granted this capacity by the USPTO. The provisional patent most likely will never be examined by the USPTO, but will be referenced by your nonprovisional patent one year from now. You do not need to know legalize; although, I would suggest having claims. However, you must present concepts in detail that are close enough to your nonprovisional filing. If you file a nonprovisional patent, it will eventually be published by the patent office. Meaning, this will be viewable in the future; thus, you need to be careful about adding additional information that you do not want published. Once you start the provisional patent, you need to be aware the clock is ticking. In fact, this influenced my timing for seeking funding because this process will soon become expensive.

There are multiple patent strategies and it's important to have one before you file your provisional patent. One year after filing your provisional you will be required to file a nonprovisional patent in order to maintain priority. Not only does your design need to be well worked out before filing the nonprovisional but if you plan to protect your device internationally you will be required to filing a PCT at the same time to enter the international phase. This can double the cost of your patent. This clock starts 1 year after your earliest priority date, which is the date of your provisional patent filing. Furthermore, if your device is not fully worked out 1 year after your priority date then you will need to choose to either abandon your provisional patent and thus lose your priority date or proceed with the nonprovisional; knowing that you may need to file additional patents or modifications to the patent, also known as a "continuation in part," in the future. Looking back, I would have delayed the provisional patent until I had a solid working prototype in order to delay my international expenditures. A final note; it is common to wait over a year before receiving any action from the USPTO; thus, anticipate remaining "patent pending" during much of your next steps.

Next steps: The 50K plan.

My early proof of concepts, were essentially mockups using bendable stylets. Nothing functional. I did try modifying some components from everyday items from the OR to try to make something functional. Essentially a low-end proof of concept as validation for my idea.

How do you take your idea to the next level

In the last 10 years there have been massive advances in rapid prototyping and computer aided design (CAD). If you can learn to draw something on the computer you can have it sitting at your front door the next day. There are several types of rapid prototyping but the most common and usually lowest cost is 3D printing. There are also many types of 3D printing but they all have one thing in common. They build your prototype one layer at a time. This is why 3D printing is also called an additive process or additive manufacturing. There are also subtractive processes such as CNC. By far, 3D printing is the lowest cost process available for rapid prototyping. You can purchase a 3D printer at a relatively low cost. I utilized both a Formlabs SLA printer and a makerbot FDM printer. Both are at extreme opposites of the type of 3D printing technology available. SLA printers have more material limitations although it is improving. This is because it's a resin material attempting to mimic an injection molded plastic such as ABS as opposed to a FDM printer that is printing in actual ABS plastic. However, the SLA printer can print at much higher quality and approach the look and feel of a manufactured product. Today they have become the mainstay of small prototype shops in many engineering firms. This way I was able to produce and test prototypes rather quickly.

When I was gearing up for a more presentable model or needed better material properties, I would send my CAD files to an online company that could print my device at a much higher resolution and with better materials. A company that does this in high volume and at rather low cost today is http://Shapeways.com. Regarding materials, there is something else that all types of 3D printing have in common. There is some element of the 3D printed materials that will always be somewhat of a compromise when compared to a manufactured injection molded part. For instance, if you want the best-looking part, it's mechanical properties may be somewhat compromised; you want a strong unbreakable part, well then the texture may feel different, the object may not be water tight (i.e., hold pressure), etc. . . .

Back to FirstLook . . . In my early days as a mechanical engineer CAD was difficult to learn. 3D CAD was unthinkable. We actually typed in our coordinates for each line. Fast forward 10 years later and CAD has become relatively easy to learn. For my initial attempt at the FirstLook LTA device I started with a simple piece of software called SketchUp. Within a couple of days, I had my design from a paper sketch to computer. I sent my file off to a local 3D printing company and waiting for 1 week with excitement. Expectation then met reality when I assembled the device. You will learn a lot from your first prototype. Nothing beats holding it in person and don't expect to get it right on your first try.

I quickly realized multiple aspects of my design needed a drastic redesign. In use, anything that is located above the uppermost portion of the ETT will feel exaggerated in weight because of the moment arm created from the location you are holding the ETT. The nozzle at the distal tip created another issue. Anything extending past the ETT will make the device difficult to maneuver. In conventional LTA you have an entire shaft dedicated to the nozzle. My nozzle was fractionally smaller but I now knew I had to reduce this to mere millimeters! Somehow, I needed something that could provide a similar radial spray pattern to conventional LTA but had to shrunken down to a few millimeters. My invention has now become 2 inventions! I needed to develop a special nozzle. From a patent perspective this is going to become expensive!

From my initial prototype I learned several things

- The device has to be much smaller and weighted differently.
- I needed a special miniature nozzle.
- This is going to be a lot more expensive to protect then initially thought.

Back to the drawing board

With a focus on size reduction, I designed a new prototype. Unfortunately, the SketchUp software was not going to cut it because of the fine detail this next iteration will require. I started looking into more professional CAD programs. Ultimately, I began working with a program called Fusion by AutoDesk. AutoDesk has been around for a long time and is a safe bet for continued support. Their fusion software made CAD approachable to nonengineers. Although it has a steeper learning curve, it is a very robust software that can assist in manufacturing and also has rendering capability.

The results were much better this time. The revised nozzle only needed to clear the ETT tip by a few millimeters; in fact the nozzle tip was almost bougie like in positive way for intubation. Hand holding the device and going thru the process of intubation on an airway mannequin was proof of concept. However, the materials were too weak to provide reliable function.

Pitch process

Something I decided early in the process, is I didn't want to be the guy with a suitcase fumbling around with a bunch of prototypes expecting it to work out perfectly. The chances of everything going as planned are slim to none. A benefit of using CAD software is the ability to provide renderings to tell your story (Fig. 34.1).

I received my first chance to pitch the device before I was quite ready. Thru a contact I set up a meeting with Medline in 2 weeks; the issue is I had no working prototype. As I mentioned prior material limitations were preventing me from housing a spring that could provide enough force for the required spray pattern. I went thru several revisions; however, 3D printed materials require compromise. The size requirements despite being of major importance was a significant limitation as well. With an unlimited budget, this would not be a limitation as there are alternatives to 3D printing that can deliver prototypes with material properties closer to a manufactured product. Depending on the complexity and size of the project, sometimes this is a first option. One technology is polyurethane casting. Polyurethane properties are better at mimicking injection molded parts. In fact, it can be very difficult to distinguish a production part from a polyurethane part. However, it's more expensive than 3D printing, especially at low quantities. A positive is that you can re-use the casting to make many identical parts. After several parts this can become a lower cost than 3D printing. At a higher cost is temporary injection molds which will provide true injection molded parts. These molds have a limited life-span when compared to a permanent mold.

It's important to maintain some degree of flexibility. Ideally, you maintain the capability to regroup before too much time and resources are spent in a direction that is fruitless. This is an additional benefit of performing an analysis at an early stage. In my earlier analysis I considered an alternative second design that was pneumatic. This had the added advantage of being able to accept a syringe in addition to a vial. Feedback, suggested this benefit was a welcome addition. Some, iterations on my Formlabs printer and we had a device that was reliably working within the limitations of 3D printing. Furthermore, I was able to utilize materials that gave a professional finish.

My pitch went well and the devices performed well on stage. Medline wanted to gather feedback at the American Society of Anesthesiologist annual meeting on display with their other devices.

This bought up a minor issue that I had never thought about. How do you travel with prototypes? Will it get thru TSA? Not wanting to take chances, I shipped off a few prototypes and took the rest on the airplane. The prototypes that were to go thru TSA I put in a box and labeled "Medical Prototypes." Yes, it went thru X-ray a couple times but it got me on to the next stage.

At the ASA, I was able to utilize the opportunity to walk the device around the showroom floor as well to gather interest from other airway device companies as well. This did result in multiple follow up meetings and getting my foot in the door at other airway device companies in addition to Medline. Simultaneously, I was introduced to the CEO of a private equity firm, Thomas Harter Sr., and I chose the route of building a company around this device and other concepts I had in my portfolio.

Regulatory

When evaluating your plan of action, it is important to understand your products FDA classification and its impact on your funding requirements. Class I is the least stringent with not much required beyond registration. Many anesthesia devices fall into the Class II category like FirstLook LTA. Class II is very broad and comprises a wide spectrum of very little to a fair number of requirements. Some airway devices are also exempt from filing a 510K; significantly decreasing your time and cost expenditures. If you are filing a 510K in the airway space; the good news is that many aspects of your design may be considered substantially equivalent to an existing device and may not need stringent testing. You can look up these 510K submissions online along with device exemptions by searching the FDA site.

Several companies will offload this process from you for a small fortune. Alternatively, you can seek the advice of an independent consultant that has written many 510K's. With FirstLook I chose the latter approach. Many of these consultants are also engineers and maybe able to advise on you other aspects of your device as well. I learned so much from these engagements I would advise you to at least look into the later route.

Assuming you will need funding to progress further, any pitch will require you to know your estimated timeline and cost to meet FDA regulatory approval. If your goal is to license the product at some point, then you will need to factor in the cost to get to various of stages of development. I would consider FDA certification a major milestone that may decrease the threshold for a company to take the chance on bringing your product to market.

YEAR	1	2	3	4	5	6	7	8	9	10
Net Sales *(Rounded)(1)*	$ 430,000	$ 1,910,000	$ 7,650,000	$ 26,780,000	$ 40,160,000	$ 46,184,000	$ 48,493,000	$ 50,918,000	$ 53,464,000	$ 56,137,000
Net Sales Growth (Rounded)		*344%*	*301%*	*250%*	*50%*	*15%*	*5%*	*5%*	*5%*	*5%*
Obsolescence Factor	100%	100%	100%	100%	100%	100%	100%	100%	100%	100%
Adjusted Sales	430,000	1,910,000	7,650,000	26,780,000	40,160,000	46,184,000	48,493,000	50,918,000	53,464,000	56,137,000
Royalty Savings	21,500	95,500	382,500	1,339,000	2,008,000	2,309,200	2,424,650	2,545,900	2,673,200	2,806,850
Less: Income Tax	(7,525)	(33,425)	(133,875)	(468,650)	(702,800)	(808,220)	(848,628)	(891,065)	(935,620)	(982,398)
After-Tax Royalty Savings	13,975	62,075	248,625	870,350	1,305,200	1,500,980	1,576,023	1,654,835	1,737,580	1,824,453

Notes:

(1) Assumes a price per unit of $15 to $25 percent and volume growth as illustrated.

		Remaining Useful Life				
		15	20	25	30	35
Royalty Rate	3.0%	3,580,000	3,770,000	3,910,000	4,020,000	4,100,000
	4.0%	4,770,000	5,030,000	5,220,000	5,360,000	5,470,000
	5.0%	5,970,000	6,290,000	6,520,000	6,700,000	6,830,000
	6.0%	7,160,000	7,550,000	7,830,000	8,040,000	8,200,000
	7.0%	8,350,000	8,800,000	9,130,000	9,380,000	9,570,000

		Discount Rate				
		12.0%	14.0%	16.0%	18.0%	20.0%
Long Term Growth Rate	2.0%	9,020,000	7,570,000	6,420,000	5,510,000	4,770,000
	3.0%	9,220,000	7,710,000	6,520,000	5,580,000	4,830,000
	4.0%	9,440,000	7,860,000	6,630,000	5,660,000	4,880,000
	5.0%	9,680,000	8,020,000	6,750,000	5,750,000	4,940,000
	6.0%	9,930,000	8,200,000	6,880,000	5,840,000	5,010,000

*Note: values modified and are for illustrative purposes only

Figure 34.3 Expected sales growth and royalty income depicted as part of a valuation generated for investors.

Valuation

Finally, no matter which path you take you must provide some valuation your invention. It is difficult to value a novel technology with no sales. In simplest terms for my initial valuation, I gathered research on the primary competitors to determine the size of the opportunity. I turned to experts in accounting to determine what portion of the market I could reasonably capture utilizing complicated algorithms (Fig. 34.3). Having a friend as a lawyer is great as well; thank you David Seidman. Finally, by the time you enter negotiations it is likely you will still be patent pending; anything you can do at this stage to de-risk the investment for your investors will lower the threshold for your obtaining funding to move your product forward.

I hope you have found this chapter insightful. Because of the amount of information proposed I wanted to keep it brief and focus on the early stages of product development.

Index

Note: Page numbers followed by "*f*" and "*t*" refer to figures and tables, respectively.

C

D

Q

R

T

[illegible]d in the United States
[illegible]r & Taylor Publisher Services